Date Due

DEC 1 7 2008			
	DISCARDED		

BRODART, CO. Cat. No. 23-233-003 Printed in U.S.A.

Analysis of Neurophysiological Brain Functioning

Springer
Berlin
Heidelberg
New York
Barcelona
Hong Kong
London
Milan
Paris
Singapore
Tokyo

Springer Series in Synergetics

Editor: Hermann Haken

An ever increasing number of scientific disciplines deal with complex systems. These are systems that are composed of many parts which interact with one another in a more or less complicated manner. One of the most striking features of many such systems is their ability to spontaneously form spatial or temporal structures. A great variety of these structures are found, in both the inanimate and the living world. In the inanimate world of physics and chemistry, examples include the growth of crystals, coherent oscillations of laser light, and the spiral structures formed in fluids and chemical reactions. In biology we encounter the growth of plants and animals (morphogenesis) and the evolution of species. In medicine we observe, for instance, the electromagnetic activity of the brain with its pronounced spatio-temporal structures. Psychology deals with characteristic features of human behavior ranging from simple pattern recognition tasks to complex patterns of social behavior. Examples from sociology include the formation of public opinion and cooperation or competition between social groups.

In recent decades, it has become increasingly evident that all these seemingly quite different kinds of structure formation have a number of important features in common. The task of studying analogies as well as differences between structure formation in these different fields has proved to be an ambitious but highly rewarding endeavor. The Springer Series in Synergetics provides a forum for interdisciplinary research and discussions on this fascinating new scientific challenge. It deals with both experimental and theoretical aspects. The scientific community and the interested layman are becoming ever more conscious of concepts such as self-organization, instabilities, deterministic chaos, nonlinearity, dynamical systems, stochastic processes, and complexity. All of these concepts are facets of a field that tackles complex systems, namely synergetics. Students, research workers, university teachers, and interested laymen can find the details and latest developments in the Springer Series in Synergetics, which publishes textbooks, monographs and, occasionally, proceedings. As witnessed by the previously published volumes, this series has always been at the forefront of modern research in the above mentioned fields. It includes textbooks on all aspects of this rapidly growing field, books which provide a sound basis for the study of complex systems.

A selection of volumes in the Springer Series in Synergetics:

Synergetics An Introduction 3rd Edition
By H. Haken

Chemical Oscillations, Waves, and Turbulence
By Y. Kuramoto

Temporal Disorder in Human Oscillatory Systems
Editors: L. Rensing, U. an der Heiden, M. C. Mackey

Computational Systems – Natural and Artificial
Editor: H. Haken

From Chemical to Biological Organization Editors: M. Markus, S. C. Müller, G. Nicolis

Information and Self-Organization
A Macroscopic Approach to Complex Systems
By H. Haken

Neural and Synergetic Computers
Editor: H. Haken

Synergetics of Cognition
Editors: H. Haken, M. Stadler

Theories of Immune Networks
Editors: H. Atlan, I. R. Cohen

Neuronal Cooperativity
Editor: J. Krüger

Synergetic Computers and Cognition
A Top-Down Approach to Neural Nets
By H. Haken

Rhythms in Physiological Systems
Editors: H. Haken, H. P. Koepchen

Self-organization and Clinical Psychology Empirical Approaches to Synergetics in Psychology
Editors: W. Tschacher, G. Schiepek, E. J. Brunner

Principles of Brain Functioning
A Synergetic Approach to Brain Activity, Behavior and Cognition
By H. Haken

Brain Function and Oscillations
Volume I: Brain Oscillations.
Principles and Approaches
Volume II: Integrative Brain Function.
Neurophysiology and Cognitive Processes
By E. Başar

Christian Uhl (Ed.)

Analysis of Neurophysiological Brain Functioning

With 120 Figures, 35 in Colour

Dr. Christian Uhl
Max-Planck-Institute of Cognitive Neuroscience
Stephanstrasse 1a
D-04103 Leipzig, Germany

Series Editor:

Professor Dr. Dr. h.c.mult. Hermann Haken
Institut für Theoretische Physik und Synergetik der Universität Stuttgart
D-70550 Stuttgart, Germany
and
Center for Complex Systems, Florida Atlantic University
Boca Raton, FL 33431, USA

ISSN 0172-7389

ISBN 3-540-65065-2 Springer-Verlag Berlin Heidelberg New York

Library of Congress Cataloging-in-Publication Data
Analysis of neurophysiological brain functioning / Christian Uhl (ed.). p.cm. -- (Springer series in synergetics, ISSN 0172-7389) Includes bibliographical references. ISBN 3-540-65065-2 (hardcover : alk. paper). 1. Neurophysiology -- Research -- Methodology. 2. Brain -- Mathematical models. 3. Cognitive neuroscience -- methodology. 4. System theory. I. Uhl, Christian, 1965- . II. Series.
[DNLM: 1. Brain -- physiology congresses. WL 300A533 1999] QP356.A69 1999 612.8'2–dc21 DNML/DLC for Library of Congress 98-41692

Typesetting: Data conversion by Katharina Steingraeber, Heidelberg, using a Springer TEX macro package
Cover production: *design & production*, Heidelberg
SPIN: 10677794 55/3144 - 5 4 3 2 1 0 – Printed on acid-free paper

Preface

The analysis of neurophysiological brain functioning is a highly interdisciplinary field of research. In addition to the traditional areas of psychology and neurobiology, various other scientific disciplines, such as physics, mathematics, computer science and engineering are involved. This is on the one hand due to the sophisticated technical equipment and the enormous amount of data "produced" by modern EEG/MEG labs. On the other hand, new concepts stemming from physics and related fields influence the field of Neuroscience. In order to learn about different models of brain functioning, we have organized a workshop at the Max-Planck-Institute of Cognitive Neuroscience, (Leipzig, December 1997) to gather researchers from various fields dealing with human brain functioning. To collect the different perspectives of an understanding of brain functioning and to review models and concepts as well as methods and applications, I have asked the participants to contribute to the present book. To avoid a collection of isolated contributions, some of the contributers have written a joint chapter dealing with a broader topic. As a result the book reviews a wide spectrum of model-based analyses of neurophysiological brain functioning. It addresses students and scientists from all the above mentioned fields.

After the "Introductory Remarks" by D. Yves von Cramon, Director at the Max-Planck-Institute of Cognitive Neuroscience – the host institution of the above mentioned workshop – the present book splits into two parts: Part I dealing with "Concepts & Models" and Part II with "Methods & Applications".

In the first part, physical and physiological models and synergetic concepts are presented: Hermann Haken reviews basic concepts of Synergetics, dealing with the emergence of macroscopic patterns due to microscopic interactions, and its applications to the understanding of brain functioning. In the chapter by Paul Nunez neocortical dynamics are discussed aiming at establishing a foundation for more physiologically based cognitive theories. Scott Kelso and coworkers present in a chapter of three parts their strategy adopted toward the brain–behavior relation: experiments, methods to analyze and visualize the spatio-temporal brain activity and a neurophysiological model to connect neural and behavioral levels of description. Finally in the last chapter of this part, Roger Cerf and coworkers review the concept of low-dimensional cortical activity observed by analyzing scaled structures that emerge in short episodes of EEG signals.

The second part of the present book focuses on methods and applications of analyzing EEG/MEG data sets. In the first chapter of this part, different approaches dealing with spatial analyses are summarized: so-called spatio-temporal dipole models, distributed sources and statistical properties of the spatial distribution of spontaneous EEG/MEG data. The following chapter then deals with temporal aspects: two methods to analyze the time evolution of synchronization processes are presented based on different aspects: coherences and phase locking. Finally, in the last chapter, a new concept of analyzing EEG/MEG data is reviewed with links to both source modeling based on "components" and stochastic descriptions of temporal fluctuations.

I would like to use this opportunity for acknowledgments: My warmest appreciations go to Hermann Haken for his support to publish the present book in *Springer Series in Synergetics* and for the great opportunity to work at his institute in a stimulating and creative atmosphere for nearly 5 years. I am also very grateful to Rudolf Friedrich for valuable support and helpful discussions in Stuttgart and for his still ongoing cooperation. From the Max-Planck-Institute of Cognitive Neuroscience I would like to thank D. Yves von Cramon and Frithjof Kruggel for long and stimulating discussions and their support to organize the workshop. Concerning the organization of the workshop I especially thank Axel Hutt and Marjan Admiraal for their help as well as all colleagues from the Max-Planck-Institute.

Finally I would like to thank the staff of Springer-Verlag, in particular Prof. W. Beiglböck, Mrs. Sabine Lehr, Mrs. Edwina Pfendbach and Mrs. Brigitte Reichel–Mayer for their cooperation.

Leipzig, July 1998 *Christian Uhl*

Table of Contents

Part II Methods & Applications

Model-Based Analysis of Neurophysiological Brain Function: Introductory Remarks

D. Yves von Cramon

Max-Planck-Institute of Cognitive Neuroscience, Stephanstrasse 1a,
D-04103 Leipzig, Germany

The analysis of neurophysiological brain functioning has become a highly interdisciplinary task: A large number of research professionals coming from outside the field of neuroscience (physicists, engineers, computer scientists, etc.) have to get involved, in order to design and perform accurate experiments, to develop and maintain sophisticated technical equipment, and, last but not least, to analyze and visualize the recorded signals. Since successful EEG and MEG studies, especially in the cognitive domain, require such a manifold expertise, as well as a significant financial investment, it is clearly necessary to take care to identify the most appropriate concepts, models and tools.

The last few decades, among them the "Decade of the Brain", have provided us with innumerable anatomical, physiological, and chemical data relating to brain function at a microscopic (submillimeter to millimeter) scale. However, despite the tremendous progress basic neurosciences have made, disappointingly little is known of the macroscopic functionality of the cortex. That is to say, the role in macroscopic neural function played by larger cortical compartments has remained more or less a mystery. We have and will continue to learn many important details from examinations at the microscopic level. However, I fear that at this scale we will not be able to obtain significant information relating to the neural correlates which are of real interest to the cognitive neuroscientists. That is for instance, data which will enable us to identify the functional components of working memory or the mechanisms involved in language comprehension. These primarily cortical functions depend upon highly integrated complex systems which are far more readily investigated at the macroscopic level. It can, however, be assumed that EEG/MEG measures at the macroscopic scale are influenced by the microscopic processes in a "bottom–up" manner. This means, in other words, that large-scale models in humans may be improved by incorporating knowledge relating to microscopical function.

The spatial resolution of conventional EEG differentiates activity between larger cortical areas with a linear scale of about 5 cm. More sophisticated EEG and MEG methods, however, seem able to obtain 1 cm resolution, which is close to the theoretical limits set by the physical separation of sensors and brain current sources. These high-resolution methods may enable us to separate functionally and morphologically meaningful subunits of those larger cortical areas.

From the point of view of neuroanatomy, an intermediate spatial scale ranging from a few millimeters up to 1–2 centimeters needs further clarification. Anatomists have identified, in hierarchical order, a microscopic world of 'minicolumns', 'cortico-cortical columns' and 'macrocolumns' – all structural units in the submillimeter and millimeter range. Somewhat surprisingly, the next larger anatomically defined structures are cortical areas, defined for instance by cytoarchitectonic features such as the classical Brodmann areas or by patterns of intracortical myelinated fibers or intracortical vasculature.

The intermediate dimension of brain organization which, in analogy to synergetics, could be denoted as the 'mesoscopic' scale, seems to correspond with what is rather vaguely conceptualized as neuronal cell assemblies binding together a multitude of functionally associated macrocolumns. These mesoscopic subunits, although not yet properly defined, do indeed seem to exist and can be described in terms of the neural correlates of 'activation areas' identified in neuroimaging (PET, fMRI) studies. These activation areas are obviously larger than single macrocolumns but in most cases substantially smaller than Brodmann areas. It is most likely that the so-called 'convergence zones', which exist as synaptic patterns within multi-layered neuron ensembles in association cortices, correspond to the mesoscopic brain structures.

I propose that a Brodmann area is composed of several mesoscopic areas (including convergence zones) representing function units or 'building stones' for a diversity of cognitive (and other cerebral) functions. It seems plausible to assume that various activation areas within a large cortical area of the Brodmann type constitute a certain class or category of cortical functions whatever dimensions or features they may have in common.

Simultaneous measurements of EEG and fMRI, which now have become technically feasible, will allow us to study the electrophysiological (neural) dimension of haemodynamically-defined mesoscopic cortical areas. They may provide us with completely new insights into the nature of spatio-temporal cooperation between cortical modules and their behavioral correlates at the mesoscopic level.

I would like to touch briefly on a second issue which will play a major role in this book – that is, the debate concerning the validity of the various mapping techniques, e.g. amplitude mapping (the most widespread method so far) and the mapping of spectral parameters, such as spectral power and coherence mappings. In the case of amplitude mapping, averaged event-related potentials or fields represent the basis for either statistical procedures with latency, amplitude and topography as considered factors, or for source localization techniques and dynamical systems based on so-called spatio-temporal models. Our own work here in Leipzig has made use of both techniques. Although not at all an expert in this field of research, I have, following many discussions, the feeling that the theories of 'dynamical and self-organizing complex systems' may enlarge our stock of knowledge on brain–behaviour relations and yield innovative experimental paradigms particularly to study cognitive processes.

Power and coherence analysis, on the other hand, take raw EEG and MEG data of event-related experiments. Averaging is performed of the obtained spectral parameters. Again, these parameters yield a basis for both statistical tests and mapping techniques. Clearly, coherence analysis provides us with important information concerning brain function, complementing data obtained by power spectral analysis, since a change of power does not necessarily mean that also the functional relationship or coupling between two signals changes. It has been argued that coherence between electrophysiological signals from different parts of the cortex may depend on structural connectivities. In anatomical terms, EEG coherence is said to be higher between cortical areas which are interconnected by dense fiber projections. Thus, coherence analysis might become a tool to characterise anatomical connectivities between activation areas at the mesoscopic scale.

Discussions in our own group have led to the conclusion that a dynamical systems approach might be used to combine the mapping of both amplitudes and spectral parameters by consideration of the so-called Fokker–Planck and comparable equations. The 'drift term', that is the deterministic component of the Fokker-Planck equation, represents a spatio-temporal model of averaged ERP or ERF data sets, whereas the 'diffusion term' in this equation deals with spectral parameters.

We all are looking forward to this comprehensive survey of the different approaches to the analysis of EEG and MEG data dealing with the various themes of source localisation, serial versus parallel processing, synchronisation phenomena, and spatio-temporal models.

Part I

Models & Concepts

What Can Synergetics Contribute to the Understanding of Brain Functioning?

Hermann Haken

Institute for Theoretical Physics and Synergetics,
Pfaffenwaldring 57/4, D-70550 Stuttgart, Germany

1 The Need to Model Brain Functioning

At present we witness the growth of the number of institutions that deal with EEG and MEG measurements. This chapter, and moreover the whole book is concerned with the analysis of EEG and MEG data. In these experiments, electric and/or magnetic fields are measured by means of arrays of sensors. The measurements consist in recording time-series of the corresponding fields. The spatial resolution is given by the separation of the sensors on the scalp. Because the temporal resolution is high and modern equipment uses numerous sensors, an enormous amount of data becomes available. To reduce the amount of information, methods such as filtering of frequency bands and/or data averaging are frequently used. More recently it has been questioned (e.g. by P. Tass, cf. his chapter in this book) whether averaging is permissible, because important correlations might be obscured. This problem probably does not occur, however, if external pacemakers are used, as in the Kelso experiments (cf. his contribution to this book).

At any rate, to return to the main line of thought, we are concerned with physical quantities that are macroscopic manifestations of the functioning of the brain in the sense of physics. Myriads of microscopic electro-chemical processes of neurons must cooperate to produce observed electric and magnetic fields. Therefore one important field of research will it be to link the physico-chemical properties of neurons with the macroscopic fields. Hereby the cooperation of neurons plays a crucial role.

In addition to macroscopic electric and magnetic fields, there are, of course, still other macroscopic manifestations of brain activity that may concur with these fields. Important examples are movements of limbs, movement coordination, or locomotion. Further examples are provided by different kinds of sensori-motor coordination.

All these events can be physically measured and thus objectively observed. In a way, the brain could be treated as a black box - were it not for the electric and magnetic fields that provide us with some insight into its physical functioning.

Our next example brings us to more fundamental questions, namely epileptic seizures. They become manifest by specific symptoms, but are also accompanied by a specific subjective experience.

A still more pronounced example of *internal* brain states is perception, e.g. visual perception. A certain bridge between internal and external can be built by means of psycho-physics, for instance by use of ambiguous figures. The goal of neuroscience is, however, far more ambitious. Here a typical question is: How are mental states, such as thoughts, feelings, etc. connected with measured electric and magnetic fields? This requires the discussion of such delicate questions that concern introspection and, perhaps, even consciousness, or the relation between subjective–objective, or inside–outside. Quite evidently, we are dragged – nolens volens – into a discussion of the philosophical aspects of the mind-body problem. For our present purpose it will be sufficient, however, to adopt the idea of a correlation between mental states (internal psychological experience) and physical brain states. Thus we do not discuss the problem of causation (cf. in this context: H. Haken [1]). In this sense, EEGs and MEGs are considered as indicators, for instance of healthy or pathological states. Or in other words, EEGs and MEGs are used to make things "objective".

As it may transpire from what I said above, at least two fundamental problems in the context of EEGs and MEGs arise:

1. How do macroscopic physical properties, such as electric and magnetic fields, arise through the cooperation of many neurons? Thus we have to study the relationship between the micro- and macro-level in the brain as a *physical* system.
2. How can we represent the relation between physical, microscopic or macroscopic brain states and psychological states?

In both cases 1 and 2 we are dealing with one common phenomenon, namely the emergence of new qualities, because of the cooperation of many individual parts: the neurons (and perhaps glia-cells). At this instance, it may be useful to mention the main tools of synergetics to answer the questions 1 and 2. These will be the concepts of order parameters and the slaving principle, which we shall discuss below. This will allow us to extract and simultaneously compress the information inherent in EEGs and MEGs to few "descriptors" of brain states. In addition, the description of mental states and their changes become accessible in a number of cases. I hope that the usefulness of synergetics will also transpire from some of the following articles.

2 What Is Synergetics About?

The interdisciplinary field of synergetics [2], [3] (in English *science of cooperation*) deals with complex systems. Such systems are composed of many elements that may produce macroscopic structures or functions by means of their cooperation. Since these functions or structures are not imposed from the outside, we deal with the phenomenon of *self-organization*. Synergetics asks whether there are general principles for self-organization irrespective of the detailed nature of the individual elements or parts. In the past, such general principles could, indeed, be found, provided we focus our attention on qualitative changes on macroscopic scales, concepts that will be elucidated in more detail below. In my article I will not be so much concerned with the unifying mathematical approach that we developed over the past decades (cf. [2], [3]), but rather with basic concepts. Readers who are interested in more details are referred to a number of volumes of the Springer Series in Synergetics (see the references). A few general remarks may be in order. The kind of macroscopic structures we have in mind may be exemplified by an example from chemistry. When we bring chemical reactants in a test tube together, they may spontaneously form spatial structures, such as stripe patterns, they may form temporal structures, such as oscillations continuously going on between colours, such as red and blue, or they may form rotating spirals (Fig. 1).

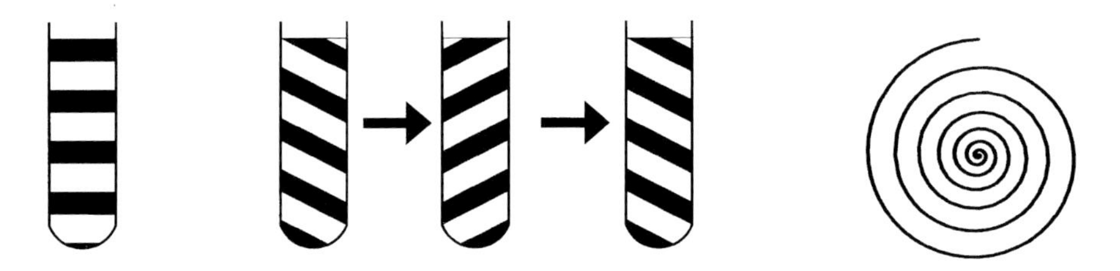

Fig. 1. Examples of macroscopic pattern formation in chemical reactions. From *left* to *right*: Spatial pattern in form of stripes, oscillations of molecular concentrations, rotating spirals

At more complex levels, such as in the brain, we may think of movement coordination as the formation of functional structures. An important issue in all these problems is the relationship between the macroscopic and microscopic levels. Such a distinction is only relative and can be made on all levels, for instance in a human being (compare Table 1).

Table 1

"macroscopic"	...	"microscopic"
molecules	...	atoms
cell	...	membranes, microtubuli, molecules
organs	...	cells
brain	mesoscopic	neurons, glia-cells
human body	...	organs, cells

There may be a spatial separation between the individual parts that is more or less static. In the present article we are interested in functional relationships between elements at the microscopic level and in particular in the resulting structures at the macroscopic level.

3 The Bottom-Up Approach or a Lesson from Physics

In many fields of science an attempt is made to understand the properties of a macroscopic system as a result of the properties of its constituents. There are fundamental limitations to this approach, however, as is witnessed by the field of physics that is in many ways far simpler than any biological science. As is well known, a gas is composed of individual molecules. But a sound wave cannot be respresented by means of the concepts of molecules alone. A sound wave is characterized by its wave length and frequency, concepts that are alien to the individual molecule. Similarly atoms may form crystals, but the occurrence of ferromagnetism and superconductivity can not be directly derived from the properties of the individual atoms. As is well known in physics, new concepts are needed that are appropriate for the macroscopic level. We strongly suspect that the same situation holds for biological systems. Indeed, any macroscopic property of a biological system, say of a human being, requires the cooperation of myriads of cells. This holds for breathing, blood circulation, heart beat, but still more for perception, thinking, a.s.o. Therefore two essential problems arise (Fig. 2):

1. What are these concepts concerning the macroscopic level?
2. How are they connected with those of the microscopic level?

Here we have applications to the brain sciences in mind.

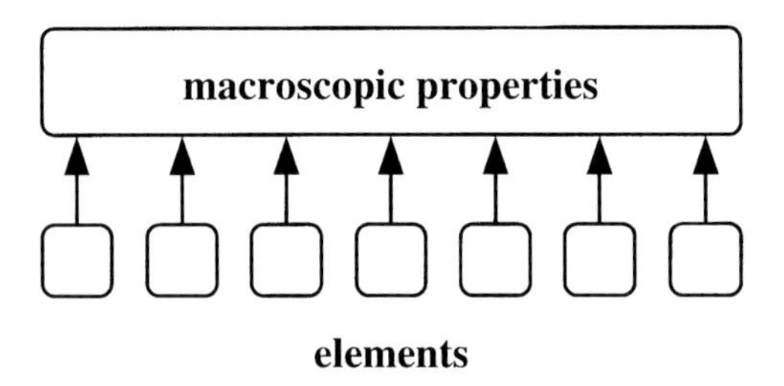

Fig. 2. The macroscopic and microscopic levels

4 The Approaches of Synergetics – 1st Approach: Bottom-Up

In this approach we assume that the properties of the individual elements or subsystems are known. By means of their interaction and of the impact of the environment on the system, in a number of cases the system forms macroscopic patterns. The impact of the environment on the system is described by control parameters. In quite a number of cases it is known that by change of control parameters an instability is caused. Such an instability means a switch between macroscopic states, for instance between a homogeneous chemical pattern to a stripe pattern. In this case the control parameter is a chemical concentration. In neuro-sciences such control parameters can be coffein, haloperidol, acoustic, optical, tactile signals, a.s.o. In generalization of the control parameter concept we may also include concentrations of neurotransmitters, hormons, etc. as such quantities, though they are not controlled from the outside, but produced by the body itself.

Synergetics considers situations close to instabilities. There it can be shown, on purely mathematical grounds, that the following happens: The many elements produce one or several order parameters. These order parameters act on the individual parts of the system and generate ordered states. This latter feature is called the slaving principle of synergetics (Fig. 3).

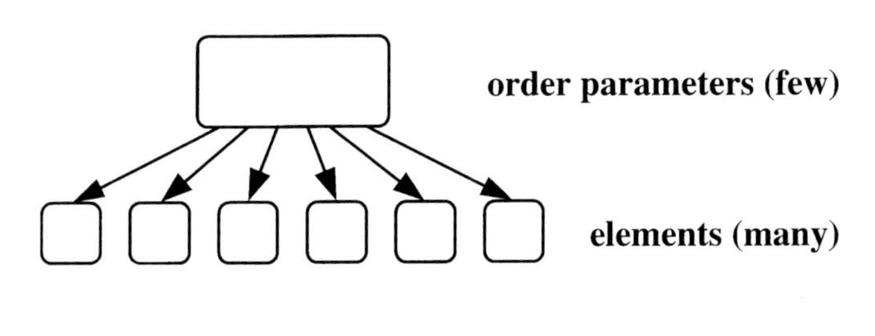

Fig. 3. The slaving principle

Because the elements may produce order parameters that in turn act on the individual elements, we speak of circular causality (Fig. 4). Because of this property, we may ignore the individual parts and describe the behavior of the system in terms of the order parameters alone (Fig. 5). Because the number of order parameters is much smaller than that of the elements, we achieve an enormous information or data compression. On the other hand we may eliminate the order parameters and deal with individual parts alone. Then we observe that they show certain kinds of ordered behavior. The procedure we just prescribed requires that we know the equations for the behavior of the individual parts. That is true in many cases in physics but hardly in biological cases. But we can capitalize on these general mathematical results which lead to the second approach of synergetics.

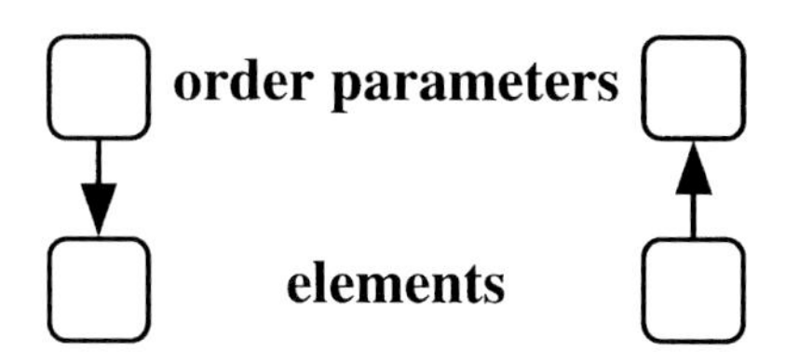

Fig. 4. Circular causality: the elements (parts) generate order parameters, that in turn enslave the parts

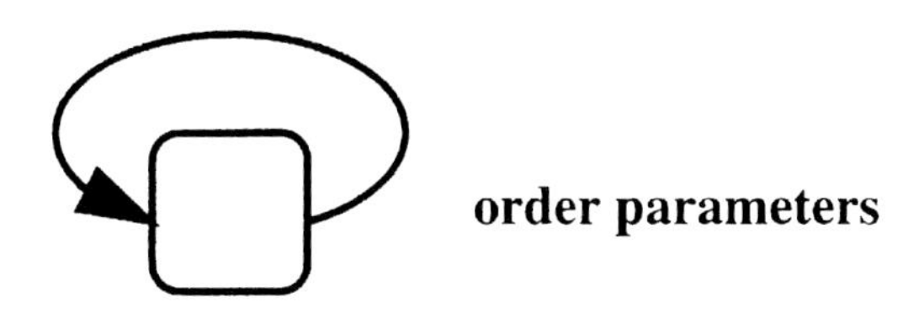

Fig. 5. The order parameters interact among each other

5 2nd Approach: Phenomenological

In this approach we start from the fact that by a change of control parameters qualitative changes at the macroscopic level can be caused and we expect that the order parameters become visible. Thus, in order to model such transitions, we have to identify the appropriate control and order parameters. Then in a second step, we may guess order parameter equations. In a number of cases, the behavior of the order parameters can be easily visualized. For instance, if there is only one order parameter, it behaves like a ball in a landscape (Fig. 6). In a certain control parameter range there is only one equilibrium point; once the order parameter has been disturbed, it will roll back to the bottom of the valley. At a critical value of the control parameter, the valley becomes very flat. As is well established in synergetics, the ball is always subject to kicks that act on it all the time. If the valley is very flat, these kicks can become very efficient and we observe the phenomenon of *critical fluctuations* and *critical slowing down*. When the control parameter is further changed, the original landscape is replaced by one with two minima, which in a number of cases may be symmetric. Since the ball can sit only in one of the two valleys so that the symmetry is no more kept, we speak of symmetry breaking.

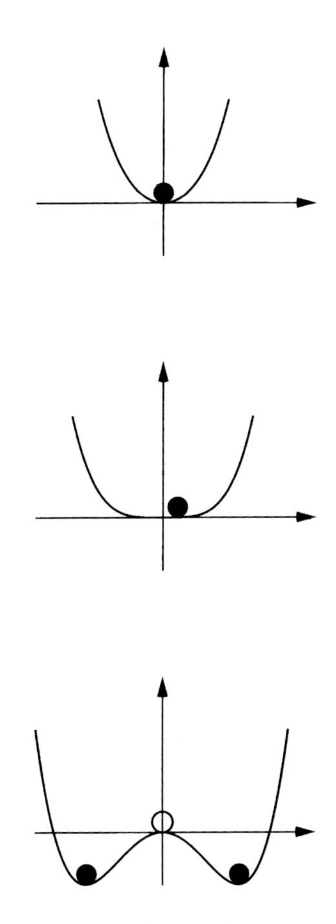

Fig. 6. Visualization of the behavior of one order parameter by the motion of a ball in a hilly landscape for different control parameter values. *Upper part*: Below a critical control parameter value, there is only one valley and thus only one stable state of the order parameter. *Middle part*: At the critical value, i.e. at the instability, the valley becomes very flat: critical fluctuations and critical slowing down occur. *Lower part*: Above the critical value, two valleys appear. The system becomes bistable

These concepts could be applied to the modelling of the by now famous Kelso experiment in which he asked test persons to move their fingers in parallel. At a certain elevated frequency of the finger movement there was a sudden change to an antiparallel, i.e. symmetric movement (Fig. 7). In this case, the control parameter is the frequency of the finger movement, whereas the relative phase between the two fingers can be identified with the order parameter. An order parameter landscape that changes during the experiment with increasing frequency is shown in Fig. 8. In the middle part of the figure again the upper valley becomes very flat and leads to the phenomena of critical fluctuations and slowing down that were indeed observed in the experiments. As it has turned out in the meantime, this so-called HKB-model is of universal nature and can describe the transition between various kinds of movements of fingers, legs, and so on.

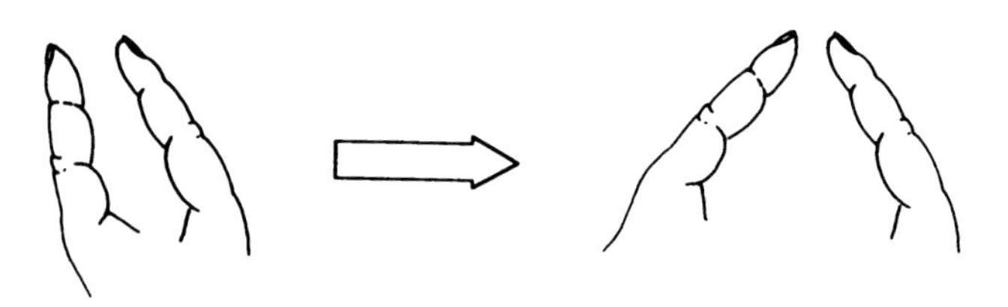

Fig. 7. Spontaneous change of finger coordination in the Kelso experiment

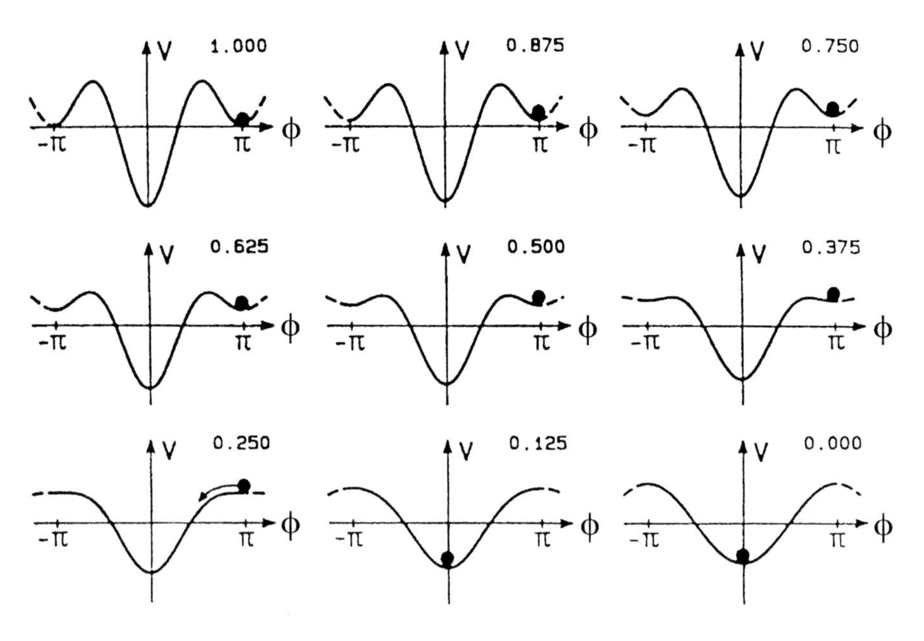

Fig. 8. The order parameter landscapes of the Haken–Kelso–Bunz model. With increasing frequency of the finger movements, first the upper valley becomes very flat and eventually vanishes

6 An Application of the Second Approach to Perception

In order to convince the reader of the usefulness of the order parameter concept, we quote a few simple examples from perception. To this end we identify the individual parts or elements with neurons and the order parameters with percepts (Fig. 9). Figure 10 illustrates the phenomenon of bistability. We may either recognize a face or fruits and vegetables. Figures 11 and 12 illustrate the effect of hysteresis in visual perception. Figure 13 finally illustrates the effect of oscillations between order parameters, in the present case between vase and faces. These phenomena are, of course, well-known in psychophysics and have been dealt with by Gestalt theory. Synergetics allows us, however, to shed new light on these phenomena at the quantitative level. Indeed, it has become possible to model oscillatory effects by means of order parameter equations, which contain parameters to which a specific psychological meaning can be attached and which allow us to make detailed comparisons with experimental results. I hope that these examples may provide the reader with some hints at the usefulness of the order parameter concept and give him or her a feeling that at the level of order parameters a quantitative modelling becomes possible.

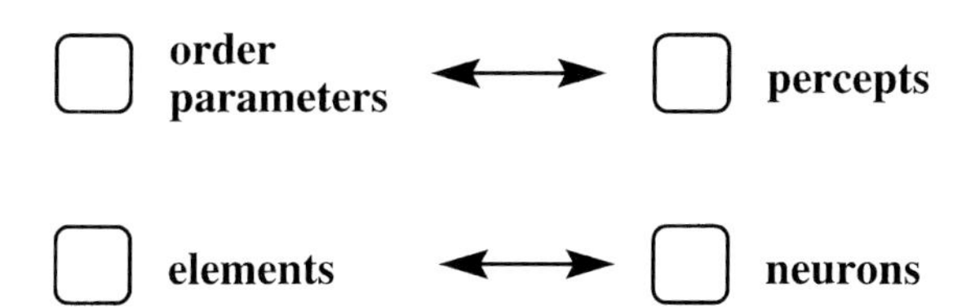

Fig. 9. The relationship between elements (parts) and order parameters is interpreted as one between neurons and percepts

Fig. 10. Bistability in perception. Painting by Arcimboldo

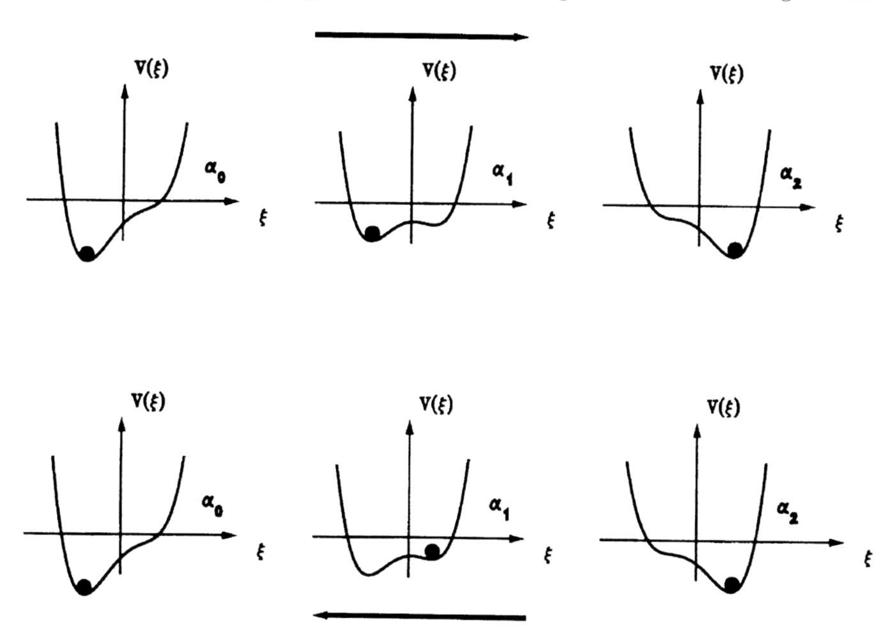

Fig. 11. The effect of hysteresis. The value of an order parameter is visualized by the position of a ball in a hilly landscape. Upper row, from left to right: When we change a control parameter, the landscape may be deformed in the way indicated. Evidently the ball rolls down to the right valley only in the third landscape. Lower row: When we now change the control parameter in the opposite direction, we recover the former landscapes. But in spite of the same landscape in the middle in both rows, the ball occupies two different positions. Thus the value of the order parameter depends on the history

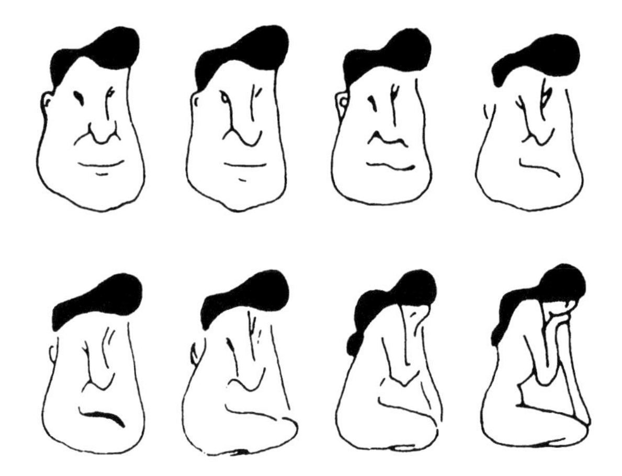

Fig. 12. Consider the first row from left to right and then the second row from left to right. You will observe a switch between two percepts in the second row. Now start in the second row from right to left and then proceed in the first row from right to left. The switch of your perception occurs in the first row

Fig. 13. Oscillations between two percepts: You may recognize a vase in the middle, then two black faces, then the vase again, and so on

7 3rd Approach: Top-Down

In this section we shall be concerned with the derivation of order parameters and enslaved modes from experimental data. These data are, for instance, electric or magnetic fields measured at a space point $\mathbf{x}$ on the scull and at time t. After some filtering, averaging etc. they are available as some function $q(\mathbf{x}, t)$, that we call the state-vector. Our starting point is the decomposition of this state vector q into a superposition of spatial modes v_j, w_k in the form

$$q(\mathbf{x}, t) = \sum_j \xi_j(t) v_j(\mathbf{x}) + \sum_k \eta_k(t) w_k(\mathbf{x}). \tag{1}$$

ξ_j are the time-dependent order parameters, η_k are time-dependent enslaved mode amplitudes. In order to derive $\xi_j, \eta_k, v_j(\mathbf{x}), w_k(\mathbf{x})$ in the context of EEG and MEG analysis two methods have been applied, namely

1) the Karhunen-Loève method (also called *principal component decomposition*) (cf. [1]), which assumes the state vector decomposition in the form

$$q(\mathbf{x}, t) = \sum_j \xi_j(t) v_j(\mathbf{x}), \tag{2}$$

and where the v_j are determined by a statistical analysis. This analysis implies that the v_j are orthogonal on each other. A 2nd approach is

2) the Uhl–Friedrich–Haken method [4], [5], which uses basic results on order parameters, enslaved modes, etc. as developed by synergetics. The basic idea is to minimize expressions that are essentially of the form

$$\int \mathrm{d}t \left(\int q(\mathbf{x}, t) v_j^+(\mathbf{x}) \mathrm{d}^3 x - \xi_j \right)^2 , \tag{3}$$

where the mode vectors v_j^+, that are adjoint to v_j, must be determined appropriately. This approach was applied to an EEG analysis by Uhl and Friedrich [6]. A typical result of this analysis is outlined in Sect. 8 below.

The decompositions (1) or (2) allow us to establish relationships between space-time correlation functions and correlation functions between order parameters or enslaved mode amplitudes. The correlation function between measured quantities, such as the electric potentials or magnetic field strengths at positions $\mathbf{x}, \mathbf{x}'$ and times $t, t-\tau$ is defined by

$$K(\mathbf{x}, t; \mathbf{x}', t-\tau) = \int q(\mathbf{x}, t) q(\mathbf{x}', t-\tau) \mathrm{d}t. \tag{4}$$

Using the decomposition (3), we may express (4) by

$$K(\mathbf{x}, t; \mathbf{x}', t-\tau) = \sum_j \sum_k v_j(\mathbf{x}) v_k(\mathbf{x}') \int \xi_j(t) \xi_k(t-\tau) \mathrm{d}t. \tag{5}$$

Multiplying K by the adjoint vectors and integrating over space, we obtain

$$\int K(\mathbf{x},t;\mathbf{x}',t-\tau)v_j^+(\mathbf{x})v_k^+(\mathbf{x}')\mathrm{d}V\mathrm{d}V' = \int \xi_j(t)\xi_k(t-\tau)\mathrm{d}t = C_{jk}(t,t-\tau) \tag{6}$$

so that the space-time correlation function (4) can be transformed into a temporal correlation function between order parameters or enslaved mode amplitudes. While the listing of the numerical values of the correlation functions (4) requires a large amount of data, in general because of the order parameter concepts only few correlation functions C need to be taken into account.

8 Example of EEG Analysis

In this section I give a brief sketch of the EEG analysis of petit-mal epilepsy by the use of concepts of synergetics. This work was performed by Friedrich and Uhl [6]. Let us consider Fig. 14 that shows the electric potentials measured on the scalp. The position of each box corresponds to that of the electrode on the scalp. The abscissa is time, whereas the ordinate in each box is the measured electric potential against the reference electrode.

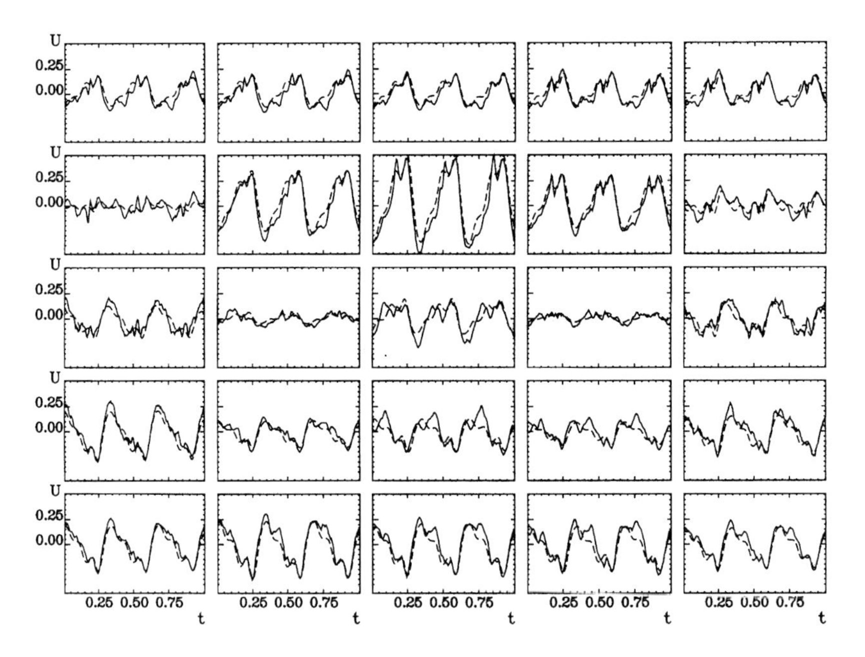

Fig. 14. Electric potentials measured on the scalp

These plots allow us to reconstruct the spatio-temporal activity pattern, which is shown in Fig. 15. Each circle corresponds to the activity at a specific instant of time. By use of the decomposition (1), Fig. 16 shows the leading spatial modes on the left-hand side. Their amplitudes are the order parameters and are plotted on the right-hand side against time. If the three amplitudes are plotted in a single three-dimensional space, a trajectory emerges that, as was shown by Friedrich and Uhl, corresponds to so-called Shilnikov chaos (Fig. 17). The modes shown in Fig. 16 cover by means of their superposition 97% of the whole electric activity measured. In this way one can say that the brain dynamics of petit-mal epilepsy is of low dimension and the order parameters show a specific kind of dynamics.

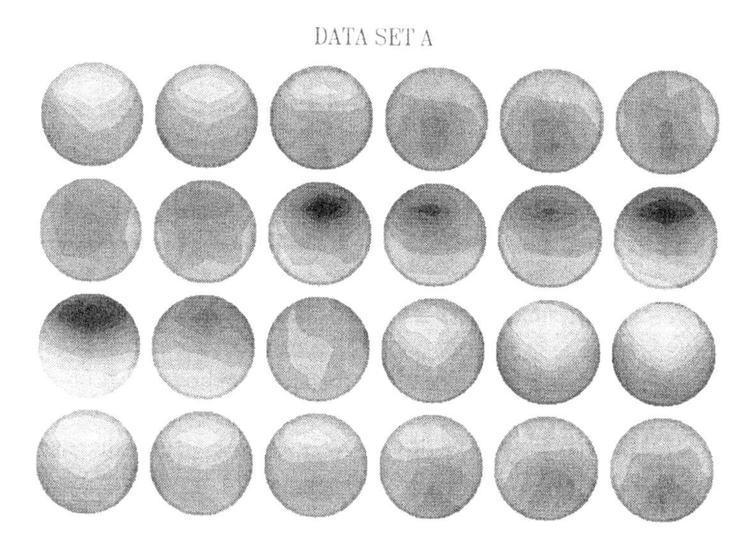

Fig. 15. Reconstruction of the spatio-temporal electric potentials by means of Fig. 14 (after Friedrich and Uhl [6])

9 Towards a Microscopic Approach

After the specific example of the foregoing section, let us discuss some general properties of the top-down and bottom-up approaches. In the top-down approach we start from macroscopic properties and want to derive microscopic models. As it turns out in many cases, the macroscopic properties alone do not allow us to derive microscopic models uniquely. But they may serve as a guideline. As it also turns out, microscopic models are strongly parameter dependent, i.e. the solutions vary greatly because of different parameters that

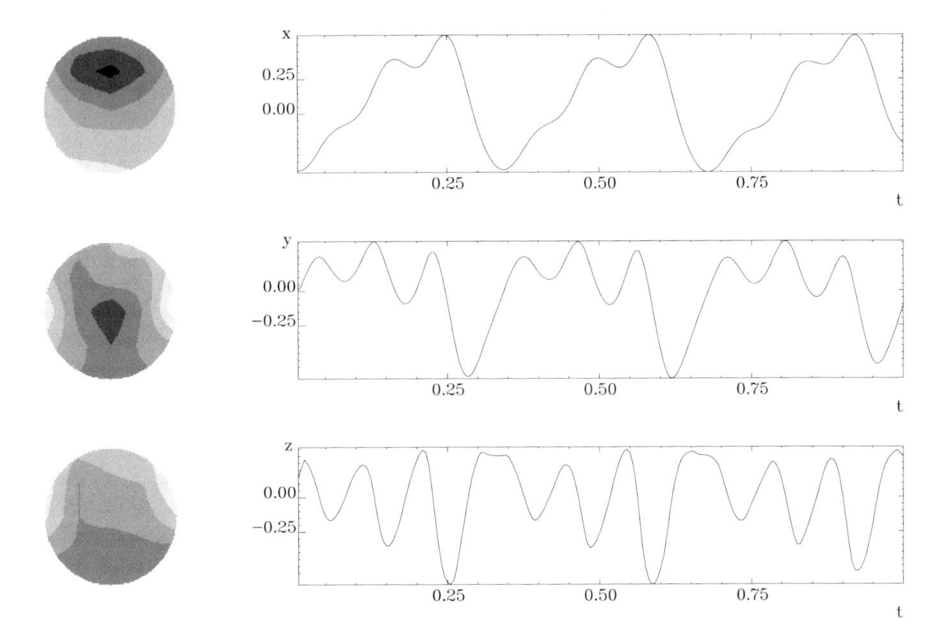

Fig. 16. Mode decomposition of the spatio-temporal patterns of Fig. 15. On the left-hand side, the leading spatial modes are shown. On the right-hand side the temporal evolution of the corresponding order parameters is shown

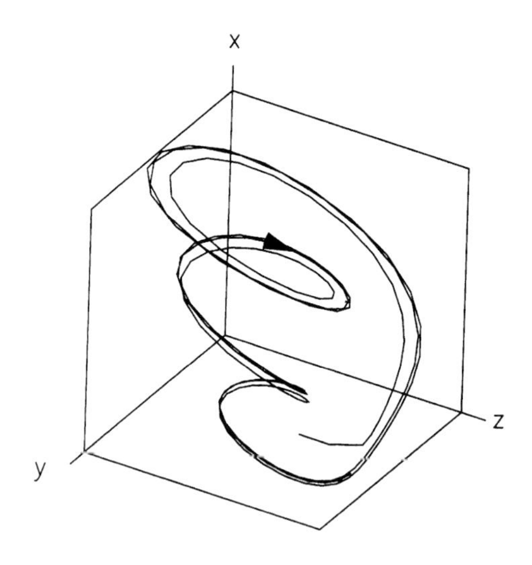

Fig. 17. The trajectory of the order parameters shows some kind of Shilnikov chaos (after Friedrich and Uhl)

occur in microscopic models. Therefore, an ideal method consists in a combination of the top-down and bottom-up approaches. Such a program has been followed up in the analysis of MEG measurements according to Table 2 (cf. also Sect. 12).

Table 2

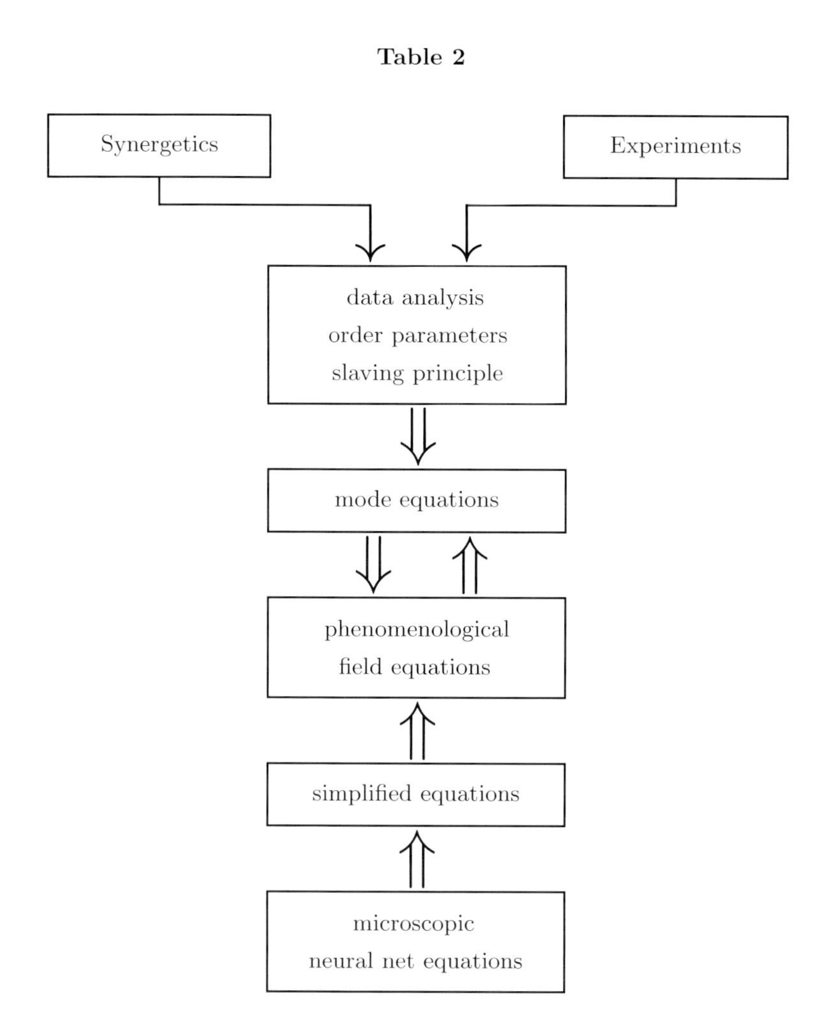

With respect to microscopic brain theories that are based on properties of neurons, we may distinguish between two lines of approach. The first line is based on averaged spike intensities, i.e. individual spikes with phase relations are neglected. The corresponding equations will be treated below in Sects. 10 and 11. The other line is based on spike trains, where phases and phase locking are taken into account. Corresponding theories were developed by Kuramoto [7], Tass [8], and others. Experimental evidence of phase locking is provided by the experiments of Singer, Gray and their co-workers [9] and by Eckhorn and co-workers [10]. Later in my article I will propose a new theoretical approach to the treatment of spike trains.

10 Microscopic Description of the Neural Network of the Brain. Phase-Averaged Equations

In this section I wish to present a microscopic description, which elucidates various aspects of neural networks. Actually, the term "microscopic" needs some discussion. In the context of my article, I consider the dendrites as well as the somas and axons as the elementary parts. In this sense they are the micro-elements. There are, of course, still more microscopic levels below that level, for instance the detailed structure and function of dendrites and axons, or the dynamics of vesicles, or the dynamics of receptors. These fine structures are not the target of our models. Furthermore, there is a somewhat higher level of consideration, where we consider populations of neurons of similar behavior. Such point of view was adopted by Beurle [11], Griffith [12], and others and may serve as an interpretation of the present approach when I shall speak below of "averages". At least in the latter cases, I should speak of "mesoscopic" rather than "microscopic".

Our general approach will provide us with a number of special cases known in the literature, such as the Wilson-Cowan model [13], [14], the Nunez model [15], [16], [17] and the McCulloch-Pitts model [18]. This general model will, I believe, shed new light on the interplay between various brain functions at various space and time scales. It will show that the brain can act, on the one hand, as a parallel computer, but that it may, on the other hand, form spatio-temporal patterns that are directly related to EEG and MEG measurements. Our approach starts from well-known physiological data and their description, in particular by the excellent tutorial of Freeman [19], [20]. To this end consider Fig. 18, which we read from left to right. We start from inputs from other neurons that are converted at the synapses into electric currents in the dendrites. The electric currents are summed up (or are integrated) in the soma and converted into an action potential that leads to pulses running along the axon. The pulses then serve as inputs to other neurons so that the picture has to be continued periodically to the right of Fig. 18. We refer to the literature with respect to time constants, propagation velocities of nerve pulses, and so on [21], [22], [23], [16], [17]. These data lead us to consider long range excitation and short range inhibition. There are two characteristic state variables of the neurons we shall consider:

1. Dendritic currents, whose size will be described by so-called wave amplitudes.
2. The axonal pulse frequency.

We shall consider two conversion processes that are visible from Fig. 18 and are now split up into two processes according to Figs. 19 and 20.

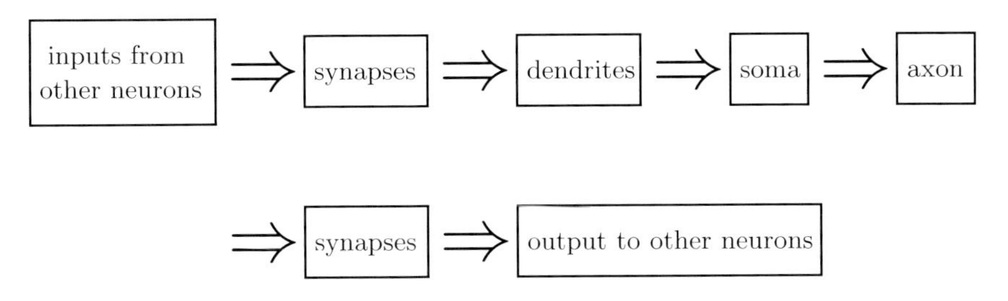

Fig. 18. Block diagram of neuronal connections

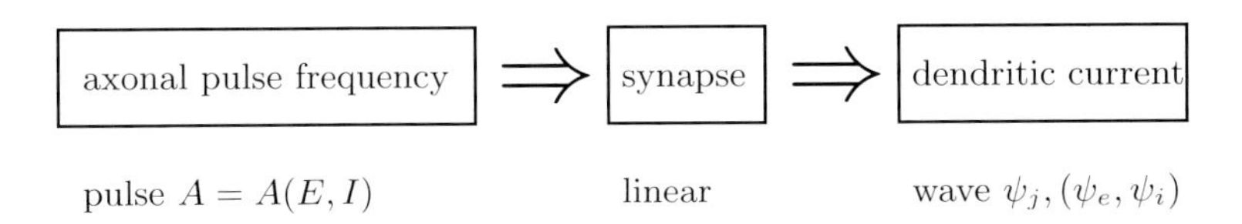

Fig. 19. Conversion of axonal pulse into dendritic wave

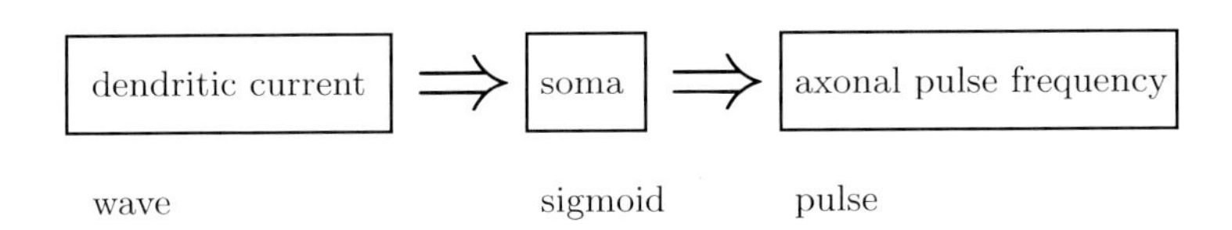

Fig. 20. Conversion of dendritic wave into axonal pulse

Figure 19 shows the conversion of axonal pulses by synapses into dendritic currents. In the following approximation of our model we shall assume that this transformation is linear which is certainly a rough approximation only. The second conversion takes place at the soma, where the dendritic currents are summed up and converted into the axonal pulse frequency by means of a sigmoid function. The wave amplitudes of the dendritic currents will be denoted by ψ_e, ψ_i, corresponding to excitatory and inhibitory currents, respectively, whereas the axonal pulse rates will be denoted by E, I, corresponding to excitatory and inhibitory neurons, respectively. A sigmoid function is shown in Fig. 21. The conversion of pulse into wave is assumed to be practically linear as mentioned above.

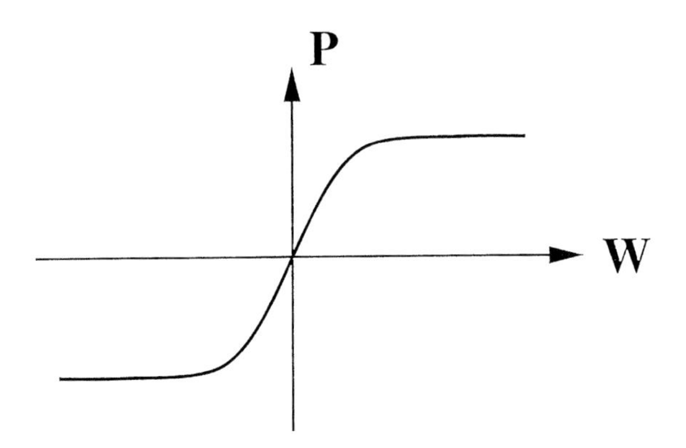

Fig. 21. Sigmoid function

In order to allow a general description that covers a number of special cases, we shall adopt a rather general notation. We denote the axonal pulse rates E, I at time t by

$$A_{jk}(t), \tag{7}$$

where the first index j refers to excitatory or inhibitory, and the second index enumerates the neurons. Later we will adopt a continuous space variable, and we shall then make the identifications

$$\begin{array}{llll} j = e & \text{(excitatory)}, & k = x & : \quad E(x,t) \\ j = i & \text{(inhibitory)}, & k = x & : \quad I(x,t). \end{array} \tag{8}$$

The dendritic wave amplitudes will be denoted by

$$\psi_{\ell,m}(t), \tag{9}$$

where in analogy to (8) we can specify the indices ℓ, m according to

$$\begin{array}{llll} \ell = e & \text{(excitatory)}, & m = x & : \quad \psi_e(x,t) \\ \ell = i & \text{(inhibitory)}, & m = x & : \quad \psi_i(x,t). \end{array} \tag{10}$$

The conversion of the axonal pulse rates into dendritic current amplitudes is described by

$$\psi_{\ell,m}(t) = \sum_k f_{\ell;m,k} S_\ell \left(A_{\ell,k} \left(t - t_{\ell,mk} \right) \right) + F_{\ell,m}(t). \tag{11}$$

$f_{\ell;m,k}$ describes the synaptic strength between neurons m and k, where we assume that inhibitory (excitatory) neurons have only inhibitory (excitatory) connections. $t_{\ell,mk}$ describes the time delay according to final propagation velocities v_e, v_i. S_ℓ is a sigmoid function, $F_{\ell,m}$ is a fluctuating force that we will discuss in Sect. 14. The conversion from dendritic amplitudes to axonal firing rates is described by the equations

$$\begin{aligned} A_{jk}(t) = & \sum_m \hat{f}_{j;k,m} \hat{S}_j \Big(\psi_{e,m} \left(t - t_{e,km} \right) - \psi_{i,m} \left(t - t_{i,km} \right) - \Theta_{j,m} \\ & + p_{j,m} \left(t - t_{km} \right) \Big) + \hat{F}_{j,k}(t), \end{aligned} \tag{12}$$

where the notation is similar to that of (11). $\Theta_{j,m}$ is a threshold, $p_{j,m}$ is an external signal of kind j at neuron m and $\hat{F}$ are again fluctuating forces. In the following we shall make two basic assumptions, namely the linearity of the conversion from axonal pulse rates to dendritic currents, i.e. we assume

$$S_\ell(Q) \approx a_\ell Q, \tag{13}$$

and we assume that the conversion from dendritic currents to axonal pulse rates is local so that

$$A_{jk}(t) = \hat{S}_j \left(\psi_{\ell,k}(t) - \psi_{i,k}(t) + p_{j,k}(t) \right) + \hat{F}_{j,k}(t). \tag{14}$$

Using the approximations (13) and (14), we can simply eliminate either A from (11) and (14) or the ψs. When we eliminate the ψs, we obtain

$$\begin{aligned} A_{jk}(t) = \ & \hat{S}_j \Big(\sum_n f_{e;k,n} a_e A_{e,n} \left(t - t_{e,nk} \right) - \sum_n f_{i;k,n} a_i A_{i,n} \left(t - t_{i,nk} \right) \\ & - \Theta_{j,k} + p_{j,k}(t) + F_{e,k}(t) - F_{i,k}(t) \Big) + \hat{F}_{j,k}(t). \end{aligned} \tag{15}$$

When we ignore the fluctuating forces and use an ensemble average that allows us to introduce connectivities that are smooth functions of the space variable, (15) are essentially the Wilson–Cowan [13], [14] equations provided the membrane decay time is short compared to the propagation times. We note that the Wilson–Cowan model has found applications to the modelling of hallucinations as was done by Cowan and Ermentrout [24], [25] and more recently in a very extensive study by Tass [26], [27].

The formulation (15) allows us immediately to make contact with approaches in neural nets. We consider only one kind of neurons, namely excitatory, and make the replacement

$$A_{ek}(t) \to q_k(t), \tag{16}$$
$$A_{ik} \to 0, \tag{17}$$

and ignore the fluctuating forces, i.e. we put

$$F = \hat{F} = 0. \tag{18}$$

Furthermore we put the delay times equal to a single time according to

$$t_{e,nk} = \tau. \tag{19}$$

Without loss of generality, we may put

$$a_e = 1. \tag{20}$$

With these assumptions, (15) is converted into

$$q_k(t+\tau) = \hat{S}\left(\sum_n f_{kn} q_n(t) - \Theta_k + p_k(t)\right). \tag{21}$$

(21) is the conventional network model in the sense of McCulloch and Pitts [18]. (For a physicist the spin glass analogy by Hopfield [28] is most appealing, but we shall not dwell on it here.) f_{kn} serve as synaptic strengths. When the sigmoid function is a step function, having the values 0 or 1, with a threshold Θ_k, (21) describes a discrete time process during which the variables q_k that may be interpreted as neuronal activity, adopt the values 0 or 1. One may easily supplement (21) by a damping term

$$-\alpha q(t) \tag{22}$$

on the right-hand side. The reader hardly needs to be reminded of the contents of (21). The network consists of neurons that can occupy the states

$$q_k = 0, 1. \tag{23}$$

The inputs into neuron k coming from neurons n are summed up by means of synaptic weights f_{kn}. Depending on the size of the total sum, the neuron with index k may change its state or remain in the former one. It is important to note that in this model the geometrical form of the network is entirely unimportant. What counts is the connections. Kohonen's idea of feature maps [29] brings (21) closer to specific spatial arrangements. However, there is, at least at present, no link with the spatio-temporal patterns found in EEG or MEG measurements.

To elucidate this case, we again start from equations (11)–(14) and eliminate the axonal pulse rates so that we obtain equations for the dendritic wave amplitudes. A simple calculation shows that these equations have the form

$$\psi_{\ell,m}(t) = \sum_k a_\ell f_{\ell;m,k} \hat{S}_\ell \left(\psi_{e,k}\left(t - t_{\ell,mk}\right) - \psi_{i,k}\left(t - t_{\ell,mk}\right) + p_{\ell,k}\left(t - t_{m,k}\right)\right), \tag{24}$$

where we omit the fluctuating forces. We can easily include thresholds Θ, but in accordance with our previous work (see below) we assume that the variables are shifted in such a way that Θ formally vanishes. As we shall see immediately, equations (23) are a good starting point for deriving field equations of brain activity. In order to do so, we identify the indices m or k with the space point x, which may be one-, two-, or three-dimensional. Perhaps one may even think of a fractional dimension, but we leave this question open for further study. It may well be that the synaptic strengths $f_{\ell;m,k}$ are extremely complicated functions of the space coordinates m, k. In order to proceed further, we shall assume that we average over ensembles of neurons so that the functions f become smoothed.

In the following then we shall make simple assumptions on the form of these smooth functions. We shall assume that the dendrite at position x receives inputs from dendrites X more strongly, if the distance is small, or, in other words, we shall assume that f decreases with increasing distance. In order to make the following approach as transparent as possible, we shall assume that we may put $\psi_i = 0$, which actually is a good approximation and its inclusion would lead only to some kind of changes of numerical factors [30–32]. Then equations (24) can be specified as equation

$$\psi_e(x,t) = a_e \int \mathrm{d}X f_e(x,X) S_e \left[\psi_e\left(X, t - \frac{\mid x - X \mid}{v_e}\right) \right. \\ \left. + p_e\left(X, t - \frac{\mid x - X \mid}{v_e}\right)\right], \tag{25}$$

where we replaced the sum over k by an integral over X, and we made the replacement

$$f_{e,m,k} \to f_e(x,X) = \beta(x - X). \tag{26}$$

Equation (25) had been derived previously [30–32] in a way where continuous space variables were used from the very beginning and the explicit form (26) of β was used. In its linearized form this integral equation corresponds to the model equation by Nunez [15] and is discussed in [17]. (Nunez included the equation for ψ_i !) Equation (25) is the starting point for our further considerations (see also [30], [31], [32]). It will be our goal to convert the integral equation (25) into a local field equation. As we shall see, the result depends sensitively on the choice of the function β. By use of the δ-function

$$\delta\left(t - T - \mid x - X \mid / v_e\right), \tag{27}$$

we can write equation (25) in the form of equation

$$\psi_e(x,t) = \int \mathrm{d}X \int_{-\infty}^{+\infty} G(x-X, t-T)\rho(X,T)\mathrm{d}T, \tag{28}$$

where

$$G(x-X, t-T) = \delta\left(t - T - \frac{\mid x - X \mid}{v_e}\right)\beta(x-X) \tag{29}$$

can be interpreted as a Green's function, and

$$\rho(X,T) = a_e S\left[(\psi_e(X,T) + p_e(X,T))\right] \tag{30}$$

as a density that occurs as an inhomogeneity in a wave or field equation. In order to derive such an equation, we proceed in a conventional manner by making Fourier transformations of the following form

$$\psi_e(x,t) = \frac{1}{(2\pi)^2} \int_{-\infty}^{+\infty}\int_{-\infty}^{+\infty} \mathrm{e}^{\mathrm{i}kx - \mathrm{i}\omega t}\psi_e(k,\omega)\mathrm{d}k\mathrm{d}\omega, \tag{31}$$

$$\rho(x,t) \leftrightarrow \rho(k,\omega), \tag{32}$$

$$G(\xi, t_0) = \frac{1}{(2\pi)^2} \int_{-\infty}^{+\infty}\int_{-\infty}^{+\infty} \mathrm{e}^{\mathrm{i}k\xi - \mathrm{i}\omega t_0} g(k,\omega)\mathrm{d}k\mathrm{d}\omega \tag{33}$$

with

$$\xi = x - X, \quad t_0 = t - T. \tag{34}$$

Then we may write the Fourier transform of ψ_e as

$$\psi_e(k,\omega) = g(k,\omega)\rho(k,\omega). \tag{35}$$

The explicit form of $g(k,\omega)$ depends on both the dimension and the specific form of β in (29). To elucidate the general procedure, we proceed with a simple example in which case we assume β in the form

$$f_e(x,X) = \beta(x-X) = \frac{1}{2\sigma_e}\mathrm{e}^{-|x-X|/\sigma_e}. \tag{36}$$

The Fourier transform g can be obtained by means of

$$g(k,\omega) = \int_{-\infty}^{+\infty}\int_{-\infty}^{+\infty} \mathrm{e}^{-\mathrm{i}kx + \mathrm{i}\omega t} G(x,t)\mathrm{d}x\mathrm{d}t. \tag{37}$$

Using (29) and (36), we can easily evaluate the integral (37), which leads to

$$g(k,\omega) = \left(\omega_0^2 - \mathrm{i}\omega_0\omega\right)\left(v_e^2k^2 + (\omega_0 - \mathrm{i}\omega)^2\right)^{-1}, \tag{38}$$

where we use the abbreviation

$$\omega_0 = \frac{v_e}{\sigma_e}. \tag{39}$$

The form of (3) suggests to multiply both sides of the Fourier transform of the equation (28) by

$$\left(v_e^2k^2 + (\omega_0 - i\omega)^2\right) \tag{40}$$

and to backtransform. This leads immediately to the equation

$$\ddot{\psi}_e + \left(\omega_0^2 - v_e^2\frac{\mathrm{d}^2}{\mathrm{d}x^2}\right)\psi_e + 2\omega_0\dot{\psi}_e = \left(\omega_0^2 + \omega_0\frac{\partial}{\partial t}\right)\rho(x,t), \tag{41}$$

where we used the abbreviation (39). (41) is of the form of a damped wave equation with an inhomogeneity that is defined by (30) and was previously derived in [30], [31], [32]. It must be stressed that these evaluations were done in one dimension.

In the following we wish to study the impact of the dimensionality and of the form of the function

$$f_e(x,X) = \beta(x - X) \tag{42}$$

on the resulting wave or field equation.

11 How Universal Are Field Equations of the Brain?

In this section, that is rather technical, I wish to elucidate the role played by the distribution function f (26) and to study the question of universality of a field equation of the brain. In all cases, we have to deal with the evaluation of the Fourier transform of the Green's function, where, eventually, we have to replace the quantities k and ω in the function g by differential operators $\mathrm{d}/\mathrm{d}x$ and $\mathrm{d}/\mathrm{d}t$, respectively, as is well known from Fourier transform theory.

1. The One-Dimensional Case: We have to evaluate the integral

$$\int_{-\infty}^{+\infty}\int_{-\infty}^{+\infty} \mathrm{e}^{-\alpha|x|}\delta\left(t - \frac{|\,x\,|}{v}\right)\mathrm{e}^{-\mathrm{i}kx-\mathrm{i}\omega t}\mathrm{d}x\mathrm{d}t. \tag{43}$$

This integral can be easily evaluated and leads to

$$(43) = \frac{2(\alpha + \mathrm{i}\omega/v)}{(\alpha + \mathrm{i}\omega/v)^2 + k^2}, \tag{44}$$

which is identical with (38) when we make a suitable identification of the constants. As a second example for a distribution function, we treat

$$f = \mathrm{e}^{-\alpha|x|} \mid x \mid^n, \quad n \quad \text{integer}, \tag{45}$$

so that we have to evaluate the integral

$$\int_{-\infty}^{+\infty}\int_{-\infty}^{+\infty} \mathrm{e}^{-\alpha|x|} \mid x \mid^n \delta\left(t - \frac{\mid x \mid}{v}\right) \mathrm{e}^{-\mathrm{i}kx - \mathrm{i}\omega t} \mathrm{d}x\mathrm{d}t. \tag{46}$$

We obtain

$$(46) = \frac{(\alpha + \mathrm{i}\omega/v + \mathrm{i}k)^{n+1} + (\alpha + \mathrm{i}k/v - \mathrm{i}k)^{n+1}}{(\alpha + \mathrm{i}\omega/v - \mathrm{i}k)^{n+1}(\alpha + \mathrm{i}\omega/v + \mathrm{i}k)^{n+1}}. \tag{47}$$

By means of these results, we can easily evaluate also the Green's function for distribution functions f of the general form

$$\mathrm{e}^{-\alpha|x|} \sum_{\nu=0}^{m} a_\nu \mid x \mid^\nu . \tag{48}$$

In order to evaluate the integrals containing (48), we just have to multiply (46) by a_ν and to sum up over ν from zero till m. This can then be done with respect to (47) also. Quite evidently, already the result (47) is quite different from that of the case (43) that has led to (44). Any of these brain models therefore cannot claim to be universal. For this reason, we shall study under which circumstances we can obtain universality. One may argue that the local form of the distribution function is not important for slowly varying waves in space and time. To this end we shall assume in the following that the following inequalities are fulfilled

$$\mid \omega/v \mid \ll \alpha, \quad \mid k \mid \ll \alpha. \tag{49}$$

We may make the following approximations in (47)

$$(\alpha + \mathrm{i}\omega/v \pm \mathrm{i}k)^{n+1} \approx \alpha^{n+1} + (n+1)\alpha^n(\mathrm{i}\omega/v \pm \mathrm{i}k) + ..., \tag{50}$$

which lead to a numerator of the form

$$2\alpha^n(\alpha + (n+1)\mathrm{i}\omega/v) \tag{51}$$

and a denominator of the form

$$\alpha^{2n}(\alpha + (n+1)(\mathrm{i}\omega/v - \mathrm{i}k))(\alpha + (n+1)(\mathrm{i}\omega/v + \mathrm{i}k)) \tag{52}$$

so that the Fourier transform of the Green's function becomes proportional to

$$(46) = \frac{2(\alpha + (n+1)\mathrm{i}\omega/v)}{\alpha^n(\alpha + (n+1)(\mathrm{i}\omega/v - \mathrm{i}k))(\alpha + (n+1)(\mathrm{i}\omega/v + \mathrm{i}k))}. \tag{53}$$

This is the same form as (44) with the exception of other numerical values of the velocity v, where v in (44) is replaced by $v/(n+1)$ in (53). Surprisingly the same result holds even for the more complicated distribution function (48), where again the general form of (44) is recovered, but the effective velocity appears now as a certain value in which the amplitudes a_ν in (48) enter. Our results are a strong indication for the idea that a universal field equation holds for brain activity provided we are dealing with sufficiently small wave vectors and frequencies.

2. The Two-Dimensional Case: We again start with the distribution function (36), where we now have to evaluate the following integral

$$\int_{-\infty}^{+\infty} \mathrm{e}^{-\mathrm{i}k_x x} \int_{-\infty}^{+\infty} \mathrm{e}^{-\mathrm{i}k_y y} \mathrm{e}^{-\alpha r} \int_{-\infty}^{+\infty} \mathrm{e}^{-\mathrm{i}\omega t} \delta(t - r/v) \mathrm{d}x \mathrm{d}y \mathrm{d}t. \tag{54}$$

Making use of the δ-function and introducing polar coordinates, whereby we put

$$(\mathbf{k}, \mathbf{r}) = kr\cos\theta, \tag{55}$$

we may cast (54) into the form

$$\int_0^{2\pi} \int_0^{\infty} \mathrm{e}^{-ikr\cos\theta} \mathrm{e}^{-\alpha r - ir\omega/v} r \mathrm{d}r \mathrm{d}\theta. \tag{56}$$

(56) can be conceived as the derivative of the integral

$$\int_0^{2\pi} \frac{1}{ik\cos\theta + \alpha + \mathrm{i}\omega/v} \mathrm{d}\theta \equiv I \tag{57}$$

with respect to α multiplied by -1. The use of complex integration makes it a simple matter to evaluate (57), which leads to

$$I = -2\pi \frac{1}{\sqrt{k^2 + (\alpha + \mathrm{i}\omega/v)^2}}. \tag{58}$$

This allows us also to evaluate (54) in the form of

$$-\frac{\mathrm{d}}{\mathrm{d}\alpha} I = 2\pi \frac{(\alpha + \mathrm{i}\omega/v)}{(k^2 + (\alpha + \mathrm{i}\omega/v)^2)^{3/2}}. \tag{59}$$

In a similar way, we may evaluate two-dimensional Fourier transforms of the distribution function

$$r^n \mathrm{e}^{-\alpha r} \tag{60}$$

by means of

$$(-1)^{n+1} \frac{\mathrm{d}^{n+1} I}{\mathrm{d}\alpha^{n+1}}. \tag{61}$$

If we use the distribution function

$$\frac{1}{r} \mathrm{e}^{-\alpha r}, \tag{62}$$

we can immediately deal with (57) so that in this case the result of the Fourier transform is given by (58). Again some kind of universality can be shown in the case of small k and ω, but since the exposition of these results is more complicated, we shall not show them here.

3. The Three-Dimensional Case: In this case we have to evaluate integrals of the form

$$\int_{-\infty}^{+\infty} \int_{-\infty}^{+\infty} \int_{-\infty}^{+\infty} \mathrm{e}^{-\mathrm{i}\mathbf{kr}} \mathrm{e}^{-\alpha r} \mathrm{d}V \int_{-\infty}^{+\infty} \mathrm{e}^{-\mathrm{i}\omega t} \delta\left(t - \frac{r}{v}\right) \mathrm{d}t. \tag{63}$$

By use of spherical polar coordinates, (63) can be cast into the form

$$\int_0^{2\pi} \mathrm{d}\phi \int_0^{\pi} \mathrm{d}\theta \int_0^{\infty} \mathrm{e}^{-ikr\cos\theta} \mathrm{e}^{-\alpha r} e^{-\mathrm{i}\omega r/v} r^2 \sin\theta \mathrm{d}\theta \mathrm{d}\phi. \tag{64}$$

It can easily be evaluated; the result reads

$$(64) = \frac{8\pi(\alpha + \mathrm{i}\omega/v)}{(k^2 + (\alpha + \mathrm{i}\omega/v)^2)^2}. \tag{65}$$

If the distribution function is given by

$$\frac{1}{r}, \tag{66}$$

the Fourier transformation reads

$$g(k,\omega) = 4\pi \frac{1}{k^2 - \omega^2/v^2}, \tag{67}$$

or if the distribution function has the form

$$\frac{\mathrm{e}^{-\alpha r}}{r}, \tag{68}$$

the result reads

$$g(k,\omega) = 4\pi \frac{1}{k^2 + (\alpha + \mathrm{i}\omega/v)^2}. \tag{69}$$

Let us conclude this section with some general remarks. We studied the impact of the distribution function of the neuronal connections on the form of the resulting field equations. In the case of one dimension, an excellent universality can be found in the case of small ω and k. Universality can also be found, at least to some extent, in the two-dimensional case. In the three-dimensional case and under the assumption that ω and k are much smaller than α, the denominator of (65) can be approximated by

$$\alpha^2 \left(\alpha^2 - \omega^2/v^2 + 2k^2 + 4\mathrm{i}\omega\alpha/v\right) \tag{70}$$

that has the same dependence on k and ω as the denominator of (69) except for the constant coefficients of k and ω. In this way, one can show in the present case, as well as in more general cases, in which powers of r in the distribution function occur, that there is some kind of universality even in the three-dimensional case. In conclusion I may state that our results are a strong indication that a field equation of a universal type holds in brain networks provided we are confining our analysis to space and time scales that are large compared to the extension of the neuronal distribution and to frequencies derived thereof by means of the velocities v_e. I believe that the above results provide us with fundamentally new insights into brain functioning. At small scales, the brain is likely to act as a parallel computer described in a way by neural network equations of the form (21). On large scales, however, the brain acts like a physical system that generates spatio-temporal patterns. If the density of the connections and/or the local velocities v_e change, we must expect that the coefficients of the derivatives as well as the other constants in the wave equation become slowly varying functions of space.

12 Bringing Neural Field Equations Closer to Application

In order to apply our field equations to experimental situations, a number of further steps must be taken (and partly has been taken). Since the observed magnetic fields must be due to electric currents, we tentatively put the magnetic field strength proportional to the dendritic currents. Certainly further theoretical work is needed that takes into account the vector character of currents and fields, the folding of the cortical sheets, a.s.o. Furthermore the coupling p_e, that occurs via (30) of the neural field equation (41), to the world *external* to the brain must be considered. This implies a study of the role of efferent and afferent nerve fibers. A brief indication must suffice here. In a beautiful experiment, Kelso and his co-workers [33] measured the MEG while the subject was reacting to an acoustic signal by finger tapping. To model this experiment by means of the neural field equation, the sensory inputs p_e, i.e. the afferences, were split into an acoustic input and a motor-sensory input from the finger tapping. The efferent signal was assumed to be proportional to dendritic currents in the motor cortex. A realistic model requires to take into account the finger movement and the feedback loop neural net → motor output → motor sensory input → neural net. Here, quite naturally, delays come into play. For further details we must refer the reader to the original publications [31], [32]. A few further remarks should be made, however. The resulting model equations are still far too general. In order to make contact with the experimental data, contact with the top-down approach must be made. In the special case of the Kelso-experiment it could be established that the brain dynamics can be described by few degrees of freedom [34], whereby the analysis was based on a Karhunen–Loéve decomposition (cf. Sect. 7). Later work, partly based on the Uhl–Friedrich–Haken method [4], [5], showed that the brain dynamics is governed by only two order parameters and their dynamic equations were derived [35], [36]. This served as an important guideline to make the appropriate approximations in the neural field equations [31], [32]. I believe that in this way it was shown how we can bridge the gap between microscopic and macroscopic approaches. I think such well-defined "physical" experiments and their analysis are indispensable if we want to learn more about brain functioning with respect to MEGs and EEGs. Most probably such an analysis can be extended to psycho-physical experiments. How to make contact between mental states and neural field equations must be left to further studies, though the work on hallucinations mentioned above might be a first step.

13 New Basic Neural Network Equations: the Light Tower Model

In Sects. 10–12 we dealt with equations in which implicitely averages over spike trains were taken. As we know from experiments by Freeman [19] on the olfactory bulb and by those of Singer, Gray et al. [9] and Eckhorn et al. [10], on the visual cortex, there are pronounced correlations between spikes of different neurons. Therefore spikes must explicitely be treated. In this section we propose model equations to cope with the generation of spikes and their interaction. We consider the generation of the dendritic current of dendrite ℓ caused by a pulse P running along the axon of neuron ℓ. The pulse is assumed to be sharply peaked. Taking into account the generation of the dendritic current and the decay of that current, we formulate the equation

$$\dot{\psi}_\ell(t) = P_\ell(t-\tau) - \psi_\ell(t). \tag{71}$$

We can easily show that the dendrite acts as a low band pass filter. In order to connect the pulse P of the axon with the phase ϕ, we make the hypothesis

$$P(t) \sim (\sin \phi(t)/2)^{2n}, \tag{72}$$

where n is a large integer number. In this way it is guaranteed that the pulse is sharply peaked. (Actually the explicit form (72) is not essential. Other sharply peaked functions do as well.) ϕ is a phase that for a regular pulse generation is given by $\phi = \omega t$. More generally, we may state that the pulses are emitted along the axon after the dendritic currents have been processed by the nucleus. The rate of pulse generation is given by

$$\dot{\phi}_\ell(t) = F\left(\sum_m \psi_m(t-\tau') + p_e(t-\tau'')\right), \qquad (73)$$

where τ' is a delay time and p_e refers to an external signal. An example for F is provided by

$$F = \sum_m \psi_m + p_e - \Theta, \qquad (74)$$

where Θ is a threshold. Then we may formulate (73) more precisely by the relations

$$\begin{aligned} \dot{\phi} &= F \quad \text{if} \quad F > 0 \\ \phi &= 0 \quad \text{if} \quad F \leq 0. \end{aligned} \qquad (75)$$

In order to take saturation effects into account, we may replace F by a nonlinear function. The central point of the present approach is provided by the equations (71) and (73). When we visualize the effect of the phase angle ϕ as that of a rotating light beam emitted by a light tower, the name of this model becomes evident. It is not difficult to show that by averaging over spikes, we can recover previous approaches that were outlined in Sects. 10–12. But because of the explicit representation of the phase angle ϕ in (71)-(73), we can also treat phase locking effects. As it turns out, phase locking can be demonstrated and also the occurrence of spindles.

14 Noise in the Brain

We denote random fluctuations of a physical quantity as noise. An example is provided by the irregular firing of a neuron, where the time-intervals between spikes are random. To a number of neurophysiologists the occurrence of noise comes as a surprise. Noise is, however, inevitable in any physical system. This is well-known since the discovery of Brownian motion. Noise is not only present in systems that are in thermal equilibrium, but in a still more pronounced manner in open physical systems, whose state is maintained by an influx and outflux of energy. Random fluctuations are inseparably connected with these fluxes and there exists a detailed theory on this connection [37], [38]. It was experimentally verified, in particular in the by now famous example of a laser. It could be shown, both experimentally and theoretically, that noise becomes especially pronouned during a nonequilibrium phase transition (critical fluctuations) (cf. Fig. 6). The corresponding effect was predicted by

Schöner, Haken and Kelso [39] for transitions between finger movements and was found by Kelso et al. [40] on the level of finger movements and, still more significantly in MEG-measurements during such transitions [33]. In the context of my article, there are at least three basic questions:

1. How can we incorporate the occurrence of noise in order parameter equations?
2. How can we deal with noise in the neural equations?
3. How can we derive equations from experimental data?

Both in cases 1 and 2 a procedure that can be borrowed mainly from laser theory [38] consists in adding fluctuating forces to the equations of order parameters or neural fields, respectively. There are well-defined mathematical rules for the formulation of the fluctuating forces and their parameters, but lack of space does not allow me to enter this problem here. At any rate, the resulting equations are of the Langevin-type. As is known in statistical physics, to a Langevin equation a Fokker–Planck equation can be attached. This allows for an elegant access to a treatment of question 3 above. Namely, some time ago I treated the problem on how to derive a Fokker–Planck equation from experimental data. To this end I used the short-time propagator of the Fokker–Planck equation [41]. In order to make an unbiased guess on the short-time propagator, I invoked the maximum information (entropy) principle. The feasibility of this approach was convincingly demonstrated by Lisa Borland [42] in her thesis and further publications for several model systems at the order parameter level.

In view of these previous results, the approach by Friedrich and Uhl (cf. this volume) to the analysis of noisy MEGs and EEGs bears great potential. These authors also use the short-time propagator, but in their present approach circumvent the maximum information (entropy) principle. Because of the scatter of their data on the drift-coefficients, future work might make it necessary to use some estimation procedures, however. At any rate, their work will provide us with important insights into noise in the brain at least at the order parameter level.

15 Concluding Remarks

In my article I tried to give an outline of basic concepts of synergetics and of some of its applications to the understanding of brain functioning. In view of the enormous number of neurons it seems to be impossible to keep track of all their activities. Therefore we look for macroscopic descriptors in form of order parameters. A number of experimental and theoretical studies, part of which were mentioned in my article, have proven the feasibility and usefulness of this approach to EEG and MEG analysis. I also indicated that these results can serve as a guideline to select parameters and solutions of "microscopic" neural network equations. Future work will not only have to extend these results to further EEG-MEG motorsensory experiments, but also to corresponding psycho-physical experiments and, eventually, to truly psychological studies.

References

[1] Haken, H.: *Principles of Brain Functioning: A Synergetic Approach to Brain Activity, Behavior and Cognition*, Springer, Berlin (1996)
[2] Haken, H.: *Synergetics. An Introduction*, 3rd ed., Springer, Berlin (1983)
[3] Haken, H.: *Advanced Synergetics*, 2nd ed., Springer, Berlin (1987)
[4] Uhl, C., Friedrich, R., Haken, H.: Reconstruction of spatio-temporal signals of complex systems, Z. Phys. B 92, 211–219 (1993)
[5] Uhl, C., Friedrich, R., Haken, H.: Analysis of spatio-temporal signals of complex systems, Phys. Rev. E, Vol. 51, No. 5 (1995)
[6] Friedrich, R., Uhl, C.: Spatio-temporal analysis of human electroencephalograms: Petit-mal epilepsy, Physica D 98, 171–182 (1996)
[7] Kuramoto, Y., Tsusuki, T.: Prog. Theor. Phys. 52, 1399 (1974)
[8] Tass, P.: Phase and frequency shifts in a population of phase oscillators, Phys. Rev. E. 56, 2043 (1997). Synchronized oscillations in the visual cortex - a synergetic model, Biol. Cybern. 74, 31-39 (1997). Synchronization in networks of limit cycle oscillators, Z. Phys. B 100, 303–320 (1996)
[9] Singer, W., Gray, C.M., König, P.: Synchronization of oscillatory responses in visual cortex: A plausible mechanism for scene segmentation. In: *Synergetics of Cognition*, Haken, H., Stadler, M. (eds.), Springer, Berlin (1990)
[10] Eckhorn, R., Reitböck, H.J.: Stimulus-specific synchronization in cat visual cortex and its possible role in visual pattern recognition. In: *Synergetics of Cognition*, Haken, H., Stadler, M. (eds.), Springer, Berlin (1990)
[11] Beurle, R.L.: Properties of a mass of cells capable of regenerating pulses, Philos. Trans. Soc. London, Ser. A 240, 55-94 (1956)
[12] Griffith, J.S.: A field theory of neural nets: I: Derivation of field equations, Bull. Math. Biophys., Vol. 25, 111-120 (1963), and II: Properties of field equations, Bull. Math. Biophys., Vol. 27, 187–195 (1965)
[13] Wilson, H.R., Cowan, J.D.: Excitatory and inhibitory interactions in localized populations of model neurons, Biophysical Journal, Vol. 12, 1–24 (1972)
[14] Wilson, H.R., Cowan, J.D.: A mathematical theory of the functional dynamics of cortical and thalamic nervous tissue, Kybernetik 13, 55–80 (1973)
[15] Nunez, P.L.: The brain wave equation: A model for the EEG, Mathematical Biosciences 21, 279–297 (1974)
[16] Nunez, P.L.: *Electric fields of the brain*, Oxford University Press (1981)
[17] Nunez, P.L.: *Neocortical dynamics and human EEG rhythms*, Oxford University Press (1995)
[18] McCulloch, W., Pitts, W.: A logical calculus of the ideas immanent in nervous activity, Bulletin of Math. Biophysics 5, 115–133 (1943)
[19] Freeman, W.J.: Mass action in the nervous system, Academic Press, New York (1975)
[20] Freeman, W.J.: Tutorial on neurobiology: From single neurons to brain chaos, International Journal of Bifurcation and Chaos, Vol. 2, No. 3, 451–482 (1992)
[21] Abeles, M.: *Corticonics*, Cambridge University Press (1991)

[22] Braitenberg, V., Schüz, A.: *Anatomy of the cortex. Statistics and geometry*, Springer, Berlin (1991)
[23] Miller, R.: Representation of brief temporal patterns, Hebbian synapses, and the left-hemisphere dominance for phoneme recognition, Psychobiology, Vol. 15 (3), 241–247 (1987)
[24] Cowan, J.D.: Brain mechanisms underlying visual hallucinations, in: Paines, D. (ed.), *Emerging Syntheses in Science*, Addison Wesley, 123–131 (1987)
[25] Ermentrout, G.B., Cowan, J.D.: A mathematical theory of visual hallucination patterns, Biol. Cybern. 34, 137–150 (1979)
[26] Tass, P.: Cortical pattern formation during visual hallucinations, J. Biol. Phys. 21, 177–210 (1995)
[27] Tass, P.: Oscillatory cortical activity during visual hallucination, J. Biol. Phys. 23, 21–66 (1997)
[28] Hopfield, J.J.: Neural networks and physical systems with emergent collective computational abilities, Proc. Natl. Acad. Sci. 79, 2554 (1982)
[29] Kohonen, T.: *Associative memory and self-organization*, 2nd ed., Springer, Berlin (1987)
[30] Jirsa, V.K.: Thesis, Modellierung und Rekonstruktion raumzeitlicher Dynamik im Gehirn, Harri Deutsch, Frankfurt a. M. (1996)
[31] Jirsa, V.K., Haken, H.: A field theory of electromagnetic brain activity, Phys. Rev. Lett., 22 July (1996)
[32] Jirsa, V.K., Haken, H.: A derivation of a macroscopic field theory of the brain from the quasi-microscopic neural dynamics, Physica D 99, 503–526 (1997)
[33] Kelso, J.A.S., Bressler, S.L., Buchanan, S., DeGuzman, G.C., Ding, M., Fuchs, A., Holroyd, T.: A phase transition in human brain and behavior, Physics Letters A 169, 134–144 (1992)
[34] Fuchs, A., Kelso, J.A.S., Haken, H.: Phase transitions in the human brain: Spatial mode dynamics, International Journal of Bifurcation and Chaos, 2, 917–939 (1992)
[35] Jirsa, V.K., Friedrich, R., Haken, H., Kelso, J.A.S.: A theoretical model of phase transitions in the human brain, Biol. Cybern. 71, 27–35 (1994)
[36] Jirsa, V.K., Friedrich, R., Haken, H.: Reconstruction of the spatio-temporal dynamics of a human magnetoencephalogram, Physica D 89, 100–122 (1995)
[37] Haken, H.: *Laser Theory*, Springer, Berlin (1970)
[38] Haken, H.: *Light*, vol. 2, North-Holland (1985)
[39] Schöner, G., Haken, H., Kelso, J.A.S.: A stochastic theory of phase transitions in human hand movement, Biol. Cybernectics 53, 247–257 (1986)
[40] Kelso, J.A.S., Scholz, J.P., Schöner, G.: Non-equilibrium phase-transitions in coordinated biological motion: critical fluctuations, Physics Letters A 118, 279–284 (1986)
[41] Haken, H.: *Information and self-organization*, Springer, Berlin (1988)
[42] Borland, L.: Ein Verfahren zur Bestimmung der Dynamik stochastischer Prozesse, Thesis Stuttgart (1993). Learning the Dynamics of Two-Dimensional Stochastic Markov Processes, Open Systems & Information Dynamics, Vol. 1, No. 3 (1992). On the constraints necessary for macroscopic prediction of time-dependent stochastic processes, Reports on Mathematical Physics, Vol. 33, No. 1/2 (1993)

A Preliminary Physiology of Macro-Neocortical Dynamics and Brain Function

Paul L. Nunez

Brain Physics Group, Dept of Biomedical Engineering, Tulane University, New Orleans, Louisiana 70118, USA, pnunez@mailhost.tcs.tulane.edu

1 Introduction

Human brains are information processing systems of vast complexity. Over the past century and especially over the past few decades, our understanding of brain function at small scales has grown substantially. For example, we know many new properties of neurotransmitters, synapses, membranes, and neurons. Higher brain functions involving the thoughts and emotions that determine our behavior, as well as consciousness itself, evidently involve the functional intergration of many sub-systems, involving large fractions of brain tissue. However, very little is known of the large-scale physiology of cell groups and their connection to psychology.

What are the most promising approaches to large scale brain study? Established fields like cognitive neuroscience, neuropsychiatry, and neurology operate as different cultures, often using different language and methods and pursuing different goals. All neuroscientists require a personal conceptual framework to guide new experimental and theoretical work. Such framework is distinguished from genuine dynamic theory, which has been advanced by only a few neuroscientists. A comphrehensive conceptual framework must embrace neocortical dynamic behavior at multiple spatial scales, ranging from single neurons to overlapping local and regional cell groups of different sizes to global fields of synaptic action density. Interaction across these scales may be essential to the dynamics (and by implication to behavior and cognition), in a manner analogous to the importance of hierarchical interactions in human social systems. Such ideas do not necessarily contradict classical neurophysiological views of focal control (bottom up) mechanisms. Rather they suggest dynamics which is more fully integrated across spatial scales (bottom up and top down).

In this chapter, I propose a general conceptual framework for neocortical dynamic function at large scales. This framework is based mostly on past work of many neuroscientists and is consistent will classical views. However, the framework has some new intellectual structure based on recent theoretical and experimental studies in EEG. Within this general framework, a greatly over-simplified theory of macro-neocortical dynamics is outlined. The theory is physiologically based with no free parameters. While several of the approximations and control parameter ranges are in doubt, the theory makes qualitative and, in some case, semi-quantitative contact with a variety of experiments. Such theories are important for the refinements of conceptual frameworks, even when the theories are later proved wrong. Thus, application of such dynamic theoretical methods to electroencephalographic (EEG) data can help to unify the sub-fields of macro-neuroscience by providing a more detailed conceptual framework for brain function, one with more consistent and precise language and culture, leading to new interdisciplinary studies.

2 Neocortical Dynamics

The outer, infolded surface of the human brain consists of neocortex, a structure 2 to 3 mm thick with surface area in the 3000 cm^2 range and containing about 10^{10} neurons. Regions of neocortex are interconnected by about 10^{10} cortico-cortical fibers, which form most of the white matter layer just below the cortical layer and provide a system of massive positive feedback between remote cortical locations (Braitenberg 1978; Braitenberg and Schuz 1991). A few (0.000001%) of these axons of cortical pyramidal cells are shown in Fig. 1. I emphasize "neocortical" rather than "brain" dynamics here for several reasons (Nunez 1995):

1. In humans, voluminous EEG data have been recorded from the scalp. The dominant current sources of scalp potentials are cortical. Relatively little human data have been recorded in subcortical regions. Such experiments are severely limited by both ethical and technical considerations.
2. Human neocortex is much more strongly connected to itself than to midbrain regions. For example, only about 2% of fibers entering the underside of neocortex originate in the midbrain. The other 98% are mostly cortico-cortical fibers; a few are shown in Fig 1. This fraction is much lower in lower mammals; e.g., perhaps 50% in rat. Thus, it has been suggested that the large fraction of cortico-cortical fibers is a major factor in making human brains "human" (Braitenberg 1977; Katznelson 1981; Nunez 1995).
3. The cortico-cortical fibers provide for long-range (non-local) interactions at cm scales. The intracortical connections provide local interactions, mostly at sub-millimeter scales. Intracortical connections are so dense that each mm^3 of cortical tissue contains more than 1 km of axon length (Braitenberg and Schuz 1991). The relative isolation of neocortex, which is much greater in humans than lower mammals, appears to justify treating it as a separate system, in which chemical and electrical input from the midbrain provide external control of neocortical state on time scales much longer than EEG oscillation times.
4. Sensory and cognitive processing of information is critically involved with neocortical dynamic function. Much of what makes us individuals is apparently due to differences in neocortical/white matter systems. Con-

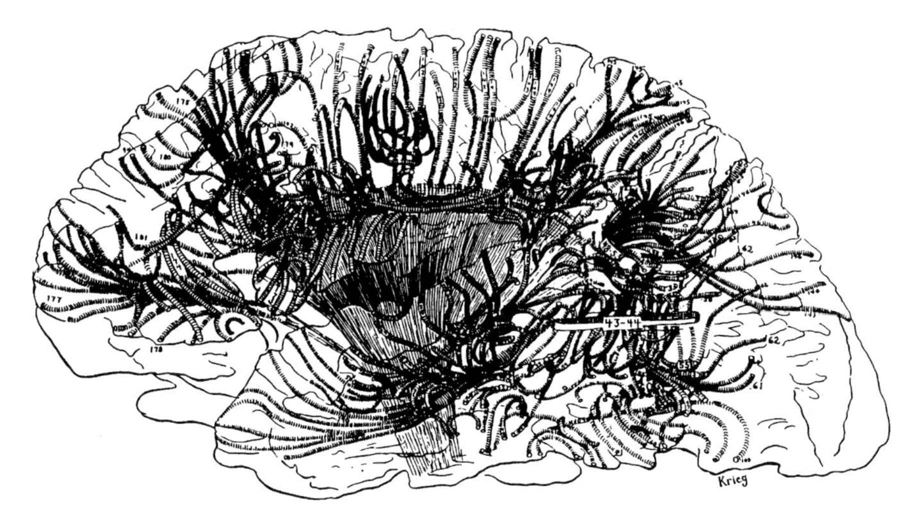

Fig. 1. Human neocortex ("gray matter", which is actually pink when alive) forms a 2 to 3 mm thick outer surface of the brain. Neocortical neurons are interconnected by short-range intracortical fibers (lengths mostly in the sub mm range), and by long range cortico-cortical fibers (1 to 20 cm range). A few of the cortico-cortical fibers that form most of the "white matter" layer, just below human neocortex, are shown here. The actual number of cortico-cortical fibers in humans is about 10^{10}. That is, for every fiber obtained here by dissection of a fresh human brain, there are another hundred million not shown. These fibers are exclusively excitatory, providing a system of massive positive (non-local) feedback between cortical regions. Propagation velocities of action potentials in the (mostly) myelinated cortico-cortical fibers are distributed with a velocity distribution function peaked in the 6 to 9 m/sec range. The "effective radius" of the unfolded human cortex, defined by its surface area, is roughly 11 to 18 cm. Reproduced with permission from Krieg (1963).

sciousness is almost certainly not possible in humans born without neocortex.

5. In theoretical models, human neocortical tissue may evidently be considered homogeneous and isotropic (to first approximation, in directions parallel to its surface) at mesoscopic and macroscopic scales (roughly larger than 0.1 mm).

Neocortical tissue has hierarchical structure. Characteristic linear scales "hard-wired" in the structure are identified by anatomists as the minicolumns (0.03 mm), cortico-cortical columns (0.3 mm), macrocolumns (1 mm), Broadman areas (5 cm), lobes (17 cm), and hemispheres (40 cm). Considering the enormous complexity of brains and our experience with successful models of relatively complex physical systems in which a range of length scales must be considered, distinct mathematical theories may be required at each neocortical scale. Such theories are directly connected only to data recorded at the same scale, although good theories strive for overlap with adjacent scales to illuminate qualitative and semi- quantitative connections across scales. For example, electrophysiological data in animals span about five orders of magnitude of spatial scale, ranging from microelectrode tip (0.001 mm) to EEG scalp electrode, which records neural source activity space-averaged over regions much larger than the 0.5 to 1 cm diameter scalp electrode (due to current spreading between sources and electrodes in the head volume conductor).

Figure 2 shows a laterial view of neocortex with distinct local regions based on different numbers and types of cells that occur in different parts of neocortex. These local differences may evidently be viewed as perturbations of the relatively homogeneous structure of neocortex. About 50 local areas were indentified by Broadman (1909); each contain about 10^8 neurons and have a linear scale of about 5 cm (including the infolded surface). This scale roughly matches the spatial resolution of conventional EEG, although high resolution EEG and MEG may be able to obtain 1 or 2 cm resolution. The conceptual framework proposed here allows for dynamic formation of overlapping cell groups which are both larger and smaller than Broadman areas. EEG and MEG measure dynamic behavior at larger scales, but this dynamics is influenced by smaller scales (bottom up), and smaller scales are influenced by larger scales (top down), in a manner perhaps somewhat analogous to the dynamics of our global weather system.

Figure 2 indicates that brain states are determined by subcortical chemical and electrical modulatory influences which change control parameters Q and B on much longer time scales than characteristic EEG oscillation periods, typically 50 ms. to 1 sec. Known time constants for cortical pyramidal cells are roughly in the 5–10 ms. range, which is a good guess for the minimum time over which large scale neocortical fields show substantial change, in approximate agreement with EEG studies. The symbols Q_n indicate control of local regions (n). Regional dynamics at higher hierarchical levels is indicated by multiple subscripts on the Qs. For example, Q_{ijk} represents the control parameter for cell groups (i, j, k), which may act as single systems in "fixed" brain states (for times much longer than EEG periods). Input cortico-cortical fibers branch out within neocortex over a tangential distance of about 0.3 mm; this structural characteristic defines the cortico-cortical column. Thus, it appears that cell groups at cortico-cortical or larger scales, forming single systems, may be contiguous, 20 cm apart, or anything in between. However, the probability that any two regions (at relatively large scales) have substantial direct (cortico-cortical) connections tends to fall off with separation distance within each hemisphere (Braitenberg and Schuz 1991; Nunez 1995). Such quasi-metric behavior at large scales coexists with specific cortico-cortical connections at submillimeter scales. Cortical tissue may also be connected through sub-cortical connections, e.g., through the thalamus. However, only a few per cent of fibers entering (or leaving)

Neocortical Control Parameters

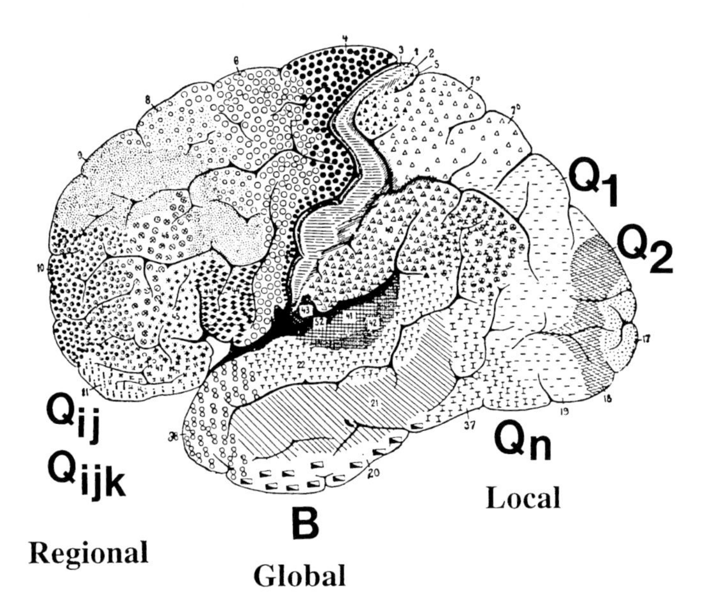

Fig. 2. A laterial view of neocortex indicating distinct local regions, based on different numbers and types of cells. These local differences may evidently be viewed as perturbations of the relatively homogeneous structure of neocortex. About 50 local areas were indentified by Broadman (1909); each contains about 10^8 neurons and has a linear scale of about 5 cm (including the infolded surface). The symbols (added by the author) suggest that brain states are determined by subcortical chemical and electrical modulatory influences which change control parameters Q and B. The symbols Q_n indicate control of local regions (n). Regional dynamics at higher hierarchical levels is indicated by multiple subscripts on the Qs. For example, Q_{ijk} represents the control parameter for cell groups (i, j, k), which may act as single systems in "fixed" brain states (for times much longer than EEG periods). Input cortico-cortical fibers branch out within neocortex over a tangential distance of about 0.3 mm; this structural characteristic defines the cortico-cortical column. Thus, it appears that cell groups at cortico-cortical or larger scales, which can evidently form single systems, may be close together or as much as 20 cm apart. In contrast to the cortico-cortical fibers, the intracortical fibers are both inhibitory and excitatory. Lateral inhibition provides a means to partly isolate patches of neocortex, which can evidently produce local dynamics, at least partly independent of global behavior. Local dynamics is believed to depend on negative and positive feedback in local "circuits". Remote regions also appear to form temporary cell groups, which may functionally connect and disconnect in typical times of ten to several hundred ms. Different brain states appear to involve more doninant local or global (e.g., spatially coherent) behavior. Global modes provide a possible mechanism to synchronize remote cell groups having no direct interconnections. Reproduced and modified with permission from Jones (1987).

the underside of human neocortex originate (or terminate) in sub-cortical regions; the remaining are cortico-cortical fibers.

Studies of the multiple hierarchy of local and regional neocortical dynamics are likely to challenge neuroscientists far into the future (Ingber 1982, 1995a,b). As a putative "introduction to 21st century neurophysics", I focus here on only a small part of this dynamic complexity, the very large scale fields of synaptic action density for which scalp EEG provides a crude measure. In this highly over-simplified description, global states of neocortex are described in terms of a single control parameter B (Fig. 2). Neocortical dynamics is imagined to involve bottom up and top down interactions of the global system with local cell groups, which are considered to have local dynamics, partly controlled by the parameter Q, which in the oversimplified mathematical formulation, is assumed to be constant over the cortical surface. Of course, in genuine brains, one expects Q to vary with location. The macroscopic fields of synaptic action, which are moderately to strongly correlated with behavior, may be both strongly influenced by and act back on smaller scale activity, which is typically weakly correlated with behavior. This general idea and other contributions to a "physiology of neural mass action" or "macroscopic physics of neocortex" at the large scales of scalp EEG may ultimately provide insight into the dynamics of smaller scales for which minimal human data are available, e.g., the cooperative workings of local and regional cell groups.

3 EEG and Mental States

Dynamic behavior of such strongly integrated, complex systems may be studied at many spatial scales. However, in humans, it is the very large scale behavior that is most easily studied experimentally. Several methods provide structural (e.g., CT or MRI, typically providing "dynamics" at time scales of weeks to years) or intermediate time scale measures (e.g., functional MRI or PET, with associated scales of seconds to minutes). However, only electroencephalography (EEG) or magnetoencephalography (MEG) provide millisecond temporal resolution required to follow directly the dynamics of brain information processing.

Consider the following experiment. Place electrodes on a human subject in one room and feed his EEG signal to a computer display in an isolated location. Monitor the subject's state of consciousness over several days and provide this information to persons following the unprocessed oscillations of scalp voltage (with axes showing amplitude and time scales). Even observers unfamiliar with EEG will recognize that the voltage record during deep sleep has larger amplitudes and contains more low frequency content. Slightly more sophisticated monitoring and training allow observers to accurately identify distinct sleep stages, depth of anesthesia, and seizures. Still more advanced methods reveal robust connections of EEG to more detailed cognitive processes.

We are now so accustomed to these EEG/brain state correlations, first studied in the 1920's, that we may forget how remarkable they are. The scalp EEG (or MEG) provides a very large scale measure of neocortical dynamic function. A single electrode pair provides estimates of synaptic action averaged over tissue masses containing something between 10 million and 1 billion neurons. Most human studies are limited to extracranial recordings, with space averaging a fortuitous data reduction process, due to passive current spread in the head volume conductor. More detailed information is sometimes obtained from intracranial recordings. However, the number of intracranial electrodes implanted in living human brains must be very small compared to anything approaching full spatial coverage, even at intermediate spatial scales. Thus, in practice, intracranial data provide different information, not more information, than is obtained from the scalp. That is, intracranial recordings provide smaller scale measures of neocortical dynamics, with scale dependent on electrode size. The small and intermediate scale data is largely independent of scalp data, and is typically weakly correlated or uncorrelated with cognitive events, which are much more easily observed at large scales. Thus, the technical and ethical limitations of human intracranial recordings force us to emphasize scalp recordings, and these methods provide data at the macroscopic scales which are most strongly correlated with cognition and behavior, a fortunate coincidence!

Although cognitive scientists make good use of raw scalp EEG data, explorations of new MEG and EEG methods to provide somewhat higher spatial resolution continue. A reasonable goal is to record averages over "only" 10 million neurons at the 1 cm scale in order to extract more details of the spatial patterns correlated with cognition and behavior. This resolution is close to theoretical limits caused by the physical separation of sensors and brain current sources. (EEG/MEG comparisons are somewhat obscured by the fact the the two systems are sensitive to different sub-sets of neural sources, refer to Nunez 1995). One approach to high resolution EEG takes advantage of the fact that skull current density is much larger in directions perpendicular to its surface than in tangential directions (because of high skull resistivity). For this reason, the two-dimensional Laplacian of surface potential (in two tangent scalp coordinates) provides robust estimates of brain surface potential which are relatively independent of head model (Nunez 1981,1995; Srinivasan et al. 1996; Nunez et al. 1997). Model-based projections of scalp potential to inner surfaces, e.g., the dura (called "cortical imaging" or "deblurring"), may eventually prove more accurate, but are limited by uncertainties in tissue boundaries and the complicated head tissue resistivity tensor, both of which can vary substantially between individuals (Nunez et al. 1994).

An example of high resolution EEG is provided in a study of data recorded at the Brain Sciences Institute in Melbourne, Australia. A comparison of coherency patterns over the scalp obtained in two distinct brain states is obtained (Nunez 1995; Nunez et al. 1997). The author alternated one minute periods of resting (slowly counting his breaths to facilitate relaxation) with summation of series like (1+2+3+...) up to sums of several hundred. Coherency is a correlation coefficient (squared); it measures phase consistency recorded at paired locations, for each frequency component in the EEG. However, raw scalp coherency between electrode sites closer than about 8 to 10 cm is typically large or moderate due only to volume conduction and reference electrode effects, even when the underlying cortical sources are uncorrelated. But, with high resolution methods, most of the erroneous high coherency may be eliminated by estimating potential at the dura surface (an unfolded layer between brain and skull) using dense scalp arrays before calculating coherences.

In the 60 channel data of this experiment, $60 \times 59/2 = 1770$ coherences were followed. The peak alpha frequency in both brain states was between 9 and 10 Hz. Coherences at 9 Hz, obtained from unprocessed scalp potentials, are plotted versus electrode separation distance in Fig. 3. These same raw potentials were passed through the New Orleans spline-Laplacian and Melbourne dura image spatial filter to improve spatial resolution, thereby obtaining estimates of brain surface (dura) potential distribution at each time point. The corresponding dura image coherences are shown in Fig. 4. The upper plots in Figs. 3 and 4 show coherences obtained during three alternating minutes of the eyes closed, resting state. The lower plots show coherences for three alternating minutes of eyes closed, mental calculation (the "cognitive state"). The solid lines are estimates of coherency fall-off with scalp electrode separation when the underlying sources are superficial, uncorrelated radial dipoles. That is, the non-zero coherency of the solid lines is due only to passive current spread in the head volume conductor.

Uncorrelated, tangential dipoles in fissures and sulci can produce longer ranged correlations in the unprocessed potentials, but have much less effect on dura images or Laplacians, which are insensitive to deep sources. Furthermore, tangential sources are generally expected produce smaller scalp potentials, due to orientation, increased depth in fissures and sulci, and the ability of radial dipoles to form large dipole layers with non-zero source correlations. For example, the maximum surface potential generated by a tangential dipole (current source) is about 1/3 that of a radial dipole with the same dipole moment, if both are at the depth of cortical gyri. More importantly, when large, correlated dipole layers are formed, radial dipoles are aligned in parallel so that corresponding scalp potentials are obtained as the linear superposition of these sources (Nunez, 1981,1995). By contrast, tangential dipoles on opposite sides of fissures and sulci produce potentials that tend to cancel. Thus, one expects radial source models to yield fairly realistic simulations, which are applicable to many EEG states in which radial sources make dominant contributions. These and earlier studies (Nunez 1981, 1995) indicate the following:

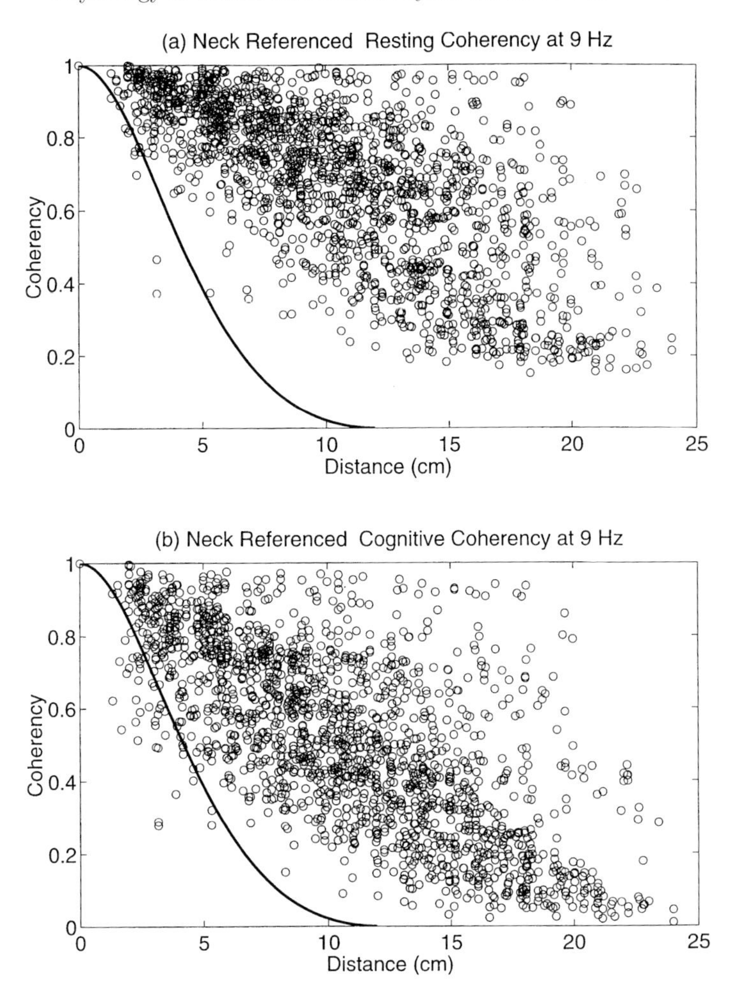

Fig. 3. Coherency at 9 Hz is plotted as a function of interelectrode distance for neck-referenced EEG data. A coherency estimate is shown for each of the 1770 electrode pairs at 9 Hz (near the peak power). The upper plot refers to data recorded during three one-minute periods when the author had closed eyes and was slowly counting his breaths to facilitate relaxation ("resting coherency"). These periods were alternated with three one-minute periods when the author summed series like (1+2+3+ ...) with eyes closed to sums of several hundred ("cognitive coherency"). Corresponding coherences are plotted in the lower row. Relative coherency reductions between resting and cognitive states were generally largest for frontal-to-frontal sites. The solid lines are the analytic estimate of scalp potential coherency expected from uncorrelated radial dipoles, uniformally distributed over neocortical gyri (the "random coherency", Nunez 1995; Nunez et al. 1997; Srinivasan et al. 1998). Deeper sources and/or tangential sources cause larger random coherency; however the radial dipole model appears appropriate for many if not most EEG phenomena, which are apparently generated by large, moderately correlated dipole layer sources alligned in parallel (Nunez 1981).

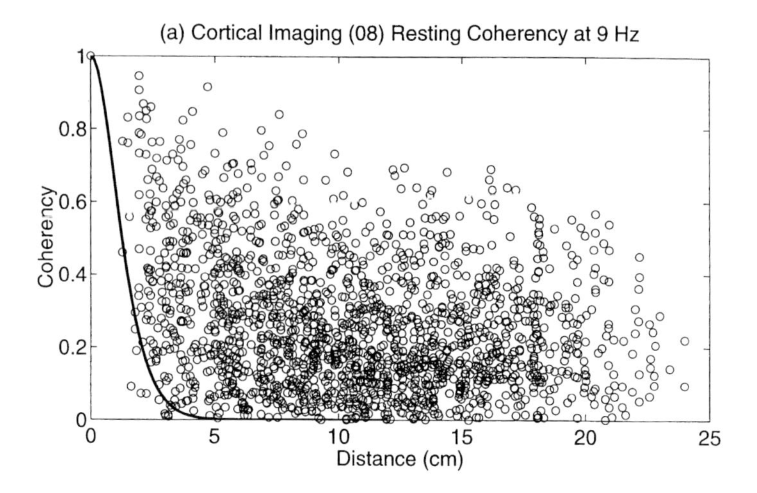

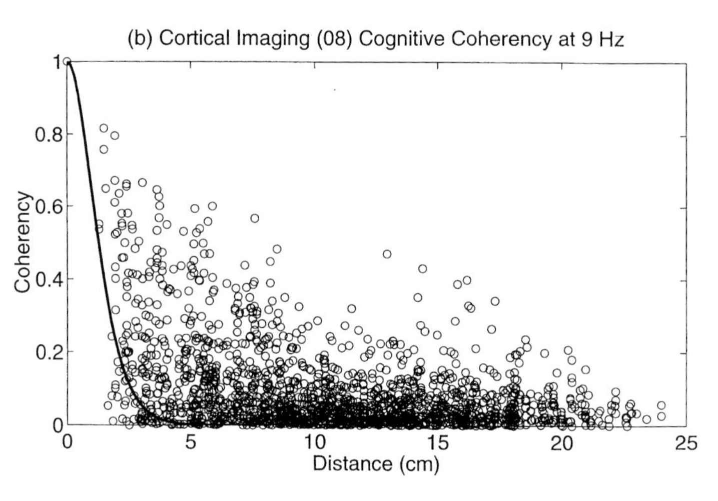

Fig. 4. The same raw data (or Fourier coefficients) used in Fig. 3 were passed through both the New Orleans spline-Laplacian and Melbourne cortical imaging algorithms to estimate potential distribution on the dura surface, thereby minimizing volume conduction distortion and eliminating reference electrode contributions to coherence estimates (Nunez 1995; Nunez et al. 1994, 1997). Dura image estimates of coherency are shown here; Laplacian coherency is very similar. The "08" label denotes the smoothing factor used by the dura imaging algorithm; coherency plots are not sensitive to this parameter. The upper plot is for the resting state; the lower plot refers to the cognitive state. The solid lines are the analytic estimate of the Laplacian or dura image coherency expected from the same uncorrelated sources described in the caption for Fig. 3. Both Laplacian and cortical image coherences at 9 Hz exhibit large and robust decreases between resting and cognitive periods.

1. Human brains produce extremes of coherent and incoherent electrical activity, depending on brain state, spatial scale, and frequency band. Coherences can be recorded with minimal volume conduction and reference electrode distortion using pairs of close bipolar electrodes (e.g., 2 to 3 cm separation) to estimate local tangential scalp electric fields. Alternately, spline-Laplacian or cortical imaging algorithms may be applied with large electrode arrays (e.g., 48 to 131 channels) to obtain local estimates of dura potential before coherences are calculated. An eyes closed, relaxed brain often produces alpha rhythm coherences in the 0.4 to 0.8 range between anterior/posterior regions separated by 20 cm or more. Such moderate to large coherences often occur only in narrow frequency bands (e.g., 9 to 10 Hz). They typically apply to averages over one to several minutes of EEG, which may include substantial amplitude (and by implication, coherency) variations. If only bursts of alpha activity are used, larger coherences are calculated.
2. Brains can produce both coherent and incoherent large-scale dynamics. Alpha coherency at large scales decreases during mental calculation. There are large, robust decreases in coherency for most paired sites between states of resting and mental calculation in roughly the 9–10 Hz alpha band. This effect is observed for both raw and high resolution coherences, but percentage changes are largest with high resolution coherences since volume conduction and reference effects are apparently minimal in the latter case.
3. Coherency changes are largest for frontal-to-frontal electrode pairs as shown by the plots of regional changes in Figs. 5 and 6. The effect shown here is much more detailed than traditional "alpha blocking". Alpha amplitude is lower during mental calculations, with the largest state changes in amplitude occurring over posterior regions. However, the alpha rhythm persists in both states, and the largest coherency state changes involve frontal-to-frontal sites, consistent with generally accepted ideas about the important role of frontal cortex for higher mental processing. Of course, these large scale coherency reductions could coexist with very specific coherency increases at scales too small to observe from the scalp, e.g., in synchronous cell groups.
4. Coherency outside the 9–10 Hz band is substantially lower. This suggests that passive current spread by the head volume conductor and reference electrode contributions to erroneous high coherency have been largely eliminated (since volume conduction is independent of frequency in this frequency range).
5. Complicated patterns of coherence and coherence state changes occur. Coherence patterns are often very specific to electrode pair and narrow bandwidths of 1 to 2 Hz, especially with high resolution coherences. Many coherences in the 4–5 Hz theta band are consistently higher during mental calculations than resting coherences ($p \ll 0.001$), although the mag-

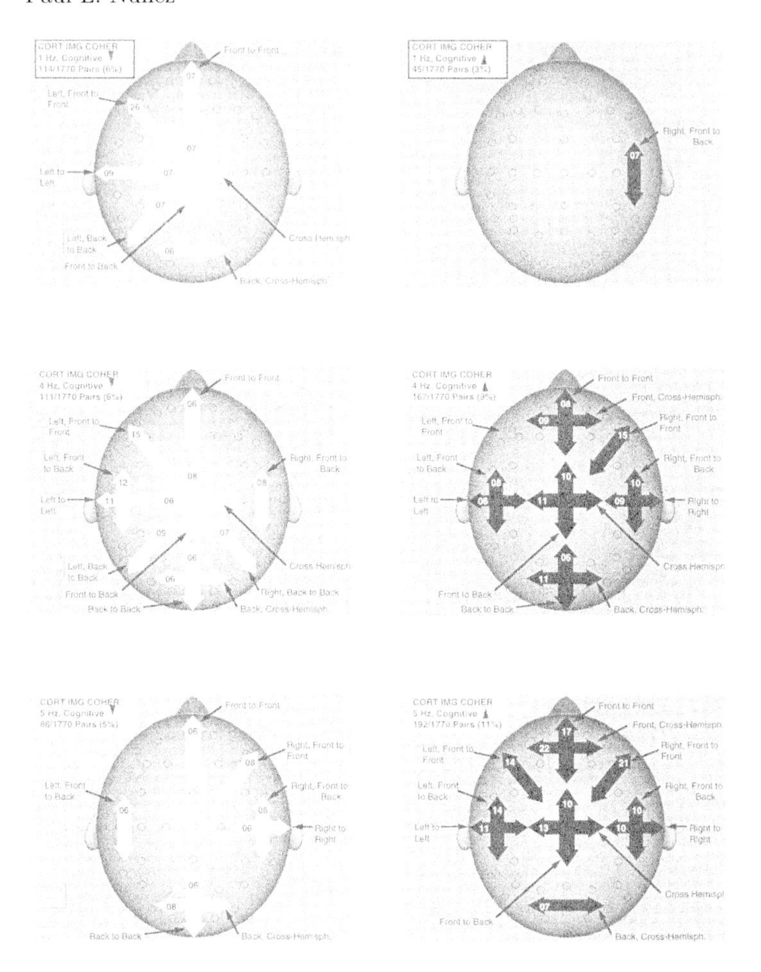

Fig. 5. Regional coherency changes between resting and cognitive states at 1, 4, and 5 Hz are shown for dura image estimates. Frequency resolution is 1 Hz. The subject alternated between three resting and three cognitive one-minute periods (six state changes). Only electrode pairs with fully consistent directions of coherency change are used to construct these plots. Percentages of consistent cortical image coherency changes between resting and cognitive states for subsets of regional electrode pairs are printed inside arrows. Patterns of "white behavior" (consistent decreases in coherency between resting and cognitive periods; white arrows) are shown in left column. Patterns of "black behavior" (consistent increases of coherency between resting and cognitive periods; black arrows) are shown in the right column. All groups of electrode pairs exhibiting more than 6% consistent behavior (roughly, $p < 0.005$) have arrows drawn. The pure chance level in each case is 1.6%. For example, the vertical arrow near the nose refers to 27 frontal electrodes (anterior to Cz). The white arrow in the plot at lower left indicates that 6% of the $27 \times 26/2 = 351$ paired sites exhibited white behavior at 5 Hz. The corresponding black arrow at the lower right shows that 17% of the same paired sites showed black behavior at 5 Hz. The remaining 77% exhibited inconsistent behavior at 5 Hz over the six state changes.

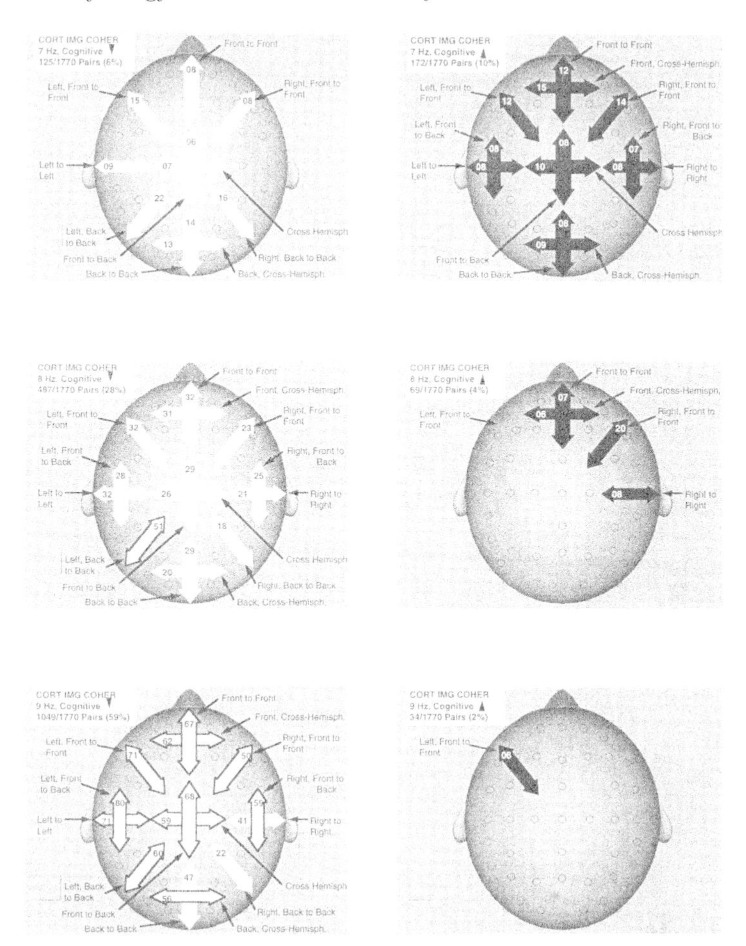

Fig. 6. Regional coherency changes between resting and cognitive states at 7, 8, and 9 Hz are shown for dura image estimates in the same format as Fig. 5. The plots in the lower row indicate that at 9 Hz, 67% of the frontal-to-frontal pairs show white behavior and none show black behavior at 9 Hz. The remaining 33% of paired frontal sites exhibit inconsistent behavior. Very large percentages of consistent behavior are emphasized by highlighted arrows.

nitudes of these changes are relatively small as indicated in Fig. 5. These data appear inconsistent with the very simple pictures often proposed, e.g., where EEG sources consist only of a few isolated "alpha generators" or "dipoles". Apparently, there are large and very detailed changes in the strength of dynamic "binding" at macroscopic scales between brain states.

6. Potential, Laplacian and dura image maps over the scalp appear to oscillate in the manner expected of standing waves, but with different spatial

structures owing to different spatial filtering. For example, instantaneous potential maps obtained at the successive extrema of a time plot of one channel are shown in Fig. 7. Corresponding dura images are plotted in Fig. 8. As expected, the Laplacian or dura image plots, which estimate potential on the dura surface, show much more detail as a result of filtering out of the very long wave scalp potentials associated with passive volume conduction distortion. But, without a perfect head model and more complete spatial sampling, we cannot distinguish genuine long wave source activity from volume conduction distortion. Laplacian or dura image filtering may remove substantial genuine long wave source activity in addition to that due only to volume conduction. Thus, Figs. 7 and 8 represent two different estimates of dura potential distributions.

7. Stable structure with apparent nodal lines occurs in both states. High resolution plots of magnitude and phase over the scalp reveal regions separated by several cm, with voltage oscillations 180 deg out of phase. Magnitude and phase patterns do not appear to change substantially between states, except for magnitude reduction in the 9-10 Hz band during the cogntive states. Nodal lines of putative standing waves are not easily observed in raw (reference) EEG data due to limited spatial sampling and volume conduction distortion. However, the spline-Laplacian and dura image algorithms improved the spatial resolution of these data to roughly 3 cm to reveal quasi-stable phase and magnitude structures in the alpha band on a second by second basis. Although the resolution is not quite good enough to fully confirm nodal lines, and the folded cortical surface confounds simple interpretations, the data are apparently consistent with standing waves of synaptic action density in the neocortical layer (Nunez et al. 1994; Nunez, 1995).

These data are also consistent with EEG and MEG studies carried out in Stuttgart, Germany and Boca Raton, United States (Fuchs et al. 1992; Friedrich and Uhl 1992; Friedrich et al. 1992; Jirsa et al. 1994, 1995; Haken 1996, 1998). For fixed brain states, spatial-temporal patterns could be accurately described in terms of two to five spatial modes (spherical harmonics), or "order parameters" in the parlance of complex physical systems theory (synergetics). The observed amplitude and phase structure support the idea that EEG and MEG dynamics are governed by macroscopic field equations for the apparent mixture of standing and traveling waves.

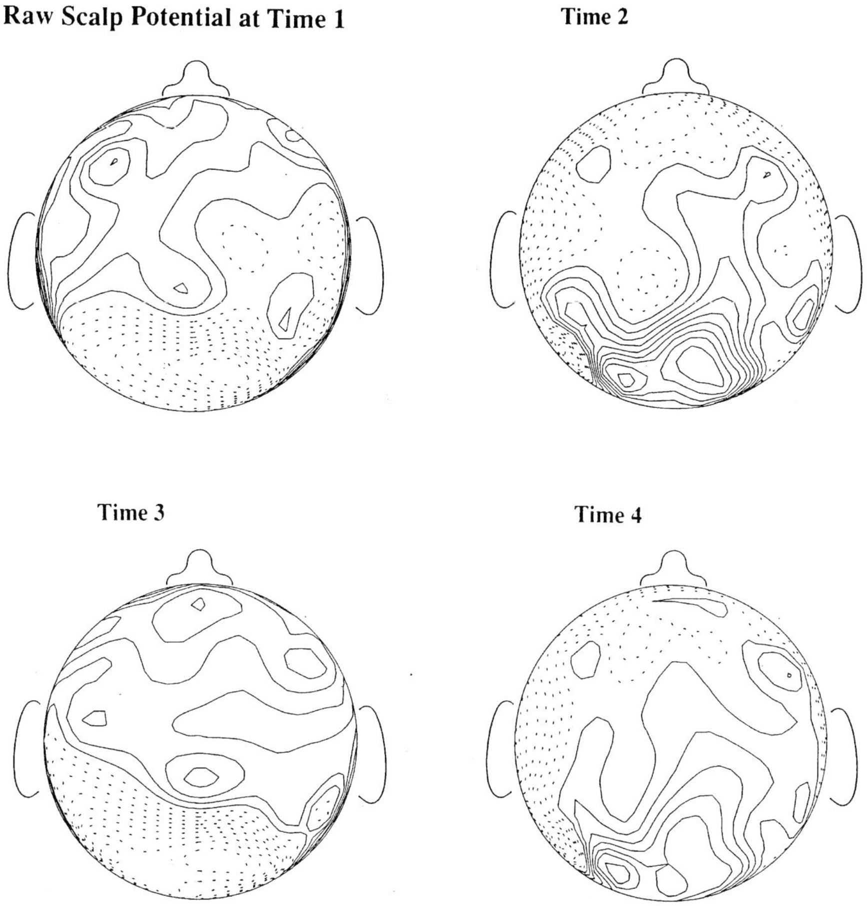

Raw Scalp Potential at Time 1

Time 2

Time 3

Time 4

Fig. 7. Instantaneous potential maps at four successive extrema of the P4 time plot were constructed using a spline function generated from potentials recorded at the 64 electrodes to estimate surface potential at about 640 locations. Solid lines indicate above average potential for the fixed time slice; maximum potential is approximately 30–40 microvolts. The large scale, anterior-posterior oscillation of potential was observed in many similar plots. Reproduced with permission from Nunez (1995).

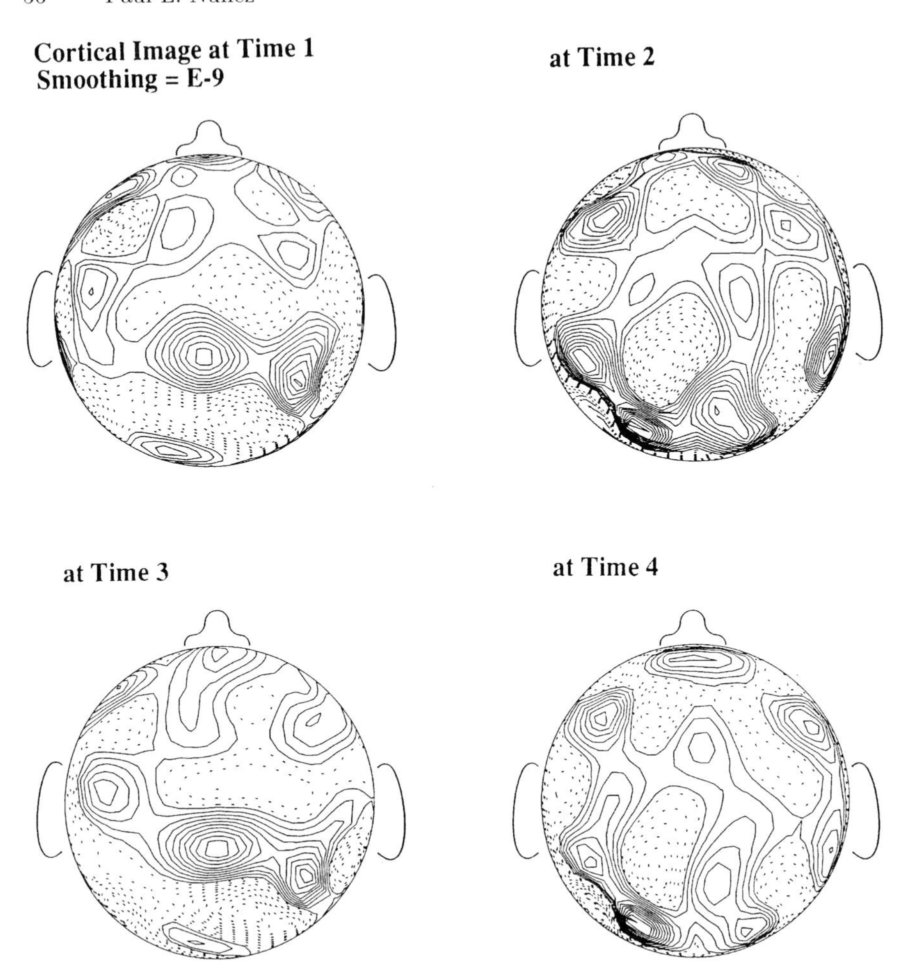

Fig. 8. Each time slice used in Fig. 7 was multiplied by the Melbourne dura imaging matrix to obtain instantaneous estimates of dura potential. The maps represent spatial band-pass versions of Fig. 7. The New Orleans spline-Laplacian maps (shown in Nunez, 1995) are similar to the Melbourne dura image maps.

4 A Physical Science Metaphor for Theoretical Development of Complex Systems

Given the enormous complexity of brains, some justification for development of (necessarily) crude neocortical dynamic theory is appropriate, especially for readers from outside the physical sciences who may question the utility of such theory. In order to address this issue and to illustrate the general philosophy motivating this chapter, I call on a metaphor from the physical sciences. Consider the fundamental idea of conservation of some substance X in a small volume of space, which is expressed quantitatively by the following equation:

$$\nabla \cdot (\rho \mathbf{v}) + \frac{\partial}{\partial t}\rho = 0. \tag{1}$$

Here ρ is the density of X (say X per cubic mm), where X might be mass, charge, thermal energy, radiation energy, probability, students, etc. The flow velocity of X is given by the vector $\mathbf{v}$; and the product $\rho\mathbf{v}$ is the flux of X. Equation (1) simply says that the net flux of X into some volume is proportional to the time rate of increase of the density of X within the volume, provided X is neither created nor distroyed in the volume. For example, if X represents students in a classroom, (1) indicates the obvious fact that by counting students going in and out all doors, we can determine the change in student density within the classroom, unless something unusual happens (e.g., a student gives birth in class). If X is electric charge, the flux $\rho\mathbf{v}$ is current density and (1) is exact. Equation (1) is also exact if X is probability; in this example (1) essentially says that the summed probabilities of all possible events equals one. If X is radiation energy, (1) requires an additional term on the right side to account for conversion of radiation energy into thermal energy (e.g., a potato in a microwave oven). The charge and radiation versions of (1) are both contained within Maxwell's equations. The probabilty version occurs in several physical systems, including classical gas kinetics, plasma physics, and all quantum systems. One of the most important lessions from these physical sciences is that one can take a very simple idea like the conservation of something, express it in precise mathematical language, and combine it with other equations and physical constraints to produce a solid foundation for a field of study.

Another example of (1) is fluid flow, e.g., water flow in pipes, air flow over airplane wings, or the global weather system. Equation (1) is then just the conservation of mass, which holds except in unusual circumstances (e.g., mass is converted into another form as in evaporation or nuclear reactions occur). These physical processes are typically very complicated, and (1) does not, by itself, allow us to predict dynamic behavior of fluids since it contains the four dependent variables, density and three velocity components. However, (1) places substantial constraints on possible dynamic behavior. These constraints are typically exploited by both experimental and theoretical scientists.

In order to predict dynamic behavior, (1) must be coupled to additional equations for the four dependent variables. In the fluid case, these are Newton's law in three directions, concerned with inertia, friction, and pressure forces. They are represented here symbolically by

$$\text{Vector Momentum Conservation } (\rho, p, \mathbf{v}) = 0, \tag{2-4}$$

where p is pressure, and the fifth equation is

$$\text{Equation of State } (\rho, p, T) = 0. \tag{5}$$

If the temperature T is constant (over relatively long times, thereby serving as a control parameter), five equations for the five dependent variables (ρ, p, and $\mathbf{v}$) are obtained. If T is a sixth variable, these five equations must be coupled to a sixth equation, the conservation of energy. In either case, fluid flow is nicely formulated as a well-posed, largely self-contained mathematical problem. If we were able to solve the general version of these five equations symbolically, predicting the dynamics of fluid motion would be a simple matter of plugging specific parameters into the general solution. However, no such general solution is known. Rather, substantial effort has been expended over the past two centuries by fluid physicists and engineers in obtaining limited solutions and using the unsolved equations directly to guide experiments. Today, giant computer programs based on these equations use input data recorded in wind tunnels to design modern airplanes or data from multiple world-wide sensors to predict global weather patterns. We should be impressed both by how far fluid mechanics has advanced in two centuries and by current limits to our understanding. For example, turbulent fluids will continue to provide substantial scientific and engineering challenges into the foreseeable future. How much we know, but how little we know!

In order to put the current state of brain dynamics in perspective, imagine ourselves as fluid engineers in the mid 19th century. We are relatively sure of the validity of (1), but are looking for plausible approximations to (2) through (5) that yield solutions approximately matching specific physical systems. A reasonable early approach is to seek simple fluid systems (e.g., quasi-linear fluid systems), or even to design such systems specifically for scientific experiments, with the goal of verifying or falsifying the approximate (perhaps linear) theory. Only after obtaining such limited success are we able to take the next step, predicting more complicated fluid dynamics using more accurate approximations to (1) through (5). In the next section, I suggest an anologous approach to the study of brain dynamics, a field which is likely to provide major challenges to neuroscientists well beyond the time when turbulence is fully mastered.

In the crude neocortical dynamic theory outlined here, three dependent variables are followed, excitatory and inhibitory "synaptic action" (number of active synapses per unit area of cortical surface, including the entire depth of cortex in the equivalent volume element) and action potential density, defined similarly. These variables are chosen in the theory because of their close connection to large scale EEG/MEG. That is, scalp data are believed to measure the modulation of synaptic action variables averaged over large surface areas and across the cortex. The three brain variables proposed here vary in time and location over the cortical surface. They describe the dynamics of model brains which can be compared with data, although in making such comparisons, spatial filtering of EEG/MEG by physical separation of sensors from sources and (for EEG) the head volume conductor must be accounted for.

The synaptic and action potential variables are then somewhat analogous to the five fluid variables $(\rho, p, \mathbf{v})$, which vary with time and location along a surface, e.g., an airplane wing. However, no basic conservation equation for the brain is proposed. Rather, the analogous equation describes a fundamental physiologic idea – action potentials fired at one cortical location produce synaptic action at other locations, generally at later times dependent on local synaptic delays and global delays due to finite propagation speeds of action potentials.

5 A Local/Global Theory of Large-Scale Neocortical Dynamics

A theory of the large scale neocortical dynamics, appropriate for verification, modification, or falsification with scalp EEG, is briefly outlined. The theory is derived from physiology and anatomy and contains no free parameters. Details may be found in Nunez (1995). The theory emphasizes a combination of local and global physiologic mechanisms. The term "local theory" indicates dynamic descriptions based on intracortical feedback "circuits", with signal delays due mostly to postsynaptic potential (PSP) rise and decay times. Feedback mechanisms at millimeter scales are critical to such theories. Local theories are most compatible with functional segregation, and global boundary conditions have no influence on predicted dynamic behavior. Such local theories have been published by Wilson and Cowan (1972, 1973), Lopes da Silva et al. (1974), Freeman (1975), van Rotterdam et al. (1981), and Zhadin (1984).

"Global theory" is based on signal delays that are mainly due to finite propagation speeds of action potentials along corticocortical fibers, providing positive feedback between multiple cortical regions. Furthermore, periodic boundary conditions due to the closed neocortical surface exert important influences on the global dynamics. This approach emphasizes functional integration in neural tissue. Global theory of EEG has been published by Nunez (1972, 1974, 1981, 1995), Katznelson (1981,1982), Srinivasan (1995), Jirsa and Haken (1997), and Robinson et al. (1997).

Both local and global mechanisms are well-established in neuroscience. However, their relative importance to neocortical dynamic behavior and EEG is controversial. Attempts to combine local and global effects and to evaluate their relative importance have been published by Nunez (1981, 1989, 1995), Nunez and Srinivasan (1993), Srinivasan and Nunez(1993), Tononi et al. (1994), Silberstein (1995a,b), Friston et al. (1995), Wright and Liley (1996), Jirsa and Haken (1996, 1997), Robinson et al. (1997), and Haken (1997,1998). Given our current meager understanding of physiological control parameter ranges, a plausible assumption is that both local and global mechanisms generally contribute to dynamic behavior, with one or the other more dominant in different brain states, or perhaps over different wave number ranges of the dynamics. With this philosophical approach, local versus global arguments of brain function and EEG reduce to issues of relative magnitudes of local and global control parameters occurring in each brain state.

The basic global equation links the excitatory "synaptic action" $h_E(\mathbf{r}, t)$ (number of active excitatory synapses per unit volume at time t and cortical location $\mathbf{r}$) to the number density of action potentials per unit volume $g(\mathbf{r}_1, t)$, fired at other times and other cortical locations $\mathbf{r}_1$, as indicated in Fig. 9. In order to define a surface distance measure, it is convenient to imagine a smooth or "inflated" cortical surface as indicated in Fig. 10. Remote cortical locations are connected by the excitatory cortico-cortical fibers which carry action potentials at speeds v. The complications of fiber density as a function of distance and distributed speeds of action potentials are lumped in the distribution function $R_E(\mathbf{r}, \mathbf{r}_1, v)$. This prescription leads to the relatively non-controversal, linear integral equation,

$$h_E(\mathbf{r}, t) = u(\mathbf{r}, t) + \int_0^\infty \mathrm{d}v \int_s R_E(\mathbf{r}, \mathbf{r}_1, v) \; g\left(\mathbf{r}_1, t - \frac{|\mathbf{r} - \mathbf{r}_1|}{v}\right) \mathrm{d}^2 r_1 . \quad (6)$$

Here $u(\mathbf{r}, t)$ is excitatory input to cortex from the midbrain, e.g., sensory input. The inner integral is over the cortical surface and the retarded time in the g parenthesis suggests possible traveling waves. A similar equation relates the inhibitory synaptic action $h_I(\mathbf{r}, t)$ to action potentials $g(\mathbf{r}, t)$. But, as a consequence of the inhibitory fibers being very short range (all intracortical), the spatial-temporal Fourier transforms of these two variables are simply related so that one variable may be eliminated from the equations at the onset. Since the cortical surface is closed, periodic boundary conditions are expected to cause wave interference and an infinite set of resonant frequencies for oscillations of the field variables $h_E(\mathbf{r}, t)$, $h_I(\mathbf{r}, t)$, and g($\mathbf{r}$,t).

Equation (6) is fundamentally linear since the number of active synapses at axon terminals is proportional to number of action potentials in cell bodies. However, (6) involves two dependent variables so a second equation is required for solution. The analogy between (6) and mass conservation in fluid mechanics (1) is noted (mass conservation is linear only for fluids that are approximately incompressible, e.g., low velocity flow). Both equations

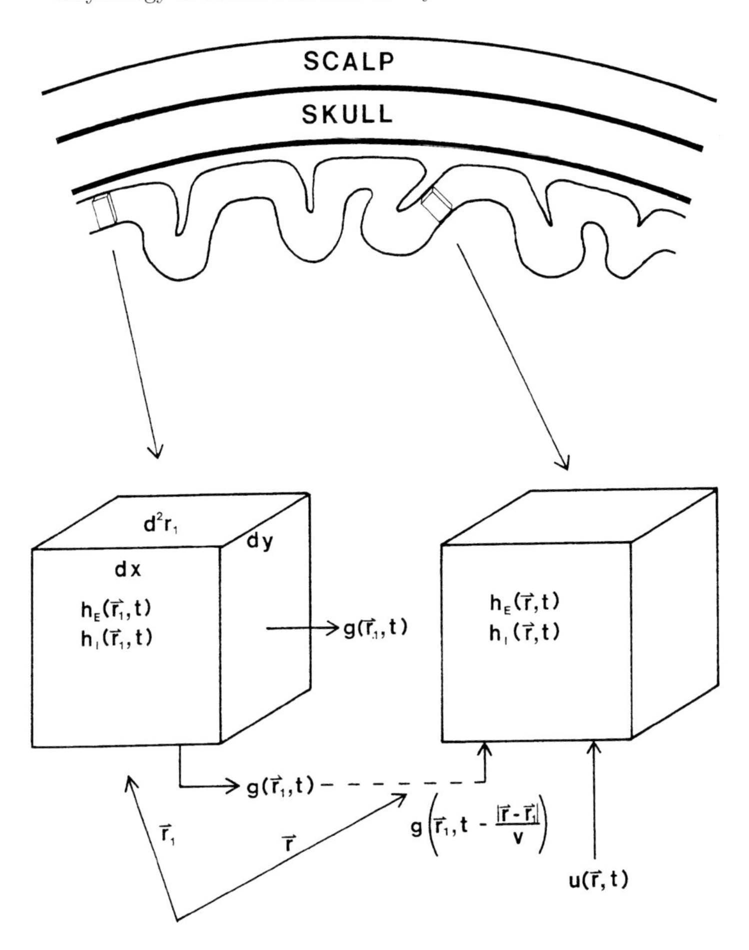

Fig. 9. A neocortical column at location $\mathbf{r}_1$ produces action potentials $g(\mathbf{r}, \mathbf{r}_1)$, thereby causing excitatory synaptic action $h_E(\mathbf{r}, t)$ at location $\mathbf{r}$ at later times, depending on separation distance between columns and action potential propagation speed v. The separation distance $|\mathbf{r} - \mathbf{r}_1|$ may be measured along a mentally inflated, smooth surface, as shown in Fig. 10. Reproduced with permission from Nunez (1995).

appear to be approximately valid in many practical cases, thereby providing relatively robust starting points for theories in both fields. In brain dynamics, a second equation connects the number density of action potentials fired in a tissue mass to excitatory and inhibitory synaptic action input. This second equation is generally nonlinear since it involves complicated, state-dependent local circuitry. But, as a very first approximation, this equation may be linearized about a fixed dynamic state (Nunez 1974, 1995), with the understanding that this is no more than a useful intermediate step towards

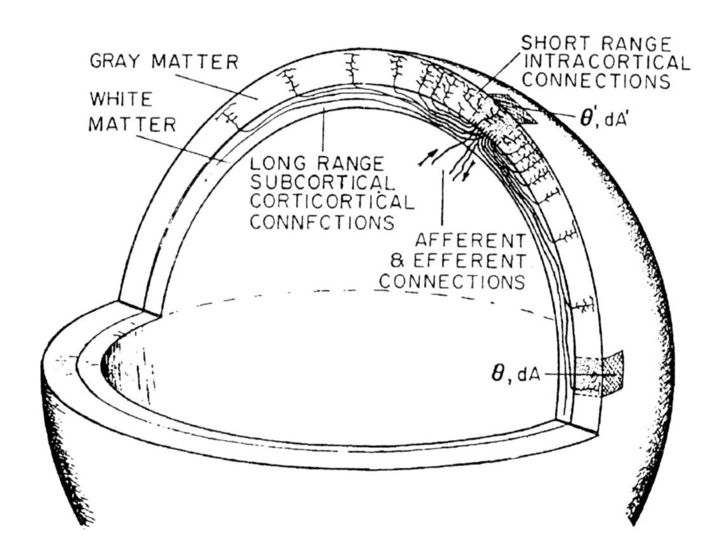

Fig. 10. This spherical structure is proposed as a crude model for neocortex, obtained by mentally inflation to smooth the fissured surface. Sphere radii, based on various estimates of neocortical surface area, are in the 11 to 18 cm range. The peak in the action potential velocity distribution function for myelinated cortico-cortical fibers is approximately 600 to 900 cm/sec. A very simple model of linear, non-dispersive standing waves in the spherical shell then predicts a fundamental resonant frequency in roughly the 7 to 18 Hz range, consistent with many EEG phenomena. More physiologically based global theories predict standing waves in the same general frequency range, but with modes dependent on at least one brain state (control) parameter (Nunez 1981, 1995). Reproduced with permission from Katznelson (1981).

more accurate theory. No claim is made that the brain is generally linear (unfortunately, a common interpretation of this theory!). That is, the linear approximation may be somewhat appropriate for very large scale dynamics and for brain states of high coherence, but it is most useful as a preliminary step to quasi-linear approximations (Srinivasan and Nunez 1993; Nunez 1995; Jirsa and Haken 1997).

To further simplify the theory, the closed neocortical surface may be replaced by a one-dimensional, closed loop of tissue (Nunez 1974, 1981) or spherical shell representing a smooth ("inflated") neocortex, as indicated in Fig. 10 (Katznelson 1981; Nunez 1995). Wave damping is predicted by the theory to be lowest in the direction of the longest cortico-cortical fibers, which run in anterior/posterior directions of each brain hemisphere, providing support for the closed loop geometry. On the other hand, the spherical model is faithful to the two dimensional cortical surface. Thus, the spherical and one-dimensional loop models may be viewed as limiting cases, representing isotropic and extreme anisotropic cortico-cortical connections. It is not now clear which model is more physiologically realistic.

At the large scales of scalp EEG, the local transfer function $L(\omega; Q)$ described here is approximately independent of spatial frequency because scalp wavelengths are much longer than intracortical fibers, which provide local length scales. $L(\omega; Q)$ depends on temporal frequency ω and the local control parameter Q, which is proportional to local (intracortical) feedback gains (van Rotterdam et al. 1982). Positive feedback in cortico-cortical fibers leads to global functions $G(k, \omega)$ or $G_I(\omega)$, where k is wave number and l is the spherical harmonic index in the loop and spherical models, respectively. The spherical symmetry assumption causes $G_l(\omega)$ to be independent of the second (m) index.

The local/global neocortical transfer function in a spherical shell $T_l(\omega; B, Q)$, which is derived from these known physiological processes may be expressed as (Nunez 1989, 1995),

$$T_l(\omega; B, Q) = \frac{L(\omega; Q)}{1 - BG_l(\omega)L(\omega; Q)}. \tag{7}$$

The global function $G_l(\omega)$ is multiplied by a global control parameter B, which is proportional to positive feedback gains established by corticocortical fibers. In the case of dynamics on a spherical surface, the (l, m) indicies label spatial frequencies of the spherical harmonics, which are functions of two surface coordinates (latitude and longitude). Nodal lines of a few of the spherical harmonic functions are plotted in Fig. 11. Equation (7) illustrates several predicted aspects of EEG dynamics that may hold, even when more accurate theories are later developed, because they have general validity, i.e., they appear to be largely independent of detailed physiological assumptions of this specific theory.

The full (local/global) dynamic transfer function T is large if the local transfer function L is large or if L is approximately equal to $(1/BG)$. This latter condition defines a "matching" of local and global resonances, which is a possible mechanism to facilitate interaction between remote cell assemblies, thereby addressing the so called "binding problem" of brain science. Regions of (l, ω) space where T is large correspond to multiple branches of dispersion relations for "brain waves". Such dynamics may exhibit large changes due to changes in the parameters (B, Q), which control the state of this simple model cortex by relatively slow chemical and electrical input from midbrain.

If long range positive feedback is negligible (e.g., there are no corticocortical fibers, or the control parameter B is small due to specific neuromodulatory influences), the term BG in (7) is small and the full transfer function is essentially the local transfer function, i.e.,

$$T_l(\omega; Q) = L_l(\omega; Q). \tag{8}$$

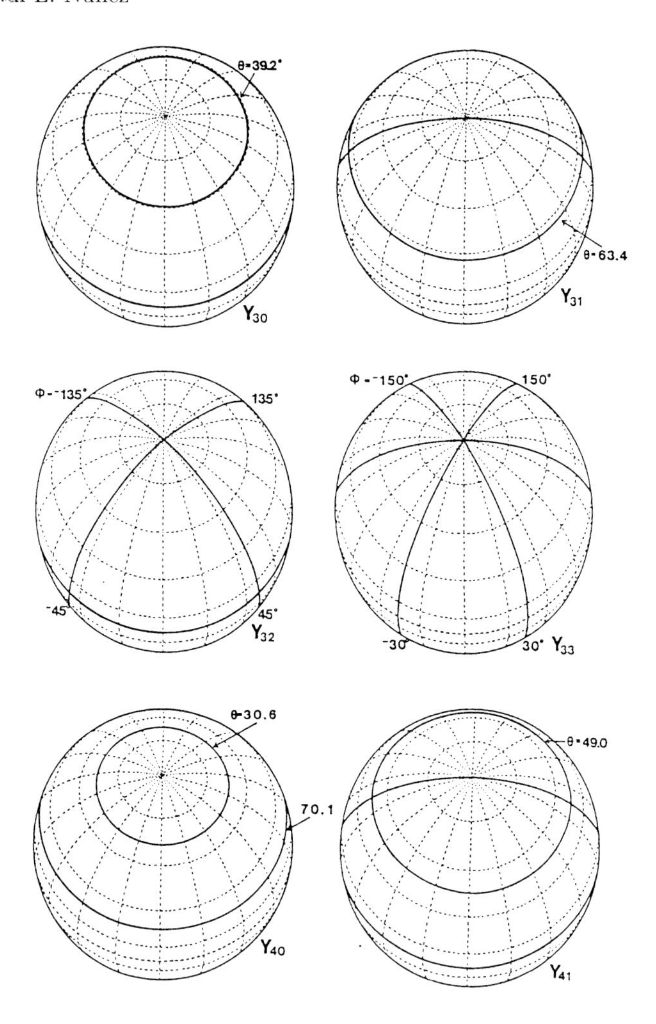

Fig. 11. Examples of the spherical harmonic functions $Y_{lm}(\theta, \phi)$ are illustrated. These functions form an orthogonal set on a sphere. Nodal lines are shown as heavy lines. Any spatial dependence of a field variable $f(\theta, \phi)$ in a spherical shell at fixed time may be expressed as a double sum over the integers (l, m), called a generalized Fourier series. A physical example is the Schumann resonances, which are electromagnetic fields in the spherical shell formed by the earth and the bottom of the ionosphere. Traveling waves are produced by ongoing lightining strikes at multiple locations. Standing waves occur in the shell as a result of interference of waves traveling in different directions around the sphere (periodic boundary conditions). A distinct resonance frequency is associated with each set of functions $Y_{lm}(\theta, \phi)$, with the index m running from $-l$ to $+l$. These modes are degenerate in a homogeneous, isotropic shell; that is, resonant frequencies are independent of the m index.

We conjecture that this is a valid approximation for some brain states, at least in some cortical regions. In this basic version of the theory, the cortical/white matter system is assumed to be homogeneous, but in more detailed versions, the parameters Q and B may be functions of cortical location. The dominant local frequency band depends on rise and decay times of PSPs, and the local feedback gain Q. The magnitude of this gain is not known for genuine tissue. But, for purposes of illustration, Q may be chosen so that stable waves with a peaked (dominant) transfer function are obtained. If EPSP and IPSP rise times (with matching decay obtained from cable theory) are about 10 and 20 ms, respectively, the dominant local frequency range is near 10 Hz (van Rotterdam et al. 1982). However, published EPSP and IPSP rise times are typically much shorter (Rall 1967), perhaps 2 and 5 ms, respectively. In this case, dominant local frequencies are in the 40 Hz range, a range sometimes associated with brain information processing.

In regions of (l, ω) space where periods are much longer than local delay times (rise and decay of PSPs), the local transfer function $L_l(\omega; Q)$ is close to one. In this limiting case, the full transfer function in (7) simplifies to

$$T_l(\omega; B) = \frac{1}{1 - BG_l(\omega)}, \tag{9}$$

and neocortical dynamics is dominated exclusively by global mechanisms in this frequency range. By setting the denominator of (9) to zero, the global dispersion relation, $\omega = \omega_l$, is obtained. Again we conjecture that this is a valid approximation for some brain states. The most likely corresponding EEG states are obtained under anesthesia, some sleep stages, and the awake alpha rhythm, e.g., states of minimal cognitive processing that exhibit widespread, spatially coherent EEG most likely associated with globally dominated dynamics, as partly illustrated in the upper parts of Figs. 4 and 5. For relatively small values of the control parameter B, dominant frequencies appear to be only weakly dependent on B. The dominant frequencies depend on corticocortical propagation velocity and allowed eigenfunctions (spherical harmonics) for the global system.

Global boundary conditions (periodic) allow oscillatory dynamics to persist only for specific wavelengths [or integer (l, m) indices on the sphere]. For example, excitatory and inhibitory synaptic action fields must be finite and single valued functions of surface coordinates. Due to the interference of propagating waves of synaptic action, only certain discrete wave lengths of standing waves can persist. In complex geometry, inhomogeneous, anisotropic, or nonlinear systems, and/or systems closed on themselves, the allowed wave lengths (or quantum numbers l, m) may be limited by complex rules; however the basic principle of wave length restriction still applies. Examples include quantum wave functions in atoms, electromagnetic radiation generated by random lightning strikes in the resonant cavity formed by the earth's surface and the bottom of the ionosphere (Schumann resonances), and as suggested here, oscillations of synaptic action (and by implication EEG) in the closed neocortical shell. When the wave length restriction is combined with the dynamic restrictions implied by the transfer function (or dispersion relation), preferred temporal frequency ranges are predicted. In other words, such systems act as band pass spatial-temporal filters. These special (or resonant) frequency ranges emerge from a combination of tissue (or other media) properties and global boundary conditions. When periodic boundary conditions for the neocortical surface are combined with the global transfer function, (9), a series of global modes (or global resonant frequencies) is predicted. When the parameter B is small to moderate, the lowest (fundamental) mode is roughly in the 10 Hz range (within a factor of perhaps three). When the parameter B is increased, mode frequencies can decrease rapidly, in a manner suggestive of transitions from the awake to sleeping states or varying depths of halothane anesthesia (Nunez 1995).

We have described extreme cases when local, (8), or global, (9), processes dominate neocortical dynamics. However, most brain states may involve both large scale functional integration (facilitating global contributions to the dynamics) and functional segregation (facilitating local effects). Examples of local/global dynamic behavior in the model system described by the transfer function (7) are discussed in Nunez (1995). Realistic delay parameters suggest that global mechanisms contribute substantially to the lower frequency range (below 10 or 15 Hz). In addition, both local and global mechanisms (with larger l indicies) may combine to provide a dominant frequency range of roughly 30 to 45 Hz, which has been associated with cognitive events. While the physiological parameters which determine the specific shapes of the transfer function are not known with sufficient accuracy for close quantitative comparison with EEG data, all parameters in these examples are in plausible physiological ranges. That is, in contrast to metaphorical approaches to modeling, the theory contains no "free" parameters, which can be adjusted to fudge qualitative agreement between theory and experiment. However, in some experiments, physiological parameters may be estimated with one set of experiments, and these parameter estimates used in the equations to predict outcomes of a second set of experiments.

6 Quasi-linear Theory

The theory presented here depends on many uncertain properties of neocortex and its white matter connections. No strong claim can be made for quantitative accuracy. However, the linear version appears to be robust in its qualitative predictions. Several dozen apparently correct, but mostly qualitative, experimental predictions of the local/global theory outlined here are described in Nunez (1995). These include connections to EEG resonant behavior, phase velocity of traveling waves, and standing waves recorded from the scalp.

Although the linear theory has clear descriptive advantages, a more accurate approximation to input-output at large scales are sigmoidal relations connecting synaptic action input to action potential output (Wilson and Cowan 1973; Freeman 1975). This is implied by the generally accepted idea that action potential output from a tissue mass is zero below some threshold and is ultimately limited by saturation effects or additional negative feedback mechanisms in the tissue. In the context of the global theory, the control parameter B of the linear version effectively becomes a function of the synaptic action variables $h_{\mathrm{E}}(\mathbf{r},t)$ and $h_{\mathrm{I}}(\mathbf{r},t)$. One very approximate approach (Nunez 1995) involved expanding B about a point on the sigmoid function. Semi-quantitative arguments were used derive differential equations for each loop mode $H(k,t)$. In one limiting case, the modes were uncoupled and Van der Pol-like equations were obtained for each spatial mode,

$$\frac{\mathrm{d}^2 H}{\mathrm{d}t^2} - 2\gamma(1-H^2)\frac{\mathrm{d}H}{\mathrm{d}t} + [\omega_0^2 - 2\lambda_{\mathrm{E}}^2 v_{\mathrm{E}}^2 H^2]H = 0, \tag{10}$$

where $\gamma, \omega_0, \lambda_{\mathrm{E}}$, and v_{E} are physiological parameters defined in the Appendix of Nunez (1995). These equations have limit cycle solutions. In other examples, the modes are coupled so that individual modes may follow point, quasi-periodic, or chaotic attractors (Srinivasan and Nunez 1993). While the details of this crude, quasi-linear approach cannot be taken too seriously at this point (partly because new control parameters are required in non-linear theory and these are mostly unknown), the quasi-linear theory suggests paths to more accurate theoretical work.

A different quasi-linear approximation to sigmoidal input/ouput in the global theory was used to describe evoked magnetic field behavior (Jirsa and Haken 1996). The 37-channel MEG recorded in Boca Raton was driven by auditory stimuli of varying frequency (Kelso et al. 1992). The subject was asked to push a button between each consecutive tone. When the stimulus rate reached a critical value (near 1.75 Hz), the subject was no longer able to perform this task, and switched from out-of-phase to in-phase (synchronous) finger motion, indicating a brain state change, or phase transition in the parlance of complex physical systems. In the brain model, this stimulus was described in terms of parametric driving. The spatial-temporal MEG dynamics was described in terms of a competition between two spatial modes, with time-dependent mode amplitudes as "order parameters" (Fuchs et al. 1992; Jirsa et al. 1995). The first order parameter dominated the pre-transition state and oscillated with the stimulus frequency; the second order parameter, with twice the stimulus frequency, dominated the post transition state. A nonlinear differential equation was derived as a an extension of the global theory, with auditory and sensory cortices considered as local circuits embedded within the neural tissue (Jirsa and Haken 1997). The theoretical model was able to reproduce the essential features of MEG dynamics, including the phase transition. Thus, a triple correspondence was obtained, relating behavior, MEG data, and physiologically based theory.

7 How Are Cell Groups Formed and What Is Their Relationship to Psychology?

My goal here has been a modest one, that of laying a tentative dynamical foundation, based on both theory and EEG data, on which more physiologically based cognitive theories may be constructed. Or, expressed another way, any physiologically-based dynamical foundation (e.g., physiology of neural mass action) is likely to provide intermediate and large scale constraints on new cognitive models, in a manner analogous to constraints imposed by established physiology at the single neuron level. For example, Edelman's (1992) "theory of neuronal group selection" suggests that selective coordination of complex patterns by "reentry" (repeated reciprocal interactions between cell groups) is the basis of behavior, and that reentry combined with memory provides the bridge between physiology and psychology. Or, Philips and Singer (1998) propose synchronized population codes, with dynamic activity "coordinated within and between regions through specialized contextual connections". These views appear to fit nicely with the dynamical picture painted here, as do other, partly overlapping ideas. One goal of dynamical theory development is to help delineate differences between cognitive theories, e.g., to pin down qualitative concepts and distinguish substantive from semantic controversies. To this physical/neuroscientist, who operates on the periphery of mainstream cognitive science, it appears that such semantic arguments and turf wars too often infect scientific exchange. In this regard, the conceptual framework summerized here is a preliminary part of the long range goal of bridging the current wide gaps between physiology and psychology.

I have argued that the development of neocortical dynamic theory and its experimental verification is essential to rapid progress in neuroscience, especially when fundamental issues of information processing and conscious experience are studied. This follows from the idea that intermediate and large scale physiology require quantitative understanding, just as we require knowledge of synapses, neurotransmitters, and action potentials at small scales. Separate theories at multiple scales, as well as efforts to cross scales, will provide formidable new challenges to theoreticians and experimentalists in the future.

We know that large scale neocortical dynamic behavior measured with EEG is correlated with internal experience. Today's established correlations can perhaps be described as moderate, but there is reason to hope for even stronger connections in the near future using more advanced experimental methods, including improved EEG spatial resolution, advanced pattern recognition methods, closer connection to dynamic theory, and more effective integration with MEG, MRI, PET, etc. (Nunez 1995; Silberstein 1995a,b).

Several theoretical studies suggest that macroscopic dynamic fields of synaptic action (estimated with EEG/MEG) are both influenced by and act back on dynamics at smaller spatial scales not easily accessible to experimental measure (Ingber 1982; Ingber and Nunez 1990; Nunez and Srinivasan 1993; Nunez 1995). It has been further suggested that such hierarchical interactions may be an essential ingredient of consciousness (Harth 1993; Ingber 1995a,b), perhaps in a manner similar to hierarchical interactions in the human global social system (Nunez 1995; Scott 1995; Freeman 1995). If brain information processing takes place at multiple spatial scales, one may conjecture that characteristic time scales (or resonant mode frequencies) of these levels occur in the same general range. This allows information processing over several minicolumns, simultaneous with mesoscopic interactions at millimeter (e.g., macrocolumn) and global scales (Ingber 1985, 1995a,b; Nunez 1989, 1995). Furthermore, in the context of the global theory outlined here, multiple global modes can drive local modes (which vary with location), thereby facilitating a top-down mechanism for establishing coherent oscillations in widely separated cell groups (Singer 1993; Bressler 1995; Phillips and Singer 1998). If this guess is correct and such multiscale information processing actually is essential to consciousness, brain evolution may have exploited synaptic plasticity at each spatial scale to effect a matching of time constants at each level. This "top-down, multiscale, neocortical dynamic plasticity" would evidently be constrained minimally at the single neuron level, but more strongly by neocortical boundary conditions (Nunez 1996). These and other speculations, based on intimate marriages of theory and experiment, will help provide a wealth of new ideas for future study. Twenty-first century neuroscience will certainly be interesting!

References

Bressler SL (1995): Large scale cortical networks and cognition. Brain Research Reviews 20: 288–304.

Braitenberg V (1977): *On the Texture of Brains.* New York: Springer-Verlag.

Braitenberg V (1978): Cortical arichitechtonics: general and areal. In: MAB Brazier and H Petsche (Eds.) *Architechtonics of the Cerebral Cortex.* New York: Raven Press, 443–465.

Braitenberg V and Schuz A (1991): *Anatomy of the Cortex: Statistics and Geometry.* New York: Springer-Verlag.

Broadman K (1909): *Vergleichende Lokalisationslehre der Grosshirnrinde in ihren Prinzipien dargestellt auf Grund des Zellenbaues.* Leipzig: Barth.

Edelman GM (1992): *Bright Air, Brilliant Fire.* New York: Basic Books.

Freeman WJ (1975): *Mass Action in the Nervous System.* New York: Academic Press.

Freeman WJ (1995): *Societies of Brains.* Hillsdale, New Jersey: Lawerence Erlbaum Associates.

Friedrich and Uhl (1992): Synergetic analysis of human electroencephalograms: petit-mal epilepsy. In: R. Friedrich and A. Wunderlin (Eds.) *Evolution of Dynamic Structures in Complex Systems.* Berlin: Springer-Verlag, pp. 249–265.

Friedrich R, Fuchs A, and Haken H (1992): Spatio-temporal EEG patterns. In: H Haken and HP Koepchen (Eds.) *Rhythms in Physiological Systems*. Berlin: Springer-Verlag, pp. 315–338.

Friston KJ, Tononi G, Sporns O, Edelman GM (1995): Characterising the complexity of neuronal interactions. Human Brain Mapping 3: 302–314.

Fuchs A, Kelso JAS, and Haken H (1992): Phase transitions in the human brain: spatial mode dynamics. International Journal of Bifurcation and Chaos 2:917–939.

Haken H (1998): What Can Synergetics Contribute to the Understanding of Brain Functioning?, this volume

Harth E (1993): *The Creative Loop*. New York: Addison-Wesley.

Ingber L (1982): Statistical mechanics of neocortical interactions. I. Basic formulation. Physica D 5:83–107.

Ingber L (1995a): Statistical mechanics of multiple scales of neocortical interactions. In: PL Nunez, *Neocortical Dynamics and Human EEG Rhythms*. New York: Oxford University Press, pp. 628–681.

Ingber L (1995b): Statistical mechanics of neocortical interactions; High resolution path-integral calculation of short-term memory. Physical Review E 51: 5074–5083.

Ingber L and Nunez PL (1990): Multiple scales of statistical physics of neocortex: Applications to electroencephalography. Mathematical and Computer Modelling 13:83–95.

Jones EG (1987): Broadman's areas. In: G. Adelman (Ed.), *Encyclopedia of Neuroscience*. Vol 1. Boston: Birkhauser, pp. 180–181.

Jirsa VK and Haken H (1996): Field theory of electromagnetic brain activity. Physical Review Letters 77: 960–963.

Jirsa VK and Haken H (1997): A derivation of a macroscopic field theory of the brain from the quasi-microscopic neural dynamics. Physica D 99:503–526.

Jirsa VK, R Friedrich, Haken H, and Kelso JAS (1994): A theoretical model of phase transitions in the human brain. Biological Cybernetics 71:27–35.

Jirsa VK, R Friedrich, and Haken H (1995): Reconstruction of the spatio-temporal dynamics of a human magnetoencephalogram. Physica D 89:100–122.

Katznelson RD (1981): Normal modes of the brain: neuroanatomic basis and a physiologic theoretical model. In: PL Nunez, *Electric Fields of the Brain: The Neurophysics of EEG*. New York: Oxford University Press, pp. 401–442.

Katznelson RD (1982): Deterministic and Stochastic Field Theoretic Models in the Neurophysics of EEG. Ph.D. Dissertation. La Jolla, CA: University of California at San Diego.

Kelso JAS, Bressler SL, Buchanan S, DeGuzman GC, Ding M, Fuchs A, and Holroyd T (1992): A phase transition in human brain and behavior. Physics Letters A 169:134–144.

Krieg WJS (1963): *Connections of the Cerebral Cortex*. Evanston, Illinois: Brain Books.

Lopes da Silva FH, Hoeks A, Smits H and Zetterberg LH (1974): Model of brain rhythmic activity. Kybernetik 15:27–37.

Nunez PL (1972): The brain wave equation: A model for the EEG. Presented to American EEG Society Meeting, Houston, Oct., 1972.

Nunez PL (1974): The brain wave equation: A model for the EEG. Mathematical Biosciences 21:279–297.

Nunez PL (1981): *Electric Fields of the Brain: The Neurophysics of EEG.* New York: Oxford University Press.

Nunez PL (1989): Generation of human EEG by a combination of long and short range neocortical interactions. Brain Topography 1: 199–215.

Nunez PL (1995): *Neocortical Dynamics and Human EEG Rhythms.* New York: Oxford University Press.

Nunez PL (1996): Multiscale neocortical dynamics, experimental EEG measures, and global facilitation of local cell assemblies. Behavioral and Brain Sciences 19: 305–306.

Nunez PL and Srinivasan R (1993): Implications of recording strategy for estimates of neocortical dynamics using EEG. Chaos: An Interdisciplinary Journal of Nonlinear Science 3: 257–266.

Nunez PL, Silberstein RB, Cadusch PJ, Wijesinghe R, Westdorp AF, and Srinivasan R. (1994): A theoretical and experimental study of high resolution EEG based on surface Laplacians and cortical imaging. Electroencephalography and Clinical Neurophysiology 90: 40–57.

Nunez PL, Srinivasan R, Westdorp AF, Wijesinghe RS, Tucker DM, Silberstein RB, and Cadusch PJ (1997): EEG coherency I: Statistics, reference electrode, volume conduction, Laplacians, cortical imaging, and interpretation at multiple scales. Electroencephalography and Clinical Neurophysiology 103: 499–515.

Phillips WA and Singer W (1998): In search of common foundations for cortical computation. Behavioral and Brain Sciences, in press.

Rall W (1967): Distinguishing theoretical synaptic potentials computed for different soma-dendritic distributions of synaptic input. Journal of Physiology 30:1138–1168.

Robinson PA, Rennie CJ, Wright JJ (1997): Propagation and stability of waves of electrical activity in the cerebral cortex. Physical Review E 56: 826–841.

Scott AC (1995): *Stairway to the Mind.* New York: Springer-Verlag.

Silberstein RB (1995a): Steady-state visually evoked potentials, brain resonances, and cognitive processes. In: PL Nunez, *Neocortical Dynamics and Human EEG Rhythms.* Oxford University Press, pp. 272–303.

Silberstein RB (1995b): Neuromodulation of neocortical dynamics. In: PL Nunez, *Neocortical Dynamics and Human EEG Rhythms.* Oxford University Press, pp. 591–627.

Singer W (1993): Synchronization of cortical activity and its putative role in information processing and learning. Annual Reviews of Physiology. 55:349–374.

Srinivasan R and Nunez PL (1993): Neocortical dynamics, EEG standing waves and chaos. In: BH Jansen and ME Brandt (Eds.) *Nonlinear Dynamical Analysis of the EEG.* London: World Scientific, pp. 310–355.

Srinivasan R (1995): A Theoretical and Experimental Study of Neocortical Dynamics. Ph.D Dissertation. New Orleans: Tulane University.

Srinivasan R, Nunez PL, Tucker DM, Silberstein RB, and Cadusch PJ (1996): Spatial sampling and filtering of EEG with spline-Laplacians to estimate cortical potentials, Brain Topography 8:355–366.

Srinivasan R, Nunez PL, Silberstein RB (1998): Spatial filtering and neocortical dynamics: estimates of EEG coherence. IEEE Transactions on Biomedical Engineering 45:814–826.

Tononi G, Sporns O, Edelman GM (1994): A measure for brain complexity: Relating functional segregation and integration in the nervous system. Proceedings of the National Acadamy of Sciences USA. 91:5033–5037.

van Rotterdam A, Lopes da Silva FH, van der Ende J, Viergever MA, and Hermans AJ (1982): A model of the spatial temporal characteristics of the alpha rhythm. Bulletin of Mathematical Biology. 44: 283–305.

Wilson HR and Cowan JD (1972): Excitatory and inhibitory interactions in localized populations of model neurons. Biophysical Journal 12: 1–23.

Wilson HR and Cowan JD (1973): A mathematical theory of the functional dynamics of cortical and thalamic nervous tissue. Kybernetik 13:55–80.

Wright JJ and Liley DTJ (1996): Dynamics of the brain at global and micoscopic scales: Neural networks and the EEG, Behavioral and Brain Sciences. 19: 285–295.

Zhadin MN (1984): Rhythmic processes in cerebral cortex. Journal of Theoretical Biology. 108: 565–595.

Traversing Scales of Brain and Behavioral Organization I: Concepts and Experiments

J.A. Scott Kelso, Armin Fuchs, and Viktor K. Jirsa

Program in Complex Systems and Brain Sciences, Center for Complex Systems, Florida Atlantic University, 777 Glades Road, Boca Raton, FL 33431, USA

1 Introduction

In this paper, and the ones following, we will present an approach to understanding behavior, brain and the relation between them. The present contribution provides a sketch of the strategy we have adopted toward the brain-behavior relation, notes its main tenets and applies them to a new and very specific experiment that uses large scale SQuID arrays to determine how the human brain times individual actions to environmental events. A second paper (Fuchs, Jirsa and Kelso this volume) will describe in more detail the various methods we and others have used to analyze and visualize the spatiotemporal activity of the brain and to extract relevant features from experimental data. Finally, in a third paper (Jirsa, Kelso and Fuchs this volume) we will spell out a theory, grounded in the neuroanatomy and neurophysiology of the cerebral cortex, that serves to connect neural and behavioral levels of description for the paradigmatic case of bimanual coordination. Our collective goal in these three papers is to set the stage for a principled move from phenomenological laws at the behavioral level to the specific neural mechanisms that underlie them. With respect to the history of science our approach is entirely conventional. Fundamentally, it begins with the identification of the macroscopic behavior of a system and attempts to derive it from a level below. Even for physical systems, however, the derivation of the "macro" from the "micro" is nontrivial. Only in the 70's, for example, was it first possible to derive the behavior of ferromagnets (as described by Landau's mean field theory) from more fundamental grounds using the so-called renormalization group method that earned Kenneth Wilson the Nobel Prize in 1982. Likewise, it took the genius of Hermann Haken to derive the behavior of a far from equilibrium system like the laser from quantum mechanics (Haken 1970). Thus, some 70 years after atoms were discovered did it become possible to derive macroscopic properties of certain materials and optical devices from a more microscopic basis, and only then using rather sophisticated mathematical techniques.

What lessons can be learned from such successes? One is that it is crucial to first have a precise description of the macroscopic behavior of a system in order to know what to derive. Another is that even in a system whose microscopic constituents are homogeneous (unlike, say the neurons and neural transmitters of the brain) special methods are needed to handle events and interactions that are occurring simultaneously on many spatial and temporal scales. For example, even in a ferromagnet, it is not possible to derive the macro from the micro in a single step. Rather the so-called block spin technique proceeds in a series of steps each of which must be repeated many times in order to calculate the overall level of magnetization (Wilson 1979). Likewise, in a heterogeneous, hierarchically organized system like the nervous system, it is necessary to proceed in a level by level fashion with an intimate interplay between theory and experiment.

2 The Strategy

Our strategy for traversing scales is shown in Fig. 1. Inspired by synergetics (Haken 1983) the basic idea is to identify relevant variables characterizing coordinated or collective states of the system and the collective variable's dynamics (i.e. equations of motion for collective variables). Note that in complex neurobehavioral systems, these are not known in advance, but have to be found. The experimental method uses transition points to clearly distinguish different coordinated behaviors. In complex systems in which many features can be measured but not all are relevant, we assume that the variable that changes qualitatively is the most important one for system function.

It is these collective variables that are mapped on to a dynamical system (see Collective level, Fig. 1). We remark that the behavioral dynamics for a given system must be understood on its own terms. Relative phase, ϕ, for example, proves to be a crucial collective variable or order parameter in a number of situations, but in others (e.g. trajectory formation of a single multijoint limb (DeGuzman, Kelso and Buchanan 1997) recruitment of additional degrees of freedom in coupled bimanual movements (Kelso, Buchanan, DeGuzman and Ding 1993) etc.) amplitudes play an important role and must be included in the collective variable dynamics. Likewise, equations of motion at the behavioral level must be (and have been) elaborated to include the influence of intention, environmental demands, handedness, learning, memory and attention (see refs. in Kelso 1995, and Treffner and Turvey 1996 for a recent example). Also, further experiments are necessary to identify the component dynamics and further theory is needed to derive the collective variable dynamics from nonlinear couplings among the components (see Component level, Fig. 1). We note that this step has been accomplished in a large number of different experimental model systems such as bimanual coordination (e.g. Haken, Kelso and Bunz 1985), multifrequency coordination (e.g. DeGuzman and Kelso 1991, Haken et al. 1996) coupled pendulum movements (see Turvey 1994 for review), trajectory formation (DeGuzman et al. 1997) with relevant experiments from laboratories in North America, Europe, and Australasia. We note also that theoretical predictions of the collective

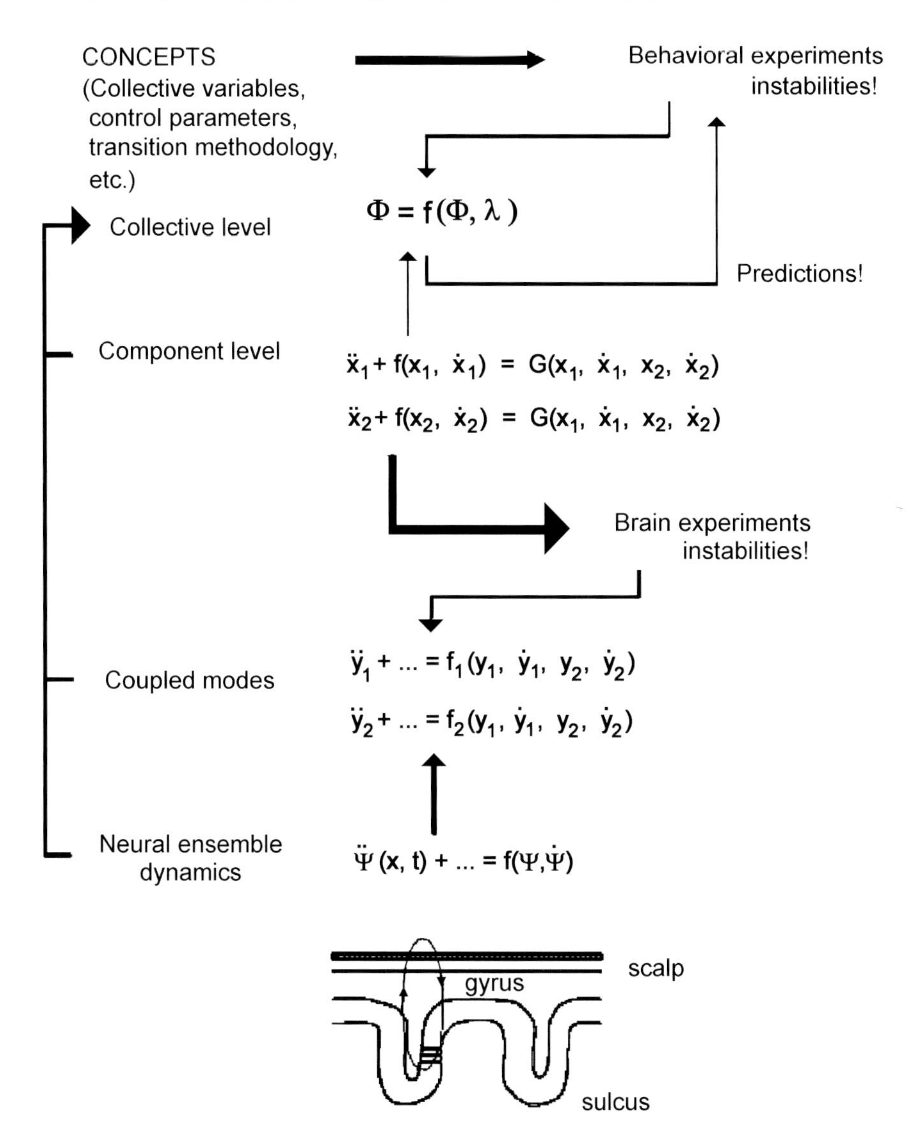

Fig. 1. The proposed level-building strategy connecting behavioral and brain dynamics (see text for details)

and component dynamics have been verified over and over again. The key notion behind our strategy, however, is to use this precise behavioral description and the methodology that allowed it, in order to probe other levels of analysis, in particular the brain. Here again, studies using large scale SQuID or multi-electrode arrays (e.g., Kelso et al. 1991, 1992, Wallenstein et al. 1995) and sophisticated analysis methods (e.g., Fuchs et al. 1992, see also Fuchs, Jirsa and Kelso this volume) have enabled us to identify relevant collective variables (such as spatial patterns and their time-dependent amplitudes). Signature features of self-organizing instabilities such as critical fluctuations and critical slowing down near transition points (Haken 1983) have been a prominent feature of these experiments. Elsewhere (Jirsa, Friedrich, Haken and Kelso 1994) we have summarized these results and presented a theoretical model that accounted for them in terms of the nonlinear interaction among spatial modes. We also showed how it was possible to derive the behavioral dynamics from this phenomenological brain mode theory (see arrow in Fig. 1 from coupled (brain) mode level back to collective behavioral level) hence providing a first glimpse of a possible connection between brain and behavioral levels. The bottom part of Fig. 1 shows a caricature of the neocortex with its sulci and gyri. Apical dendrites of pyramidal cells in columns are densely packed together and tend to discharge synchronously. A volume of cortex about 0.1mm^3 contains enough neurons to generate a magnetic field which can be measured outside the head as MEG. The term neural ensemble dynamics in Fig. 1 reflects our initial attempts to formulate a theory of the measured magnetic field generated by intracellular dendritic currents that is grounded in the neuroanatomy and neurophysiology of the cerebral cortex. Using various known facts about intra- and cortico-cortical connections, and a number of simplifying assumptions Jirsa and Haken (1996) already showed how it was possible to obtain the earlier brain mode and behavioral models, hence setting the stage for a principled move from phenomenology to brain mechanism (One man's phenomenology is another's mechanism!). In this paper, and the ones that follow, we will describe some new theoretical and empirical developments that have emerged from the strategy illustrated in Fig. 1.

3 Knowing What to Derive: Dynamics of Behavioral Function

But let's start at the beginning. The goal of experiment is to invent or discover paradigms that allow us to understand essential aspects of biology and behavior (e.g. formation of synergies, multifunctionality, stability and flexibility of function, invariance under change, pattern selection etc.). Thus we study the simplest system that contains these interesting properties. Following this approach, (1) represents an elementary equation of motion for how just two biological components are coordinated:

$$\dot{\phi} = \delta\omega - \lambda_1 \sin\phi - \lambda_2 \sin 2\phi + \sqrt{Q}\xi_t \quad (1)$$

This dynamical law was progressively established in a series of detailed experiments and theoretical steps (Kelso 1984, Haken, Kelso and Bunz 1985, Schöner, Haken and Kelso 1986, Kelso, DelColle and Schöner 1990, see also Fuchs and Kelso 1994). It constitutes a macroscopic description of the behavior of the system and contains three essential kinds of parameters:

- one that reflects whether the individual components are the same or different ($\delta\omega$). In general $\delta\omega$ may be viewed as an asymmetry parameter the sources of which are many (e.g. handedness, laterality, locus of attention, etc.);
- one that reflects external or internal factors (control parameters) that govern the strength of coupling between the components (λ_1, λ_2);
- and one that reflects the fact that all real systems contain noise or fluctuations (ξ_t) of a given strength Q.

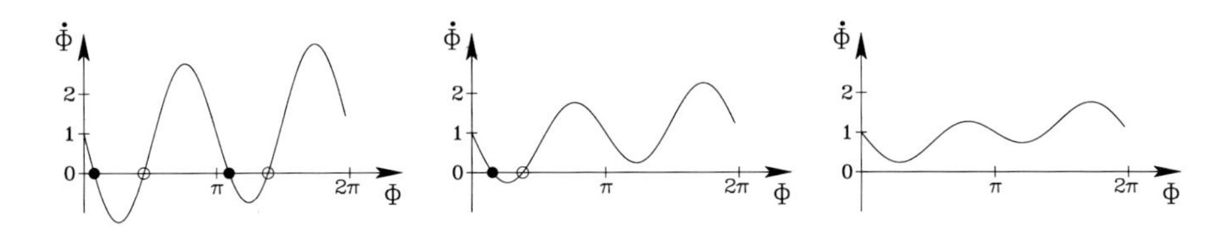

Fig. 2. An elementary law of behavioral coordination (see text for details)

Experiments showed that the relevant collective variable describing the functional synergy or spatiotemporal ordering between individual components is the relative phase, ϕ. For high values of the coupling ratio $\frac{\lambda_2}{\lambda_1}$, both modes of behavioral organization coexist, the essentially nonlinear property known as bistability (Fig. 2, left). Bistability (or, in general, multistability) confers multifunctionality on the system. That is, at least two forms of behavior are possible for exactly the same control parameter values. Notice that each is stable (negative slopes of the function cross the x-axis, denoted by solid circles, open circles mark unstable states) over a range of coupling, though the degree of stability may change. In this (bi)stable region (Fig. 2, left) the system's behavior will be restored despite any slight perturbation. As the coupling ratio is decreased, however, the system switches from one mode of behavior to another (Fig. 2, middle). Near the critical point, the slightest fluctuation will kick the system into a new form of stable organization. We may refer to this spontaneous transition or bifurcation as a form of pattern selection or decision-making which underlies the flexibility of the system's behavior. Switching is due to instability: under certain conditions, one mode of behavioral organization is less stable than another. On the right side of Fig. 2, there are no longer any stable states in the system. Due to changes in control parameters or coupling ratio, the entire function has lifted off the x-axis. Note however that the function retains its curvature; there is still attraction to, or remnants of, previously stable states (so-called metastability). This effect is entirely due to broken symmetry in the dynamics, itself due to the fact that the individual parts of the system or their properties are not the same. As a consequence of such broken symmetry, the system produces a far more flexible form of behavioral organization in which the individual components are free to express themselves yet still work together in a looser kind of harmony. Metastable dynamics may help us understand a longstanding either-or conflict in brain theory, namely how "global" integration in which parts of the brain are locked together, may be reconciled with localized, independent activity in individual brain areas. Metastable dynamics says that the brain, like other complex living systems, uses a subtle blend of both. Finally, if the direction of the control parameter values changes after the transition shown in Fig. 2 (middle), the system stays in the stable state around $\phi = 0$, i.e., it exhibits hysteresis, a primitive form of memory. In Jirsa, Fuchs and Kelso (this volume) we will derive the coordination dynamics shown in Fig. 2 from neural ensemble properties for the paradigmatic case of bimanual coordination.

4 Brain-Behavior Experiments

It is well-known that animals and humans can accomplish the same goal using different body parts and end-effector trajectories, even when the path to the goal is disrupted or perturbed (Kelso, Bateson, Tuller and Fowler 1984). Likewise, people can accomplish the same temporal rhythm using different anatomical structures. Think of the pianist whose fingers, feet, torso and head all conform to the basic beat. Or of people dancing at a rock concert. What is going on in the brain when humans produce this kind of "motor" (more properly, functional) equivalence? How is the spatiotemporal activity of the brain related to the actual behavior produced? Recent studies of single cell activity in monkeys suggest that certain parameters of voluntary movement such as direction may be specified in the motor cortex independent of the particular muscles required to execute the act (Georgopoulos 1997 for review). Here, using a full-head SQuID array to record ongoing brain activity, we demonstrate 1) a robust relationship between time-dependent activity in sensorimotor cortex and movement velocity, independent of the direction of movement and the explicit timing requirements of the task and 2) dynamic patterns of brain activation that are specific to task demands alone. Taken together, we believe these new results provide evidence of "motor" or "function" equivalence in humans at the level of cortical function.

Recently we performed two experiments (Kelso, Fuchs, Lancaster, Holroyd, Cheyne and Weinberg 1998) The main one required human volunteers to perform four different coordination tasks: Simple flexion or extension movements of the preferred index finger either on the beat of a metronome or in-between metronome beats, the frequency of which was fixed at 1Hz. One hundred cycles of continuous movement were recorded in each condition. Notice that these experimental conditions may be grouped with respect to the kinematics of motion (flexion versus extension movements), or with respect to the coordination task (synchronization or syncopation). Figure 3 shows plots of the relative phase between stimuli and movement peaks on a cycle-by-cycle basis for all four conditions. As requested, the peak of the movement is closely synchronized to the stimulus in the flexion-on-the-beat and extension-on-the-beat conditions. Likewise, subjects are able to place a movement in between stimuli in the flexion-off-the-beat and extension-off-the-beat syncopation conditions.

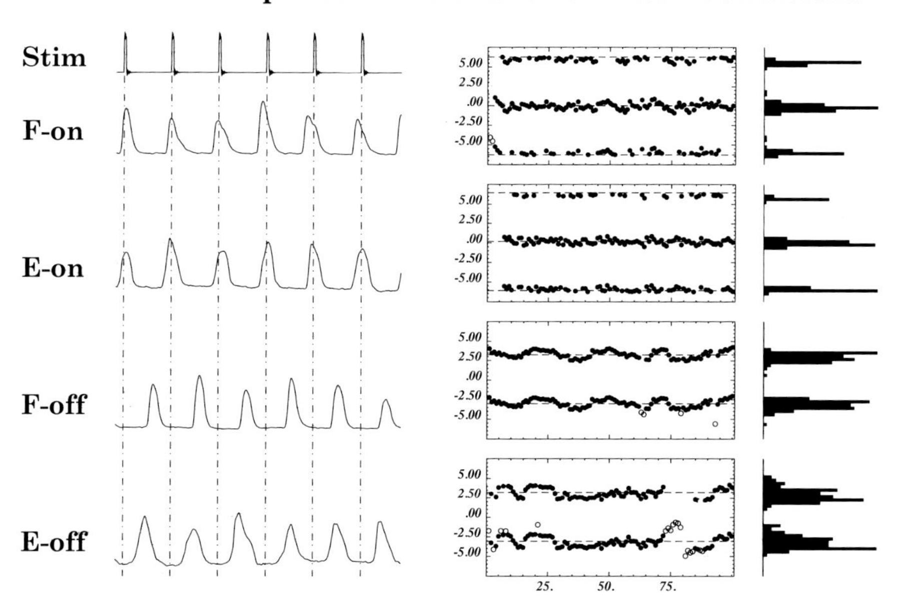

Fig. 3. Relative phase (in radians) on a cycle by cycle basis for all conditions (top rows: synchronization; bottom rows: syncopation) for a representative subject. Solid circles indicate cycles within a $\pm 60^\circ$ range of the average phase. Open circles are outside this range and were rejected from further analysis

Also shown are histograms, the width of which is a measure of the quality of performance. In general, the distributions for the syncopate conditions are broader, hence more variable than those of synchronization. Subjects tended to have more difficulty syncopating than synchronizing, which conforms to everyday experience and detailed behavioral experiments (Kelso et al. 1990).

Brain activity was recorded continuously as subjects performed these tasks using a 64 channel magnetometer (CTF Systems Inc., Vancouver) sampled at 250 Hz. This device consists of gradiometers arranged radially around the subject's head, each gradiometer consisting of a pair of parallel, axially centered, oppositely wound detection coils coupled to a SQuID. All the magnetic fields generated by the brain are due to the flow of electrical current, which is mainly ionic and is generated in the dendrites and somas of cortical neurons. About 10 000 neurons must be synchronously active to produce fields in the order of 200 fT (femtoTesla). Thus, any spatial patterning of activity within the brain is at the level of neuronal ensembles (see next section).

What is the relation between this evolving brain activity and the actual behavior produced? Figure 4 shows cortical activity patterns displayed in polar coordinates on the plane (see Fuchs, Jirsa and Kelso this volume for details) for each task averaged across subjects sampled at various points (shown by the red line) throughout the movement. Also shown (in green, inside the boxes) is the average amplitude profile of the movement. To ease visualization across conditions the movement profiles are all plotted in the same positive going fashion. Notice the presence of a strong dipolar field in the sensorimotor area of the left hemisphere during the first part of the movement, regardless of whether it involves flexion or extension. Notice also that the field reverses just after the peak movement (column 5) and then becomes much weaker and more distributed.

Decomposition of the brain's magnetic field into components corresponding to localized current sources is an ill-posed problem. Nevertheless, the spatial patterns of cortical activity shown in Fig. 4 suggest that the underlying neural ensemble is quite localized and fairly stationary during particular phases of the task. We decomposed the brain signals into spatial patterns and time-varying amplitudes using so-called Karhunen–Loève (K-L) decomposition or Principal Components Analysis (e.g. Fuchs, Jirsa and Kelso this volume). Tangential currents naturally produce spatial correlations (the field entering the scalp at one location and leaving at another), and the resulting principal components may capture this dipolar structure. Figure 5 shows that the first two spatial modes (1st and 2nd columns) capture about 80% of the variance in the brain signals. Because the top spatial mode, like the underlying neural ensembles, is under no orthogonality constraint, it is identical to the dominating spatial pattern of brain activity observed experimentally. The remarkable result shown in Fig. 5 is that its time-dependent amplitude tracks the velocity profile extremely well, especially for the initial velocity peak associated with the active phase of the coordination task. The second velocity peak constitutes the less active phase of the task and the match to the brain signal is weaker. It seems likely that the corresponding minimum in the brain signal occurs after movement onset and reflects reafferent activity from the periphery to somatosensory cortex (Cheyne and Weinberg 1989) although precentral sources may remain active throughout the movement.

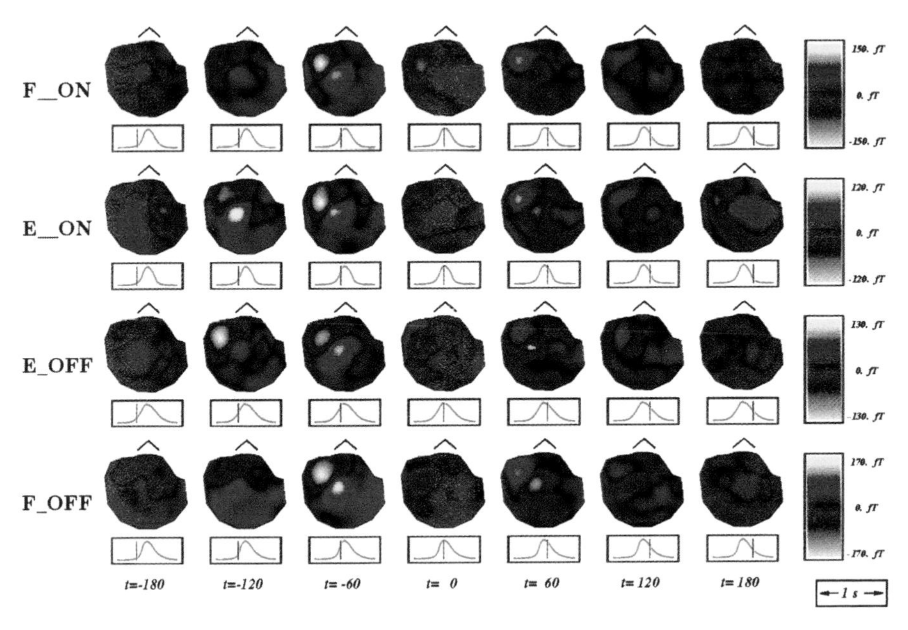

Fig. 4. Brain activity patterns averaged across cycles and subjects (N = 5) at time points t (in ms) for all conditions. The time point $t = 0$ is the peak movement amplitude. The green curve in the box shown below each pattern is the average amplitude profile for each condition. The red line indicates the time ($t = \pm 180$ms) at which brain activity is sampled during the movement cycle. Each condition has been scaled separately to highlight the peak fields. In general, the activity for flexion conditions has a slightly higher amplitude range than the extension conditions (adapted from Kelso et al. 1998)

These results are even more striking because the K-L mode decomposition simply minimizes the mean square error without regard to the movement at all. A second method calculates the spatial patterns that best fit the movement and its derivatives (Uhl, Friedrich and Haken 1995). The results of that procedure (see Fuchs, Jirsa and Kelso this volume) once again show that the largest mode corresponds to movement-related brain activity, specifically the velocity of finger flexion or extension.

Two questions arise from these results. First, is the strong relationship between average movement velocity and the time course of cortical activity a mere coincidence or is it consistent across different manipulations of the velocity profile? In particular, does this relationship hold across a number of peak velocities (with the same length timecourse) and does it hold across a number of movement rates (similar peak velocity, but shorter timecourse)? Second, what, if any, influence does the task alone contribute to patterns of brain activation? The issue of motor equivalence is double-sided, requiring an identification of both invariant and task-specific aspects of brain activity. To answer the first question we performed two different manipulations. First,

KL-expansion

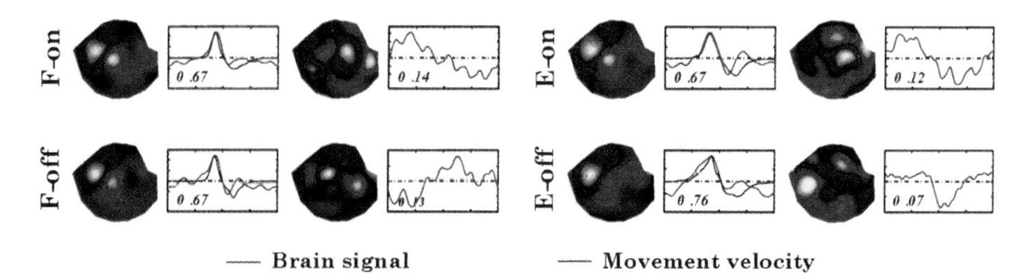

Fig. 5. Decomposition of the spatiotemporal signal into spatial patterns and corresponding time-dependent amplitudes for all conditions. The first two modes of a Karhunen-Loeve decomposition and their time-dependent amplitudes (in red). The numbers in the lower left corner of each box indicate how much of the variance of the entire signal is contained in a given mode. Notice the first two modes cover about 80% in all conditions. Overlaid in the first column is the movement velocity (in blue). The tight relationship between time-dependent neural activity and movement velocity is apparent especially for the first peak for all task conditions (F_ON and E_ON refer to flexion and extension movements on-the-beat. F_OFF and E_OFF refer to flexion and extension movements off-the-beat)

we sorted the existing data from all four original task conditions into sets representing different peak velocity ranges, from slowest to fastest. The data from four different peak velocity ranges for a representative subject are shown in Fig. 6. The brain signal (again the time-dependent amplitude of the top spatial mode) is plotted (in red) along with the velocity profiles (in blue) for the overall data (left box) and four non-overlapping bins in which velocity increases from left to right. The steepness of the displacement profile (in green) reflects the derivative, actual values of which are shown in blue. Notice that each graph in Fig. 6 is scaled individually, highlighting the correspondence between the brain signal and the velocity profile. The degree of covariance, given by the correlation value on the top right of each box, is high.

Next, we performed another experiment in which we asked the same subjects to perform the two basic syncopation tasks at six different movement rates. Beginning either in the flexion- or extension-off-the-beat conditions, subjects were instructed to syncopate with the metronome, the rate of which was increased every 10 cycles from 1.25 Hz to 2.5 Hz in 5 steps of 0.25 Hz. It is known that transitions from syncopation to synchronization occur spontaneously in both brain activity and behavior as movement rate is increased beyond a critical value (Kelso et al. 1990, Kelso et al. 1992) but the relationship of interest here has not, to our knowledge, been examined before. The results were unequivocal across subjects, initial conditions and movement rates. A representative example is shown in Fig. 7 for the flexion-off condition. Once again, the same basic dipolar-like spatial pattern was observed at all movement rates. For ease of visualization, the velocity profile (in blue)

Peak Velocity Binned Data

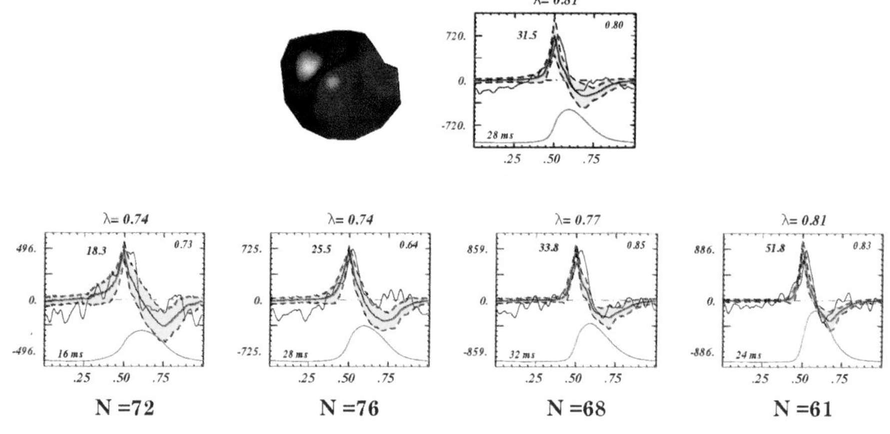

Fig. 6. The relationship between brain activity and movement velocity. The top box indicates the average data for all conditions for a representative subject. In the four boxes below these data are sorted according to different peak velocities which increase from left to right. Displacement profiles shown in green are represented on a common scale and increase in magnitude and steepness across the four bins. The time-dependent amplitude of the top spatial mode is plotted in red (eigenvalues are indicated above each box); the velocity profile and its standard deviation in blue and yellow, respectively. Labels on the y-axis (in red) indicate the range and magnitude of the brain signal, which increase from left to right. Blue numbers indicate the mean peak velocity for a given bin. From left to right, the ranges of peak velocity (in arbitrary units) in each bin are: 11.5 to 73.4 (average), 11.5 to 22.9, 23.0 to 28.5, 28.6 to 39.9, and 40.0 to 73.4. Black numbers indicate the correlation value between brain and velocity profile (top right of each box) and corresponding lag in ms (bottom left) (Adapted from Kelso et al. 1998)

is superimposed on the brain signal (here the best fit spatial pattern). The corresponding value of the correlation function at the time lag between the first velocity peak and the brain signal is also reported in the top right of the boxes. Given the predictive, rhythmical nature of the task it is not unexpected to find that the peak movement velocity leads the cortical activity by a small amount (shown in the lower left corner of Figs. 6 and 7). Yet the similarity in the neural and velocity profiles is once again striking, especially in the initial active phase of voluntary movement.

To answer the second question about task-specific brain activity, we removed the dominant velocity-related spatial pattern from the recorded brain signals (see accompanying paper by Fuchs, Jirsa and Kelso for details and results). These residual brain patterns turn out to be similar for the syncopation tasks, which in turn are very different from the corresponding brain activity patterns for synchronization. Thus, any differences attributable to

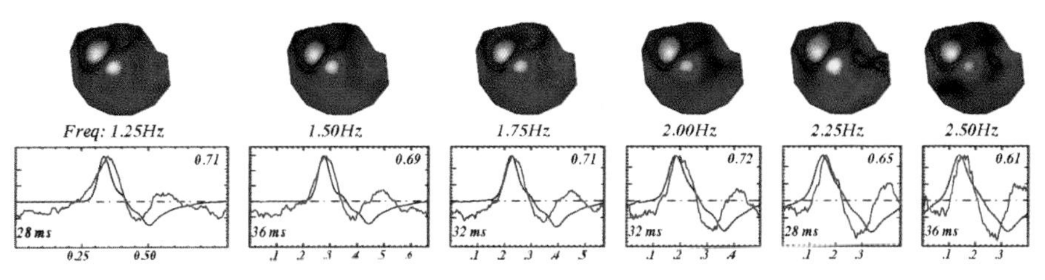

Fig. 7. The relationship between brain activity and movement rate (1.25 Hz to 2.5 Hz) for the Flexion-on-the beat condition (the other conditions are very similar). The number in the lower left corner of the boxes represents the shift between the brain signal (red) and the movement velocity (blue). In the upper right corner the correlation value for the two curves is shown. The smaller boxes as one moves from left to right reflect the number of points sampled in a given cycle which decreases with movement rate (adapted from Kelso et al. 1998)

movement direction (e.g., flexion-on and extension-off are kinematically similar) are far outweighed by differences due to the task, i.e., the goal of the subject's action.

In summary, the demonstration of a robust relation between movement velocity and the time course of cortical activity across a broad range of initial conditions, peak velocities and movement rates complements single-cell studies in monkeys which show that one of the most clearly represented parameters associated with motor cortical activity is movement direction (Georgopoulos 1997 for review). Although obtained in a rhythmic, not discrete movement context, our results are congruent with very recent findings which show that speed is directly represented in the discharge rate of cells in primary motor cortex when the directional component of the discharge pattern is removed (Moran and Schwartz in press). During the task of synchronization, the cortex appears to control the speed of movement so as to arrive or "collide" at the target (the metronome beat) at the right time. For syncopation the cortex must even plan for a virtual target in-between metronome beats. Of course, our results do not deny and even suggest that somatosensory information is used both in the planning and execution phases of the tasks. Nor do they exclude a role for other brain areas (e.g. cerebellum, putamen and thalamus) that appear to be involved in the internal generation of precisely timed movements (Rao et al. 1997). Nevertheless, our findings help resolve a longstanding question in studies of human synchronization, namely, how the brain coordinates actions in time with external events. Cortical correlates of the velocity profile hold across different movement directions, rates and task demands. At the same time, patterns of brain activation appear to be task-specific, conforming to particular modes of coordination. Taken together, these results reveal signatures of motor equivalence in dynamic patterns of cortical activity.

5 A Little Theory

Our experimental findings can be interpreted from the viewpoint of a field theoretical description that has been developed based on properties of excitatory and inhibitory neural ensembles and their corticocortical (long range) and intracortical (short range) interactions. The resulting spatiotemporal dynamics is represented by a set of retarded coupled nonlinear integro-differential equations for the excitatory and inhibitory neural activity (see Jirsa, Kelso and Fuchs of this volume for details). Due to differences in spatial and temporal scales this system can be reduced and transformed into one single nonlinear partial differential equation:

$$\ddot{\psi}(x,t) + 2\,\omega_0\,\dot{\psi}(x,t) + \{\omega_0^2 - v^2\Delta\}\psi(x,t) = \left\{\omega_0^2 + \omega_0\frac{\partial}{\partial t}\right\}\rho(x,t) \qquad (2)$$

with

$$\rho(x,t) = S[\psi(x,t) + \sum_i p_i(x,t)]\;. \qquad (3)$$

Here $\psi(x,t)$ represents the spatiotemporal neural activity, $\omega_0 = v/\sigma$ a frequency defined by the axonal propagation velocity v and the mean axon length σ, S a sigmoid function, and $p_i(x,t)$ functional input and output units. In the framework of this theory the experimental findings described above can be represented as depicted in Fig. 8. There the behavioral level may be represented by an equation of motion for the relative phase ϕ which serves as a collective variable coupling the visual stimulus to movement (see Fig. 2). At the brain level, a functional input unit $\alpha(x)$ embedded into the cortical sheet receives input signals from the visual metronome. An output unit in the motor cortex sends signals to the finger muscles which provide sensory feedback from muscle spindles and joint receptors to a second input unit located in sensormotor cortex. We seek an explicit account for the relation between finger displacement $r(t)$ and neural activity $\psi(x,t)$, represented by:

$$r(t) = \int \mathrm{d}x\,\beta(x)\int \mathrm{d}\tau\;\mathrm{e}^{-\gamma(t-\tau)}\;\psi(x,\tau) \qquad (4)$$

where $\beta(x)$ describes the localization of the output unit in the cortical sheet and is to be identified with the dipolar mode observed experimentally (see Figs. 5 and 6). Notice the finger motion $r(t)$ arises out of an integration in space as well as time. Such a mechanism of spatial and temporal integration is well-known neurophysiologically and serves to smoothen intrinsically noisy brain signals. It is well known that $r(t)$ is the particular solution of

$$\dot{r}(t) + \gamma\,r(t) = \int \mathrm{d}x\,\beta(x)\,\psi(x,t) \qquad (5)$$

where the lhs represents a linear damped system driven by the neural signal (rhs). Figure 9 shows the reconstruction of the movement profile from neural activity according to (4) for all task conditions. Note the reconstructed movement profile fits the experimentally observed movement particularly well in the active phase represented by its positive flank. The discrepancies mainly occur after peak displacement and are probably due to the influences of sensory feedback which are not accounted for by (4).

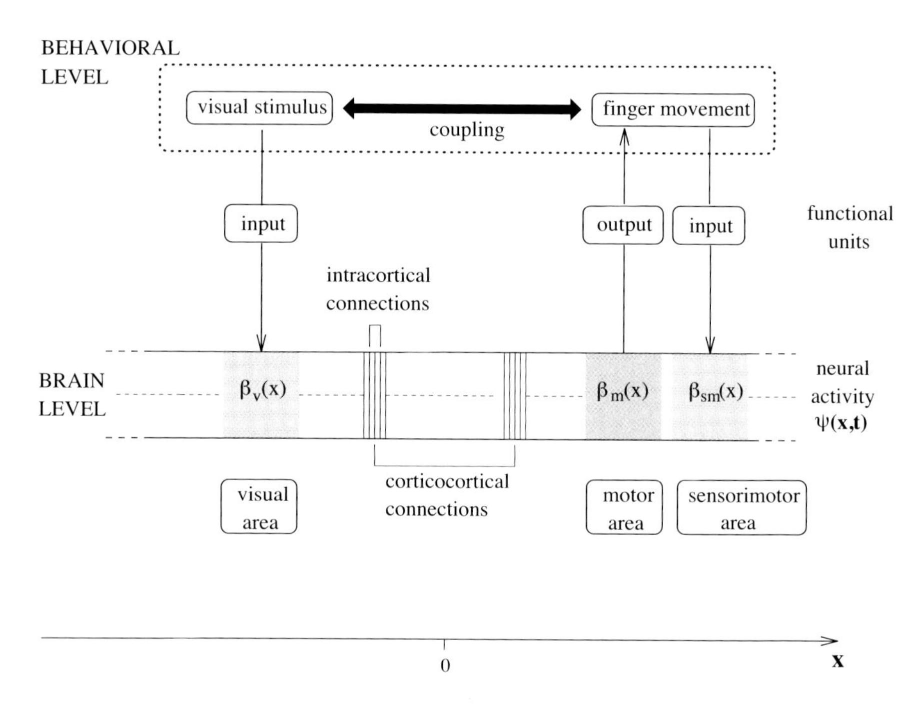

Fig. 8. Functional units embedded in a cortical sheet with long range and short range interactions (see text for details)

6 Concluding Remarks

Modern neuroscience at the end of the 20th Century is successfully reducing the brain to its physical elements, but it is becoming increasingly unclear how to put Humpty Dumpty together again, i.e., to understand how orchestrated biological functions arise from such structural complexity. New concepts and strategies are needed to handle such complex systems, and a vocabulary must be devised that is rich enough to characterize both behavioral and neural levels of analysis. A main idea behind our approach is that it is difficult, if not impossible, to integrate levels except in the context of well-defined behavioral functions. Thus, much work has gone into identifying the quite abstract, but

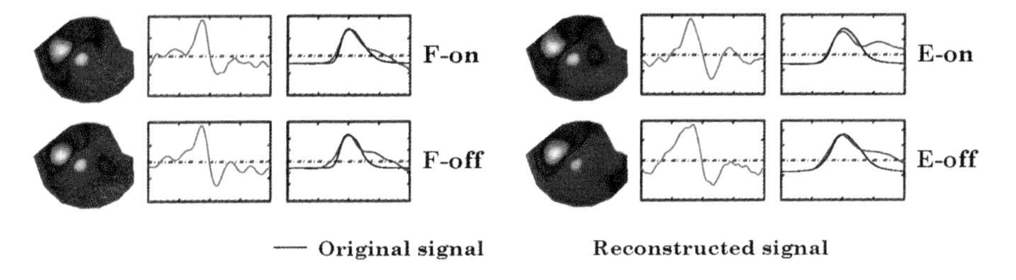

Fig. 9. Reconstruction of the movement profile using brain activity (the dipolar mode's amplitude over time, shown in green) as a driving signal for the finger oscillator

empirically-grounded phenomenological laws at behavioral and brain levels. These are then used as a necessary guide into the materially based mechanisms and principles that underlie them. Obviously the neurophysiology, neuroanatomy and neurochemistry of the central nervous system constitute a deeper basis upon which to derive phenomenological laws of behavioral function. But to start with the former without knowing the latter may be a grave mistake. Casting back to the conceptual scheme shown in Fig. 2, notice that the order parameters or collective variables are function, task or context dependent. Nothing subjective is implied by this statement. As a considerable amount of experimental research now shows, the same order parameters have been identified in a variety of contexts. Notice also in Fig. 2 that the order parameters are different at each level and that there may be mutability between order parameters and control parameters across levels. What is a collective variable at one level may be a control parameter at another, and vice-versa. The connection across levels – the traversing of scales in the title of this trilogy of papers – is by virtue of coupled spatiotemporal dynamics on all scales.

Acknowledgements

This research was supported by NIMH (Neurosciences Research Branch) Grant MH42900, KO5 MH01386 and the Human Frontiers Science Program. We are grateful to Hermann Haken for many interesting discussions and his encouragement over the years. VKJ gratefully acknowledges a fellowship from the *Deutsche Forschungsgemeinschaft*.

References

DeGuzman, G., Kelso, J.A.S. (1991): Multifrequency behavioral patterns and the phase attractive circle map, Biol. Cybern. 64, 485–495

DeGuzman, G., Kelso, J.A.S., Buchanan, J.J. (1997): Self-organization of trajectory formation. Biol. Cybern. 76, 275–284

Fuchs, A., Kelso, J.A.S., Haken, H. (1992): Phase Transitions in the Human Brain: Spatial Mode Dynamics, Inter. J. Bifurc. Chaos 2, 917–939

Fuchs, A., Kelso, J.A.S. (1994): A theoretical note on models of interlimb coordination, Journ. Exp. Psych.: Human Perception and Performance, 20, No. 5, 1088–1097

Georgopoulos, A.P. (1997): Neural networks and motor control, Neuroscientist 3, 52–60

Haken, H. (1970): *Laser theory.* Encyclopedia of Physics Vol. XXXV/2C, Springer, Berlin

Haken, H. (1983): *Synergetics. An Introduction*, 3rd ed., Springer, Berlin

Haken, H., Kelso, J.A.S., Bunz, H. (1985): A Theoretical Model of Phase Transitions in Human Hand Movements, Biol. Cybern. 51, 347–356

Haken, H., Peper, C.E., Beek, P.J., Daffertshofer, A. (1996): A model for phase transitions in human hand movements during multifrequency tapping, Physica D 90, 179–196

Jirsa, V.K., Friedrich, R., Haken, H., Kelso, J.A.S. (1994): A theoretical model of phase transitions in the human brain, Biol. Cybern. 71, 27–35

Jirsa, V.K., Haken H. (1996): Field theory of electromagnetic brain activity, Phys. Rev. Let. 77, 960–963

Kelso, J.A.S. (1984): Phase transitions and critical behavior in human bimanual coordination, Am. J. Physiol. 15, R1000–R1004

Kelso, J.A.S. (1995): *Dynamic Patterns. The Self-Organization of Brain and Behavior*, MIT Press, Cambridge, MA

Kelso, J.A.S., DelColle, J.D., Schöner, G. (1990): Action-perception as a pattern formation process, in: *Attention & Performance XIII*, Jeannerod, M., ed., 139–169, Erlbaum, Hillsdale, NJ

Kelso, J.A.S., Buchanan, J.J., DeGuzman, G.C., Ding, M. (1993): Spontaneous recruitment and annihilation of degrees of freedom in biological coordination. Phys. Let. A 179, 364–371

Kelso, J.A.S., Bressler, S.L., Buchanan, S., DeGuzman, G.C., Ding, M., Fuchs, A., Holroyd, T. (1992): A Phase Transition in Human Brain and Behavior, Phys. Let. A 169, 134–144

Kelso, J.A.S., Bressler, S.L., Buchanan, S., DeGuzman, G.C., Ding, M., Fuchs, A., Holroyd, T. (1991): Cooperative and critical phenomena in the human brain revealed by multiple SQUIDS. In: *Measuring chaos in the human brain*, Duke D., Pritcard W., eds., World Scientific, NJ

Kelso, J.A.S., Fuchs, A., Lancaster, R., Holroyd, T., Cheyne, D., Weinberg, H. (1998): Dynamic Cortical Activity in the Human Brain Reveals Motor Equivalence, NATURE 392, 814-818

Moran, D.W., Schwartz, A.B. (in press): Motor cortical representation of speed and direction during reaching, J. Neurophysiol.

Rao, S.M., Harrington, D.L., Haaland, K.Y., Bobholz, J.A., Cox, R.W., Binder, J.R. (1997): Distributed neural systems underlying the timing of movements, J. Neurosci. 17, 5528–5535

Schöner, G., Haken, H., Kelso, J.A.S. (1986): A Stochastic Theory of Phase Transitions in Human Hand Movement, Biol. Cybern. 53, 247–257

Treffner, P.J., Turvey, M.T. (1996): Symmetry, broken symmetry, and handedness in bimanual coordination dynamics. Exp. Brain Res. 107, 463–478

Turvey, M.T. (1994): From Borelli (1680) and Bell (1826) to the dynamics of action and perception. J. of Sport and Exercise Psychology 16, 2, S128–S157

Uhl, C., Friedrich, R., Haken, H. (1995): Analysis of spatiotemporal signals of complex systems, Phys. Rev. E 51, 3890–3900

Wallenstein, G.V., Kelso, J.A.S., Bressler, S.L. (1995): Phase transitions in spatiotemporal patterns of brain activity and behavior, Physica D 84, 626–634

Wilson, K.G. (1979): Problems in Physics with many scales of length. Sci. Am. 241, 158

Traversing Scales of Brain and Behavioral Organization II: Analysis and Reconstruction

Armin Fuchs, Viktor K. Jirsa, and J.A. Scott Kelso

Program in Complex Systems and Brain Sciences, Center for Complex Systems, Florida Atlantic University, 777 Glades Road, Boca Raton, FL 33431, USA

1 Introduction: Probing the Human Brain

Brain signals can be recorded from humans in a great variety of experimental setups and task conditions. As outlined in the first of our papers (Kelso, Fuchs and Jirsa this volume, referred to in the following as KFJ) our approach aims at specific situations were (in most cases) coordination tasks are used to prepare the brain into a certain state. Changing this state by either the variation of an external parameter or slight changes in the task allows us to link properties of movement behavior to ongoing neural activity. In so-called transition paradigms we have shown how the spatiotemporal patterns obtained by recordings of the electric scalp potential (EEG) (Wallenstein et al. 1995), (Meaux et al. 1996) or the magnetic field (MEG) (Fuchs et al. 1992) undergo changes when spontaneous switches in the subject's motor coordination occur. Here we describe in detail how analysis and visualization techniques can be used to show how brain activity is related to a kinematic feature of finger movement – its velocity profile, as outlined in KFJ. We are going to show which parts of the brain signals are invariant a cross different movement types, i.e. the flexion or extension of an index finger, and what differences can be found for different task conditions, i.e. syncopation or synchronization with an external stimulus.
This paper is organized as follows:

- Sect. 2 gives a brief overview about the state of the art technology for (noninvasive) recordings from the human brain and it summarizes the advantages and disadvantages of different methods. We also briefly review the visualization and analysis techniques we are going to apply to the experimental data;
- Sect. 3 presents the results; and
- Sect. 4 contains the conclusions and some future perspectives.

2 Recording Technology, Visualization, and Analysis

2.1 Technology

One of the goals in brain research is to find the spatial distribution of electric current density $\mathbf{j}(\mathbf{r}, t)$ inside the brain at each moment in time. In a perfect situation this distribution could be measured from the outside and, knowing all the inputs, could be predicted from a "brain theory". Unfortunately, the problem of calculating the current density inside a volume from measurements on a surface outside this volume (known as the "inverse problem") is ill-posed, i.e., it has no unique solution. There are essentially two methods in use to record the electric and magnetic activity of the human brain that are purely non-invasive:

Electroencephalography (EEG) records the electric potential at the scalp surface from typically a few (in clinical applications) to up to about 200 (in research labs) different locations. The electric potential is a quantity that can only be measured with respect to a reference (i.e., only potential differences can actually be measured). It has always been an issue what is a "good" reference because different locations of the reference electrode lead to different spatial patterns of the electric potential, and what is the "best" reference procedure is mostly a philosophical issue (see e.g. Nunez (1981) for a detailed discussion of the physics of this problem). Moreover, EEG signals are contaminated and smeared due to electric volume conduction inside the brain tissue and cerebral fluid before they reach the scalp leading to different activity patterns on the scalp versus the cortex. There are different techniques in use to deblur these signals. One common method is calculating spatial derivatives (Laplacians, which also solves the reference problem because this quantity is independent of the reference used) (Nunez et al. 1993). It is also possible to estimate the local conductivity of the tissue from MRI scans (Le and Gevins 1993), and calculate the current density on the surface of the cortex;

Magnetoencephalography (MEG). Most of today's magnetometers measure the radial component of the magnetic field produced by intra-cellular currents inside the brain at up to also about 200 spatial locations. It is estimated that about 10000 neurons have to be active simultaneously to create a magnetic field that can be picked up from outside the brain even though the sensors used (so-called Superconducting Quantum Interference Devices or SQuIDs) are sensitive to magnetic fields on the order of 10 fT (femto Tesla = 10^{-15}Tesla) which is less than one billionth part of the earth's magnetic field. MEG signals generally show more spatial structure than EEG signals because the magnetic field can penetrate the tissue with very little interaction. MEG is mainly sensitive to tangential current flow because this type of current creates magnetic fields that enter and exit the head whereas radial currents have magnetic fields which are confined inside. It should be mentioned that even though only the radial component of the field is measured it is possible (at least in principle) to calculate the other two components tangential to the surface defined by the sensors using Maxwell's equations (Lütkenhöner 1994). In any case, there always exist (infinitely many) current distributions inside a volume (the brain) that leave no signal at all in a set of sensors arranged on a surface around it; any one of them can be added to a current distribution that is compatible with the measured field, and therefore the inverse problem has no unique solution.

Both techniques have a high resolution in time (on the order of ms) and a low resolution in space (on the order of cm) due to inter-sensor distances on the head. "Low" here is with respect to techniques like MRI where the spatial resolution is on the scale of mm in a 3d volume. Nevertheless, the amount of data produced in experimental sessions is enormous as a simple estimate shows: If we record from 200 EEG and/or MEG sensors at a sampling frequency of 250 Hz and an accuracy of 16 bits we have $200 \times 250 \times 2 = 100\ 000$ bytes/sec or more than a GigaByte for 3 hours of recording time. Interestingly, these numbers don't seem frightening anymore because almost two hours fit on a single CD. The more important question is how to extract the relevant information from such a dataset because even after say averaging we are still left with one time series for each sensor as shown in Fig. 1.

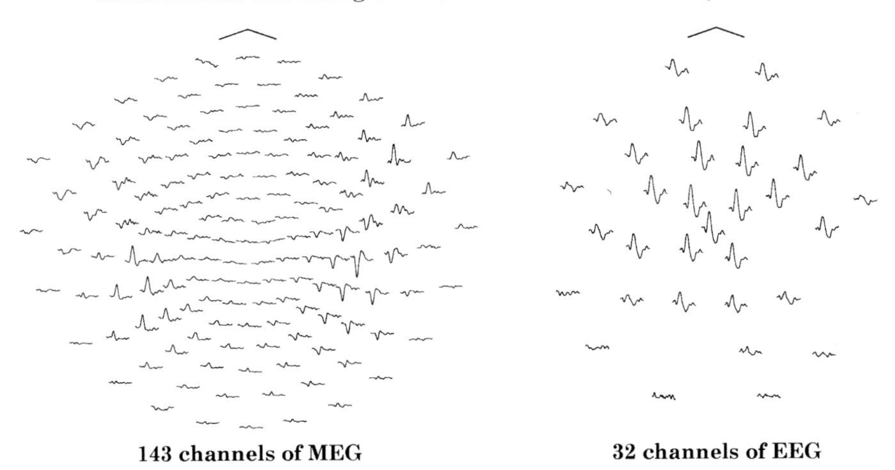

Fig. 1. Time series of a simultaneous recording of 144 channels of MEG and 32 channels of EEG for an auditory stimulus (data courtesy of CTF Inc., Vancouver)

It is evident that a great amount of the information in this dataset is redundant and plotting all these time series is certainly not the best way to visualize the spatiotemporal dynamics we are interested in. Color coded topographic maps plotted at certain timepoints can be used to show the spatial patterns for the electric potential or the magnetic field. Because the sensors are usually not arranged on a planar surface projection procedures must be used to obtain a pattern on flat paper without loosing information from hidden sensors. The topos in this article are calculated the following way: Assuming a center in the middle of the head, the 3d (x, y, z) coordinates of the sensors can be transformed into spherical coordinates (r, θ, ϕ). The 2d coordinates for the plots are then given by

$$X = s\,\theta\cos\phi, \quad Y = s\,\theta\sin\phi \tag{1}$$

where s is a scaling factor that determines the size of the figure. An example showing the sensor locations and a pattern of neural activity on a subject's head, together with the projection and a color scale that links certain colors to magnetic field amplitude is given in Fig. 2. Figure 3 shows topographic maps from the time series in Fig. 1 at two different timepoints ($t = 108$ ms, and $t = 220$ ms after stimulus onset). These points were picked to answer the question what is the "better" technology EEG or MEG (this issue is currently fought out in the BIOMAG mailing list (which can be found on CTF's web site: http://www.ctf.com). In Fig. 3 at $t = 108$ ms, clearly, the MEG pattern shows more structures, i.e. two dipolar patterns indicating current dipoles in the left and the right hemisphere pointing posterior. The pattern of the electric potential is quite structureless. At $t = 220$ ms there is virtually no magnetic field but a strong positive potential picked up by the EEG. Of course, the scaling for the two time points is the same as indicated by the color scales. This is a situation where current flow exists inside the head but in a way that leaves virtually no traces in the SQuID sensors. Therefore, to get as much information as possible about electromagnetic brain activity one would like to have measurements of both EEG and MEG, preferably recorded simultaneously.

To visualize the dynamical nature of brain activity, topographic maps can be produced for each time point and animated as movies. Examples of such animations for different conditions for EEG and MEG (in MPEG format) can be downloaded from our web site (http://www.ccs.fau.edu).

2.2 Analysis Methods

A better understanding of what is going on in large data sets can often be achieved if the data is preprocessed in certain ways that allow for an extraction of relevant information or elimination of redundancies. Here we briefly describe two linear methods that we use for this purpose.

The Karhunen–Loève Transformation. Several names are on the market for this decomposition technique including Principal Component Analysis or Singular Value Decomposition. Essentially they all follow the same basic idea: we have time series (EEG or MEG) measured at (very) many locations in space. These time series represent spatial patterns at each time point, i.e. a spatiotemporal pattern. In principle any spatiotemporal pattern $H(\mathbf{x}, t)$ can

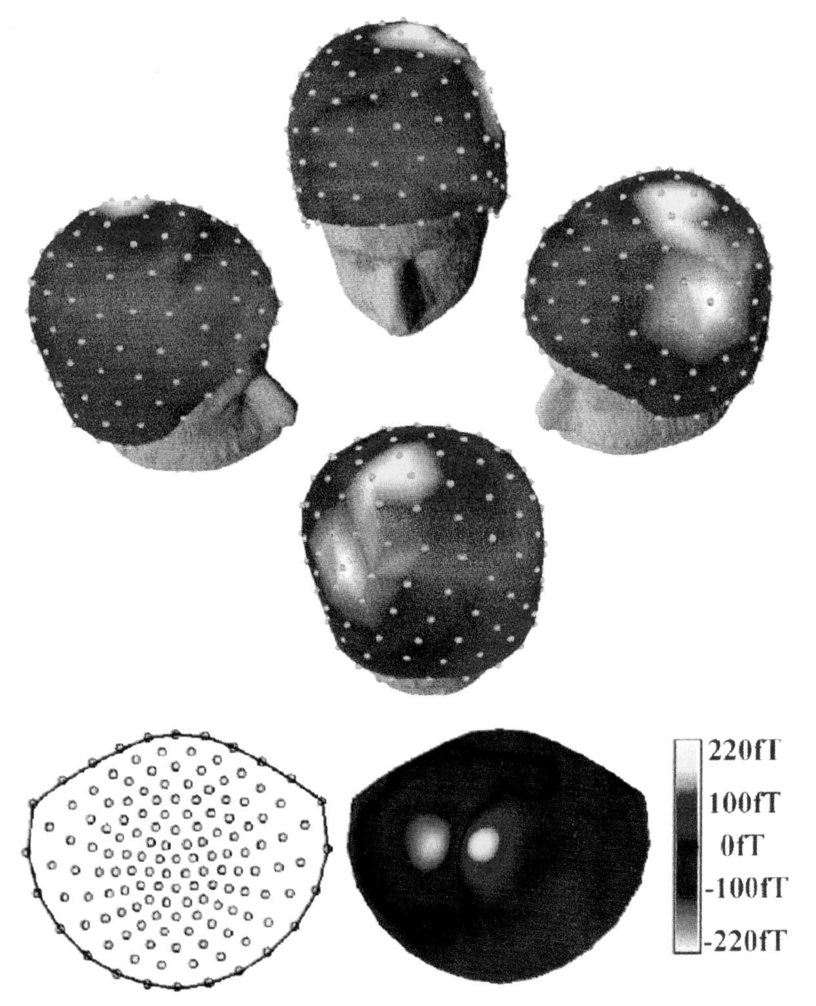

Fig. 2. A pattern of magnetic brain activity and sensor locations from four views on the subject's head (*top*), and in the projection (*bottom*). Blue indicates locations where the magnetic field is entering the head, red-yellow sites where the field is exiting the head. The color scale is in units of femtoTesla (fT)

MEG and EEG of an Auditory Stimulus: Topographic Maps

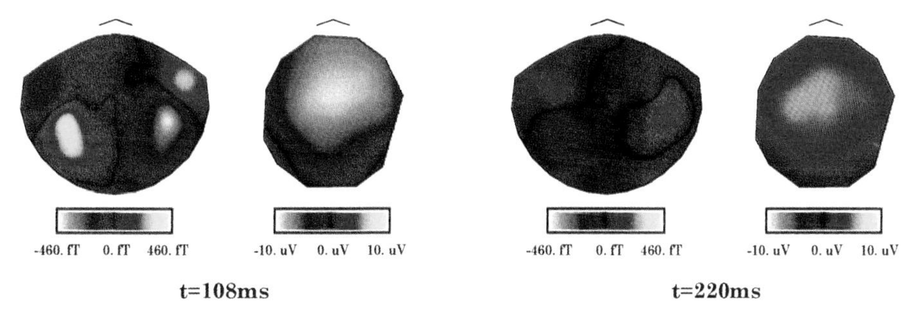

Fig. 3. Topographic maps of the data shown in Fig. 1 at two representative points in time

be written as the sum over a set of spatial patterns $\psi^{(i)}(\mathbf{x})$ multiplied by time dependent amplitudes $\xi_i(t)$:

$$H(\mathbf{x}, t) = \sum_{i=1}^{N} \xi_i(t)\, \psi^{(i)}(\mathbf{x}) \,. \tag{2}$$

In our case space and time are discrete, i.e. the function $H(\mathbf{x}, t)$ is known at certain locations k in space and at certain time points depending on the sampling frequency. However, in the following we assume that space is discrete but keep time continuous because it simplifies the notation. This way $H(\mathbf{x}, t)$ becomes a vector $\mathbf{H}(t)$ at every time point where the component $H_k(t)$ represents the amplitude (potential difference between an electrode and the reference, or magnetic field component) at sensor location k and time t. By applying the same notation to the spatial functions $\Psi^{(i)}(\mathbf{x})$ (2) reads for component k:

$$H_k(t) = \sum_{i=1}^{N} \xi_i(t)\, \psi_k^{(i)} \,. \tag{3}$$

If the vectors $\Psi^{(i)}$ form a complete set and N is equal to the number of sensors the original signal $\mathbf{H}(t)$ is reconstructed *exactly*. The more interesting question, however, is: Can we obtain a set of vectors $\Psi^{(i)}$ for which we have to sum over only a few, say $M \ll N$, and still get a decent (90–95%) reconstruction of the original signal? It can be shown that the Karhunen–Loève (KL) transformation creates such a set of vectors $\Psi^{(i)}$ such that for *any* truncation point M, the mean square error between the original and the reconstructed signal becomes a minimum:

$$E_M = \sum_{k=1}^{N} \int_0^T \mathrm{d}t \left(H_k(t) - \sum_{i=1}^{M} \xi_i(t)\, \psi_k^{(i)} \right)^2 = \mathrm{Min}\ \forall M \,. \tag{4}$$

It can also be shown that the basis vectors $\Psi^{(i)}$ in this case are the eigenvectors of the covariance matrix $\mathbf{C}$ from the time series $h_i(t)$:

$$C_{kl} = \int_0^T \mathrm{d}t \, h_k(t) \, h_l(t) \quad \text{with} \quad h_k(t) = H_k(t) - \bar{H}_k \tag{5}$$

where $\bar{H}_k$ is the mean value of the time series from sensor k. The eigenvalues of this matrix are a measure of how much the corresponding eigenvector contributes to the variance of the original signal. Notice that the covariance matrix is symmetric and therefore its eigenvalues are real numbers and its eigenvectors are orthogonal. The time dependent amplitude $\xi_i(t)$ for the vector $\Psi^{(i)}$ is given as the scalar product between the vector and the pattern at time t, $\mathbf{H}(t)$

$$\xi_i(t) = \mathbf{H}(t) \, \Psi^{(i)} = \sum_{k=1}^{N} H_k(t) \, \psi_k^{(i)} \, . \tag{6}$$

The patterns $\Psi^{(i)}$ and the time series $\xi_i(t)$ fulfill the orthogonality relations

$$\Psi^{(i)} \, \Psi^{(j)} = \sum_{k=1}^{N} \psi_k^{(i)} \, \psi_k^{(j)} = \delta_{ij} \quad \text{and} \quad \frac{1}{T} \int_0^T \mathrm{d}t \, \xi_i(t) \, \xi_j(t) = \lambda_i \, \delta_{ij} \tag{7}$$

where δ_{ij} represents the Kronecker-delta. In the sense of (4) the set of basis vectors $\Psi^{(i)}$ is optimal which means it is not possible to find a different set of M vectors that has a smaller mean square error. There may be circumstances, however, where it is of advantage to minimize other quantities (like in the bi-orthogonal decompositions discussed below).

Figure 4 shows the first three spatial patterns and their corresponding time dependent amplitudes from a KL-expansion of the dataset from Fig. 1 for the case of EEG (top row) and MEG (bottom row). For EEG the pattern representing the first eigenvector contributes already about 97% to the variance of the signal, i.e. by using only this pattern we can get a very good reconstruction of the whole dataset – an enormous compression of information compared to the 32 time series. For MEG three patterns are necessary to account for 90% of the variance. Movies in MPEG format showing the spatiotemporal dynamics of the original signal together with a reconstruction from one, two, and three spatial patterns can be downloaded from our web site.

Bi-Orthogonal Expansions. The bi-orthogonal expansions are used to obtain a bi-orthogonal set of vectors and time dependent amplitudes calculated from a spatiotemporal pattern corresponding to a given second set of time series. Say we have a set of time series $H_k(t)$ from EEG or MEG recordings where k represents the different sensors. In addition, we have a second set of time series, $r_i(t)$, for instance the movement profile of a finger and its derivative. From these two sets we want to calculate the spatial patterns $\Psi^{(i)}$ in

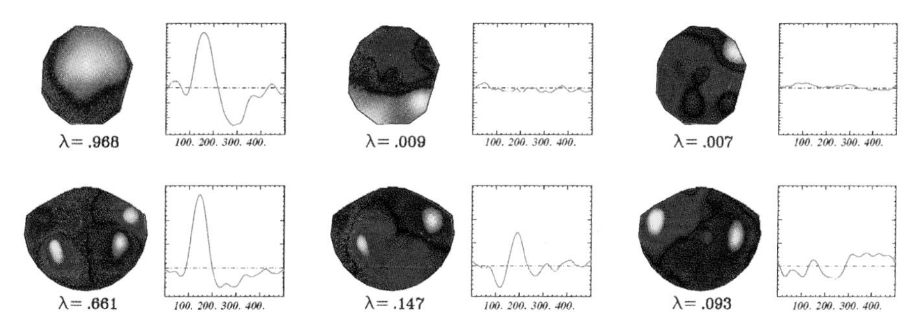

Fig. 4. Kl-expansion of the data shown in Fig. 1 (see text)

a way that if we project $\mathbf{H}(t)$ onto them we get an approximation of $r_i(t)$. There are various ways to do that:

- We can calculate the patterns according to

$$\sum_{k=1}^{N} \sum_{i=1}^{M} \int_0^T \mathrm{d}t \, \{H_k(t) - r_i(t)\, \psi_k^{(i)}\}^2 = \mathrm{Min} \tag{8}$$

  In this case the patterns can readily be calculated by taking the derivative of (8) with respect to $\psi_m^{(j)}$ and solving for $\psi_k^{(i)}$. The equations are completely decoupled and $\psi_k^{(i)}$ is given by:

$$\psi_k^{(i)} = \frac{\langle h_k(t)\, r_i(t)\rangle}{\langle r_i^2(t)\rangle} \tag{9}$$

  where $\langle ...\rangle$ is an abbreviation for the time average $\frac{1}{T}\int_0^T \mathrm{d}t \, ...$.
- A second possibility is to calculate the patterns according to:

$$\sum_{k=1}^{N} \int_0^T \mathrm{d}t \, \{H_k(t) - \sum_{i=1}^{M} r_i(t)\, \psi_k^{(i)}\}^2 = \mathrm{Min} \tag{10}$$

  Notice that this case is more like an expansion. Now, taking the derivatives with respect to $\psi_m^{(j)}$ leads to linear systems of equations of dimension M for the components $\psi_k^{(i)}$ which have to be solved for every k:

$$\langle r_j(t)\, H_k(t)\rangle = \sum_{i=1}^{M} \langle r_i(t)\, r_j(t)\rangle \, \psi_k^{(i)} \tag{11}$$

A priori it is not evident which one of these techniques is "better". In any case, the vectors $\Psi^{(i)}$ will not form an *orthogonal* basis. Therefore, a set of adjoint vectors $\Psi^{(i)+}$ needs to be calculated that fulfills the relation:

$$\Psi^{(i)+}\,\Psi^{(j)} = \delta_{ij} \quad \text{and} \quad \Psi^{(i)+} = \sum_{j=1}^{M} a_{ij}\,\Psi^{(j)} \tag{12}$$

Using these adjoint vectors the set of time series $\xi_i(t)$ corresponding to $r_i(t)$ is then obtained as:

$$\xi_i(t) = \sum_{k=1}^{N} h_k(t)\,\psi_k^{(i)+} \quad . \tag{13}$$

Friedrich and Uhl (1992, 1996) used such expansions to obtain spatial patterns that represent the dynamical properties of a data set more appropriately than the patterns obtained by the KL-expansion which are restricted by the orthogonality constraint (7). The idea is to use the time series for the dominating pattern from a KL-expansion and to calculate the other patterns in a way that they correspond to temporal derivatives of this function. This way the data can be represented by a dynamical system of first order differential equations of the form

$$\dot{\eta}_1 = \eta_2 \quad , \quad \dot{\eta}_2 = \eta_3 \quad , \quad \ldots \quad , \quad \dot{\eta}_n = f\{\eta_1, \eta_2, \ldots, \eta_n\} \tag{14}$$

where a set of two bi-orthogonal patterns relates to each variable η_i. Bi-orthogonal expansions of the second type are depicted in Fig. 5, again for the EEG (upper row) and MEG (lower row) data sets. Shown are the most dominant patterns from a KL-expansion (left most column) for EEG (top) and MEG (bottom) and the corresponding adjoint pattern (second column from left) together with the time series calculated according to (13) in red. The original time series from the KL-decomposition is also plotted in green but is not visible in this case because the two curves are virtually identical. In the fourth and fifth columns the patterns that best fit the time dependence of the derivative of the curves in column three are shown with their corresponding adjoint patterns, respectively. Again in column six the original time series (the derivatives of the curves in column 3) are plotted in green and the

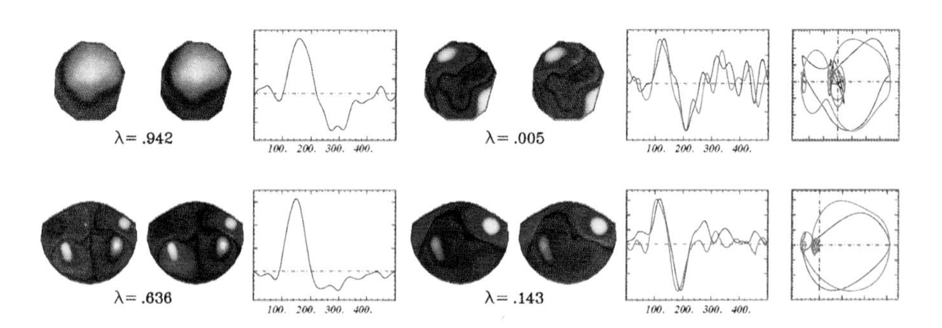

Fig. 5. Bi-orthogonal expansion of the data shown in Fig. 1 (see text)

curve reconstructed from (13) in red. In this example the patterns and their adjoints are very similar because the patterns are almost orthogonal which is not always the case. The right-most column shows phase space plots, i.e. $\xi(t)$ plotted in the x-direction and $\frac{\mathrm{d}}{\mathrm{d}t}\xi(t)$ in the y-direction. The attractors are obviously not unfolded in two dimensions since the trajectories intersect. However, the skeleton is the same for all four curves: There is one fixed point at $\xi = 0$ and there appears to be a second fixed point for a negative value of ξ. So even though the spatial patterns picked up be MEG and EEG are quite different the main features of the underlying dynamics from the viewpoint of dynamical systems are similar (for a much more sophisticated reconstruction of attractor skeletons and phase space dynamics see e.g. the contribution by Uhl et al. this volume).

3 Experimental Results

We are now going to apply the techniques outlined in the previous section to the data from the flexion-extension experiment outlined in KFJ. There we claimed that a strong correlation exists between the pattern of neural activity and the velocity profile of the finger movement for a variety of different movement directions or movement rates. In order to examine this relation further we first have a closer look at the dynamics of the movement itself (the movement amplitude as a function of time, or movement profile) and then analyze its connection to the observed patterns of brain activity in detail.

3.1 Behavioral Dynamics

The results obtained for the coordination behavior as far as timing (in terms of relative phase between stimulus and the movement) is concerned have already been discussed in KFJ (cf. Fig. 3 therein). It has been shown that the distributions of relative phase for the syncopation conditions are broader, indicating a less stable state of the system, compared to synchronization. Here we want to concentrate on the shape of the movement profile for the different conditions. Figure 6 shows the averaged movement profile together with the corresponding standard deviations for all four task conditions.
All curves are bell-shaped but show remarkable differences between the different kinematics flexion and extension. The flexion movements are shorter, steeper during the left flank when compared with the right flank, and more variable at the time of peak movement amplitude. The extension movements, on the other hand, are wider with an earlier onset in the cycle, and more symmetric with an almost constant variability. If our claim stated in KFJ about the relation between brain activity and movement velocity is valid, these differences must be visible in the MEG data, i.e. we expect that the dipolar pattern of movement activity is visible in the extension conditions before it can be seen in the flexion conditions due to the wider profile of the movement.

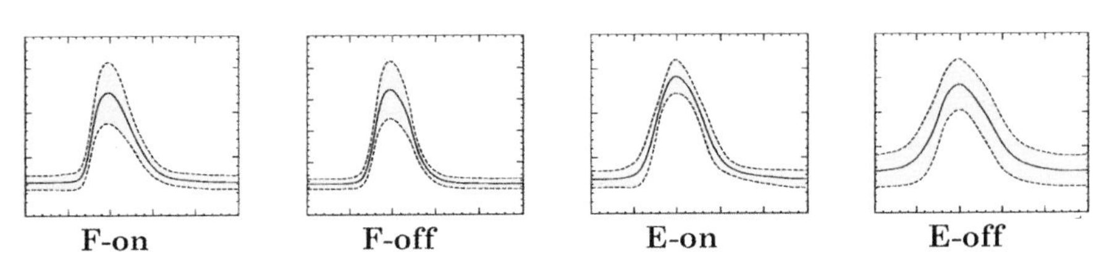

Fig. 6. Shape of the averaged movement profile with the standard deviation (shaded) for all task conditions

A movie at our web site shows this is indeed the case. It can also be seen in Fig. 4 in KFJ in which the dipolar structure becomes first visible in the row $t = -120$ ms for the extension and at $t = -60$ ms for the flexion movement.

3.2 Brain Dynamics

To examine the relation between movement velocity, $v(t)$, and the brain signal in sensor k, $H_k(t)$, the cross correlation at zero time lag, $C_k(\tau = 0)$, was calculated as

$$C_k(\tau) = \frac{\langle \{H_k(t) - \langle H_k(t)\rangle\}\{v(t) - \langle v(t)\rangle\}\rangle}{\sqrt{\langle \{H_k(t) - \langle H_k(t)\rangle\}^2\rangle \langle \{v(t) - \langle v(t)\rangle\}^2\rangle}} \; . \tag{15}$$

Figure 7 shows its value color coded as a function of space. Note that at the locations of maximum field magnitude (see Fig. 4 KFJ) the cross correlation is close to ± 1 indicating strong correlations or (anti-correlations) between movement velocity and magnetic brain activity. Figure 8 shows an overlay of the time series at sensor locations (green) and movement velocity (red; note the velocity profile is multiplied by the corresponding correlation value to correct for the fact that due to the dipolar nature of the pattern, the signal

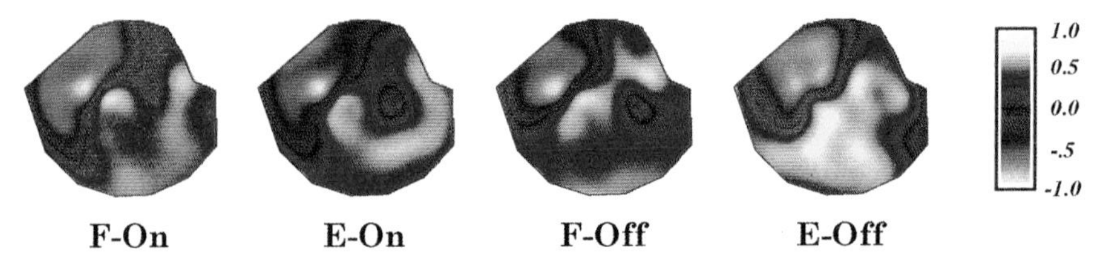

Fig. 7. Cross-correlation as a function of space for all task conditions

is reversed in some of the sensors). Within the highlighted region (where the field maximum and minimum is located), the time series of these two quantities match extremely well.

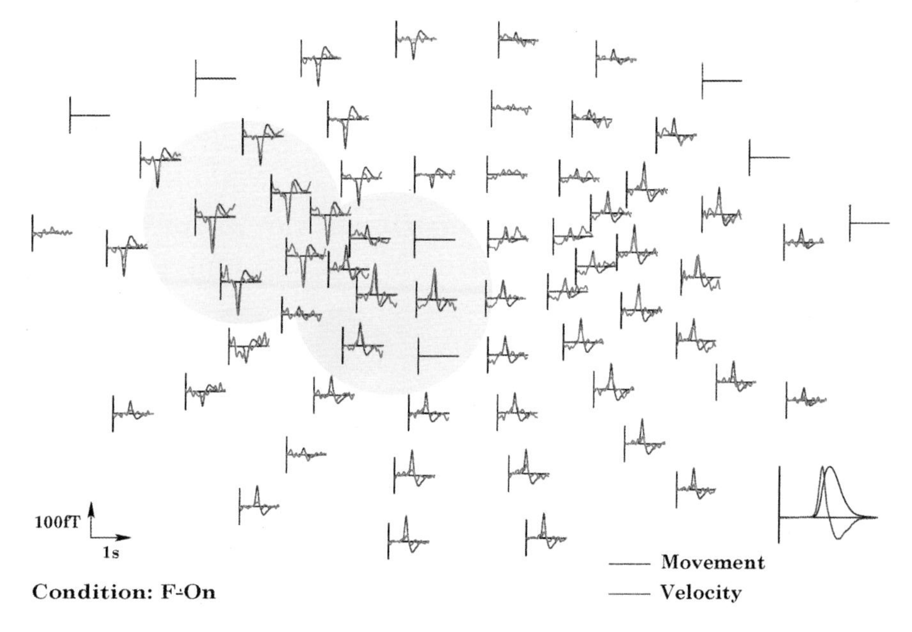

Fig. 8. Overlap between brain signals in all sensors (green) and the correlated movement velocity, i.e. the product of the velocity and the cross-correlation value $C_k(\tau = 0)$ in the sensor (red). This quantity was chosen, because due to the dipolar nature of the magnetic field patterns, the signals in some sensors are reversed versions of the signal in others. In the highlighted region these correlation values are close to ± 1 as shown in Fig. 7

As described earlier, the decomposition of a spatiotemporal pattern into a spatial part (consisting of static spatial patterns $\psi^{(i)}(\mathbf{x})$) and a temporal part (consisting of a time series, i.e. amplitudes corresponding to the different patterns at each time point) can lead to a drastic compression of information. How many patterns one has to take into account in order to get a decent reconstruction of the original signal depends on the data set under consideration. As seen above for the case of an auditory evoked response in EEG we needed only one pattern to account for over 95% of the variance in the original signal whereas for the case of MEG at least 3 patterns were necessary (if the original data set is purely random in space and time no reduction at all is achieved with these techniques).

In KFJ it has been shown that the time dependent amplitude corresponding to the dominating pattern from a KL-expansion approximates the shape of the movement velocity profile quite well for all task conditions (see Fig. 5 in KFJ). The time series for the second mode has no obvious interpretation due to the orthogonality constraint (7). Therefore, we applied the second of the bi-orthogonal expansions described in the previous section to the data and calculated the set of bi-orthogonal patterns corresponding to both the movement profile amplitude and velocity. Figure 9 shows the decompositions of the spatiotemporal brain signal for the four task conditions.

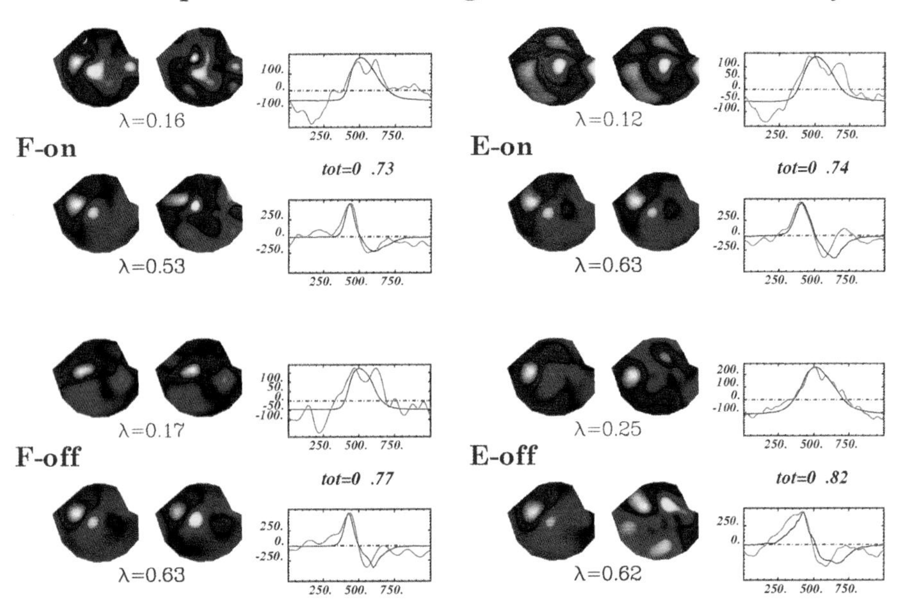

Fig. 9. Decomposition of the data using the second bi-orthogonal expansion. For each condition the upper patterns represent the modes $\Psi^{(1)}$ and $\Psi^{(1)+}$ corresponding to the finger displacements while the lower patterns $\Psi^{(2)}$ and $\Psi^{(2)+}$ relate to movement velocity. The values of λ are an estimate of the contribution of these modes to the variance of the original signal. On the right, the time series of the averaged displacement (red, top) and its temporal derivative, the finger velocity (red, bottom), are plotted for each condition. The green curves represent the projection of the original brain signal onto the adjoint vectors, i.e. the time-dependent amplitudes $\xi_1(t)$ and $\xi_2(t)$ from a reconstruction of the signal according to (3). The value of *tot* is an estimate of the quality of the reconstruction if both modes are used

Two remarks have to be made here: First, the two λs in general do not add up to the total contribution (as they do for the KL-expansion) because the patterns are no longer orthogonal and therefore some contributions of one pattern can be canceled by contributions from the other pattern. The decomposition only makes sense if the angle between the first and the second pattern (given by their scalar product) is not too small (estimates say at least 30°). Second, the quality of the reconstruction in terms of least square error can not be better than the reconstruction from the first two KL-modes (in fact, the fit is worse in all cases where the bi-orthogonal set of patterns are not the same as those obtained from the KL-expansion). In some sense we are trading here accuracy of the fit against functional relevance of the patterns, because now these vectors can be interpreted in terms of the behavior: They are the spatial patterns of the magnetic field produced by the neural activity that best follows the time course of movement amplitude and movement velocity - functionally relevant quantities created by the human motor system. The quality of the reconstruction from these two modes is still excellent as can be seen in Fig. 10 showing an overlay of the original (green) and the reconstructed brain signal (red).

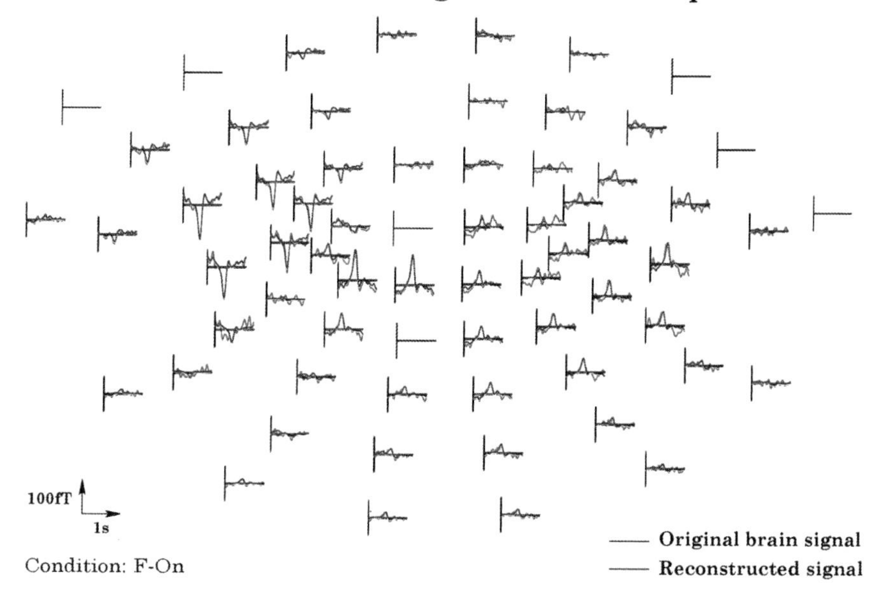

Fig. 10. Reconstruction of the signal from the two spatial patterns corresponding to movement amplitude and movement velocity

So far we have seen that the brain signals are very similar for the four different task conditions, i.e. independent of movement direction or coordination requirements. Specifically, the spatiotemporal patterns corresponding to both movement amplitude and velocity are dominated by a strong dipolar structure over motor cortex in the left hemisphere. After demonstrating these similarities we now ask the question: Can we find differences? Due to their large amplitudes, the motor-related patterns mask any differences that might exist between task conditions. Therefore, we first remove them by subtracting the contribution of both patterns at each point in time. In other words we subtract the red curves from the green curves (Fig. 10) and calculate the residual patterns for each condition separately. Figure 11 shows the results from this procedure.

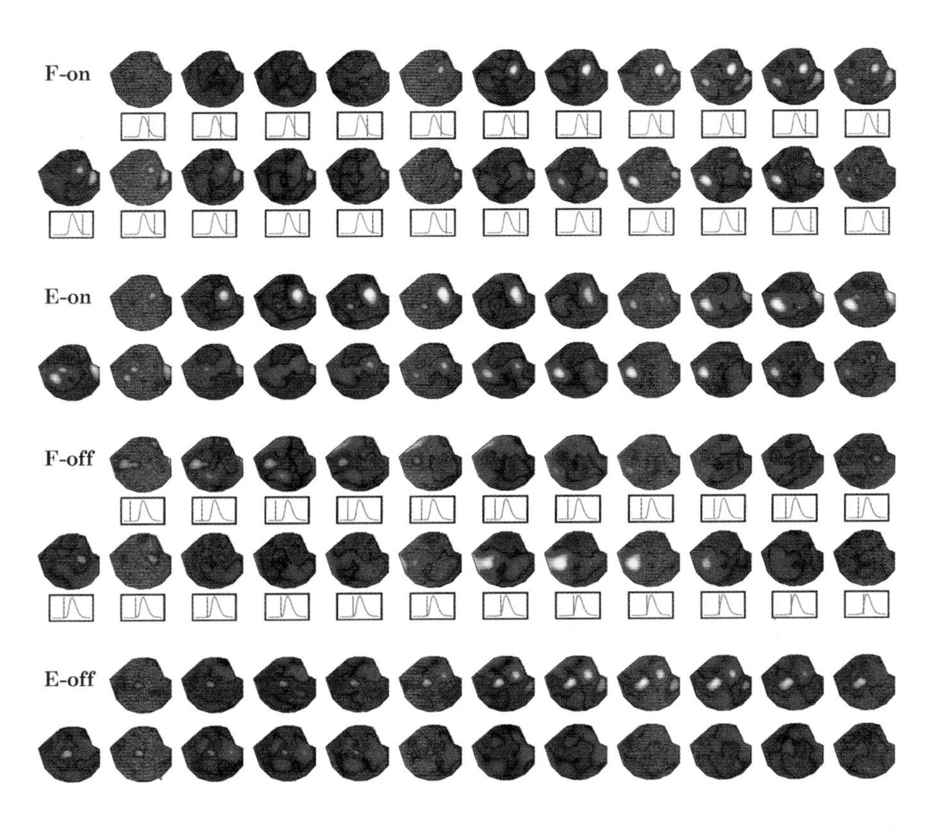

Fig. 11. Residual patterns after the dominating motor-related activity is removed for all task conditions at certain locations within a cycle (see text for details)

We are looking for dynamical structures that are similar in different conditions and occur at about the same time within the cycle. The top two rows show a sequence of topographic maps from the flexion-on condition occurring after peak movement as indicated in the boxes below each pattern. From left to right in the top row a red-yellow structure, indicating magnetic field that exits the head, appears over right frontal areas followed by activity of the same polarity over left central areas a little later in the cycle. In the second row this pattern of activity almost vanish before the left central structure reappears. A very similar scenario can be seen in the third and fourth rows which show the spatial patterns from the extension-on conditions at the same time points. The right frontal activity appears a little earlier in the cycle but the sequence is the same and the left central activity disappears at the same time in both conditions. For the off-the-beat conditions no such structure is observable after peak movement (not shown). However, if we look at time slices that are shifted by half a cycle with respect to the patterns shown for the on-the-beat conditions we find the two pattern sequences displayed in the bottom rows. Given the time shift, the left central activity in the flexion-off shows approximately the same time course observed in in the on-the-beat conditions. Similar observations can be made for the activity over right frontal areas in the extension-off condition. These similarities are probably due to the fact that the pattern sequences are now stimulus locked in all conditions. However, it is not easy to explain them because the stimulus in this experiment was a small LED and control conditions revealed no evoked visual fields. Therefore, it is likely that these sequence are related to some higher level processing of the stimulus. The most interesting point here is that in the off-the-beat conditions only parts of the sequence exists which could mean that the rest is annihilated by other ongoing activity and one may speculate that the instability of the syncopic coordination is related to these interferences.

4 Conclusions and Outlook

The main point in this paper was to show how visualization and analysis techniques can be used to extract relevant features from large sets of experimental data. Relevant here means features that link neural activity patterns on one level and human behavior, in this case motor behavior in a coordination experiment, on another level. It takes the coherent action of tenthousands of neurons to produce a signal that we can measure from outside the head or to make a finger move, but it is the movement of the finger on the other hand that leads to the coherent neural activity. This is the kind of circular causality which is present in all systems showing self-organization, and explored in the theory of synergetics founded by Hermann Haken. We can substitute "atoms" for "neurons" and "laser light" for "finger movement" because the same basic laws of self-organization govern the dynamics of entirely different systems. The hope is thus that some day we will be able to understand brain function as well as we understand the laser. To achieve this goal, both experimental and theoretical investigations are needed. Experimentally, performing the movement at different coordination frequencies should tell us whether or not the sequence of the stimulus related activity in the off-the-beat conditions is longer at lower frequencies, shrinks when the frequency is increased and eventually vanishes when syncopation becomes unstable. From theory we could learn what kind of nonlinear interactions lead to such an annihilation process and where in the cortex the sources of activity must be located to create the observed activity patterns. The high spatial resolution of functional MRI could be used to identify these regions. Once known, together with high resolution EEG and MEG measurements, a combination of experimental data and knowledge about dispersion and propagation properties of the neural tissue (including inhomogeneities) could eventually lead us to a "solution" of the inverse problem after all.

Acknowledgements

We thank Tom Holroyd for help with data collection and Justine Mayville for her help in the final stages of the manuscript. This research was supported by NIMH (Neurosciences Research Branch) Grant MH42900, KO5 MH01386 and the Human Frontiers Science Program. VKJ gratefully acknowledges a fellowship from the *Deutsche Forschungsgemeinschaft.*

References

Friedrich, R., Uhl, C. (1992): Synergetic Analysis of Human Electroencephalograms: Petit-Mal Epilepsy. In *Evolution of Dynamical Structures in Complex Systems*, Friedrich, R., Wunderlin, A., eds., Springer, Berlin, Heidelberg

Friedrich, R., Uhl, C. (1996): Spatiotemoral Analysis of Human Electroencephalograms: Petit-Mal Epilepsy. Physica D 98, 180

Fuchs, A., Kelso, J.A.S., Haken, H. (1992): Phase transitions in the Human Brain: Spatial Mode Dynamics. Int. J. Bifurc. Chaos 2, 917-939

Le, J., Gevins, A. (1993): Method to Reduce Blur Distortion from EEGs using a Realistic Head Model, IEEE Trans. on Biomed. Eng. 40, 517-528

Jirsa, V.K., Haken, H. (1996): Field theory of electromagnetic brain activity. Phys. Rev. Let. 77, 960–963

Jirsa, V.K., Haken, H. (1997): A derivation of a macroscopic field theory of the brain from the quasi-microscopic neural dynamics, Physica D 99, 503–526

Lütkenhöner, B. (1994): The magnetic field arising from current dipoles randomly distributed in a homogenous spherical volume conductor. J. Appl. Phys. 75, 7204-7210

Meaux, J.R., Wallenstein, G.V., Nash, A., Fuchs, A., Bressler, S.L., Kelso, J.A.S. (1996): Cortical Dynamics of the Human EEG Associated with Transitions in Auditory-Motor Coordination. Society for Neuroscience Abstracts Vol. 22/2, 890

Nunez, P.L. (1981): *Electric fields of the brain*, Oxford University Press

Law, S.K., Nunez, P.L., Wijesinghe, R.S. (1993): High-Resolution EEG Using Spline Generated Surface Laplacians on Spherical and Elliptical Surfaces. IEEE Trans. Biomed. Eng. 40, 145–153

Wallenstein, G.V., Kelso, J.A.S., Bressler, S.L. (1995): Phase transitions in spatiotemporal patterns of brain activity and behavior. Physica D 84, 626-634

Traversing Scales of Brain and Behavioral Organization III: Theoretical Modeling

Viktor K. Jirsa, J.A. Scott Kelso, and Armin Fuchs

Program in Complex Systems and Brain Sciences, Center for Complex Systems, Florida Atlantic University, 777 Glades Road, Boca Raton, FL 33431, USA

1 Introduction

In coordinated movements typically several states related to different behavioral patterns can be found, e.g. different gaits of horses (Collins and Stewart 1993, Schöner et al. 1990) or different configurations among the joints for trajectory formation tasks (Buchanan et al. 1997, Kelso et al. 1991). These states have different stabilities dependent on external or internal control parameters. When such control parameters are manipulated, coordination states may become unstable and the system exhibits a transition from one state to another. These phenomena have intensively been investigated experimentally and theoretically and mathematical models have been set up reproducing the experimentally observed coordination behavior as well as predicting new effects (see (Haken 1996, Kelso 1995) for reviews). On the other hand, recent MEG and EEG experiments (Kelso et al. 1992, Wallenstein et al. 1995) have investigated the spatiotemporal brain dynamics during coordination of finger movements with external periodic stimuli. To accommodate these results, a mathematical phenomenological model was developed describing the on-going brain activity (Jirsa et al. 1994). In (Jirsa and Haken 1996a, 1996b) [Jirsa and Haken 1996a, Jirsa and Haken 1996b] a neurophysiologically motivated field theory of the spatiotemporal brain dynamics was elaborated which combined properties of neural ensembles, including their short- and long-range connections in the cortex, in addition to describing the interaction of functional units embedded into the neural sheet. This approach was applied to the brain-coordination experiment (Kelso et al. 1992) where the subject's task was to coordinate rhythmic behavior of a finger with an external acoustic stimulus. During the experiment the MEG of the subject was recorded. Complex systems, such as the brain, have the general property that they perform low-dimensional behavior during transitions from one macroscopic state to another (Cross and Hohenberg 1993, Haken 1983, 1987). This type of behavior has also been found in the analyses (Fuchs et al. 1992, Jirsa et al. 1995) of the brain data from the coordination experiment in (Kelso et al. 1992). On the basis of these analyses the phenomenological model in (Jirsa et al. 1994) describing the brain activity was derived qualitatively from the neurophysiologically motivated theory in (Jirsa and Haken 1996a, 1997).

The goal of the present paper is to show how it may be possible to traverse levels of organization from the behavioral level to the brain level. For this purpose, we choose a bimanual coordination experiment (Kelso 1981,1984) in which a transition in coordinated behavior is observed between finger movements when a control parameter is changed. Along this example, we will treat the organization on the level of behavior (Sect. 2) and then make the the connection to the organization on the level of the brain (Sect. 3).

2 The Level of Behavior

Experimental studies by one of us (Kelso 1981,1984), as well as others (see e.g. (Carson et al. 1994) for a review) have shown that abrupt phase transitions occur in human finger movements under the influence of scalar changes in the cycling frequency. Below a critical cycling frequency two dynamical patterns or states are possible: An in-phase state where the finger movements are symmetric and an anti-phase state where the finger movements are antisymmetric. Starting the finger movements in the anti-phase state and increasing the cycling frequency, a spontaneous transition from anti-phase to in-phase is observed at a critical frequency. Beyond this frequency only the in-phase state is stable. Further, it is experimentally found that the amplitude of the finger movements decreases when the cycling frequency is increased.

In 1985 these phenomena were theoretically modeled by Haken, Kelso and Bunz (1985) by formulating a model system for the dynamics of the collective variable represented by the relative phase between the fingers. This model system was then used as a guide to establish a model for the dynamics of the component variables represented by the finger positions x_1 and x_2. These component variables perform an oscillatory behavior and interact nonlinearily. Two ordinary differential equations, again based on detailed experimental results, describe the dynamics of the individual fingers with the amplitudes x_1 and x_2. This model system reads

$$\ddot{x}_1 + (Ax_1^2 + B\dot{x}_1^2 - \gamma)\dot{x}_1 + \Omega^2 x_1 = (\dot{x}_1 - \dot{x}_2)\,(\alpha + \beta(x_1 - x_2)^2) \qquad (1)$$

$$\ddot{x}_2 + (Ax_2^2 + B\dot{x}_2^2 - \gamma)\dot{x}_2 + \Omega^2 x_2 = (\dot{x}_2 - \dot{x}_1)\,(\alpha + \beta(x_1 - x_2)^2) \qquad (2)$$

The left hand sides of (1), (2) describe the motion of the individual fingers, while the right hand sides describe the coupling.

With the goal of connecting behavioral coordination dynamics (1) and (2) to brain dynamics we introduce an alternative description of these phenomena in terms of symmetric and antisymmetric modes. These modes directly correspond to the behavioral states of the system, for which we seek correspondence at the brain level.

We define the following variables:

$$\tilde{\psi}_+ = x_1 + x_2 \qquad\qquad \tilde{\psi}_- = x_1 - x_2 \qquad (3)$$

These variables represent modes of behavior where $\tilde{\psi}_+$ corresponds to the symmetric (in-phase) mode and $\tilde{\psi}_-$ to the antisymmetric (anti-phase) mode. The back transformation onto the amplitudes of finger movement reads

$$x_1 = \frac{1}{2}(\tilde{\psi}_+ + \tilde{\psi}_-) \qquad x_2 = \frac{1}{2}(\tilde{\psi}_+ - \tilde{\psi}_-)\,. \tag{4}$$

In order to obtain the equations governing the dynamics of the new variables $\tilde{\psi}_+$, $\tilde{\psi}_-$ we subtract and sum (1), (2), respectively, and obtain

$$\begin{aligned} &\ddot{\tilde{\psi}}_+ - \gamma\dot{\tilde{\psi}}_+ + \Omega^2\tilde{\psi}_+ + \tfrac{A}{12}\tfrac{\partial}{\partial t}(\tilde{\psi}_+^3 + 3\tilde{\psi}_-^2\tilde{\psi}_+) + \tfrac{B}{4}(\dot{\tilde{\psi}}_+^3 + 3\dot{\tilde{\psi}}_-^2\dot{\tilde{\psi}}_+) = 0 \\ &\ddot{\tilde{\psi}}_- - \gamma\dot{\tilde{\psi}}_- + \Omega^2\tilde{\psi}_- + \tfrac{A}{12}\tfrac{\partial}{\partial t}(\tilde{\psi}_-^3 + 3\tilde{\psi}_+^2\tilde{\psi}_-) \\ &\qquad\qquad + \tfrac{B}{4}(\dot{\tilde{\psi}}_-^3 + 3\dot{\tilde{\psi}}_+^2\dot{\tilde{\psi}}_-) = 2\dot{\tilde{\psi}}_-\,(\alpha + \beta\tilde{\psi}_-^2). \end{aligned} \tag{5}$$

The left hand sides of (5) represent fully symmetric (with respect to the exchange of the indices + and −), nonlinearly coupled equations. The former coupling terms with α and β in the variables x_1, x_2 now appear only in one equation solely in terms of the antisymmetric mode $\tilde{\psi}_-$.

In order to treat the system (5) analytically we make the following ansatz:

$$\tilde{\psi}_+ = R_+ e^{i\Phi_+} e^{i\Omega t} + R_+ e^{-i\Phi_+} e^{-i\Omega t} \tag{6}$$

$$\tilde{\psi}_- = R_- e^{i\Phi_-} e^{i\Omega t} + R_- e^{-i\Phi_-} e^{-i\Omega t} \tag{7}$$

where R_+, R_- denote real time dependent amplitudes and Φ_+, Φ_- the corresponding time dependent phases. Inserting this ansatz into (5) and performing two approximations well known in nonlinear oscillator theory (rotating wave approximation, slowly varying amplitude approximation, see e.g. (Haken (1983))) we obtain the following equations for the amplitudes

$$\begin{aligned} \dot{R}_+ &= \tfrac{1}{2}\gamma R_+ - a(\Omega)\,(R_+^2 + 2R_-^2 + R_-^2\cos 2(\Phi_- - \Phi_+))\,R_+ \\ \dot{R}_- &= \tfrac{1}{2}\gamma R_- - a(\Omega)\,(R_-^2 + 2R_+^2 + R_+^2\cos 2(\Phi_+ - \Phi_-))\,R_- + \alpha R_- + \beta R_-^3 \end{aligned} \tag{8}$$

and for the phases

$$\dot{\Phi}_+ = -a(\Omega)R_-^2 \sin 2(\Phi_- - \Phi_+) \tag{9}$$

$$\dot{\Phi}_- = -a(\Omega)R_+^2 \sin 2(\Phi_+ - \Phi_-) \tag{10}$$

where $a(\Omega) = 1/8(A + 3B\Omega^2)$. Defining the new variable $\phi = \Phi_+ - \Phi_-$ we can rewrite (9), (10) as

$$\dot{\phi} = -a(\Omega)(R_+^2 + R_-^2)\sin 2\phi \tag{11}$$

which has the only stable solutions $\phi = \frac{\pi}{2}, \frac{3\pi}{2}$ for nontrivial amplitudes R_+, R_-. Thus (8) can be reduced to

$$\begin{aligned} \dot{R}_+ &= \tfrac{1}{2}\gamma R_+ - a(\Omega)\,(R_+^2 + R_-^2)\,R_+ &&= -\frac{\partial V}{\partial R_+} \\ \dot{R}_- &= \tfrac{1}{2}\gamma R_- - a(\Omega)\,(R_+^2 + R_-^2)\,R_- + \alpha R_- + \beta R_-^3 &&= -\frac{\partial V}{\partial R_-} \end{aligned} \tag{12}$$

where the dynamics of R_+, R_- can be expressed in terms of a gradient dynamics with the potential

$$V = -\frac{1}{4}\gamma(R_+^2 + R_-^2) + \frac{1}{4}a(\Omega)\,(R_+^2 + R_-^2)^2 - \frac{1}{2}\alpha R_-^2 - \frac{1}{4}\beta R_-^4 \tag{13}$$

A linear stability analysis of (12) yields the same results as in (Haken et al. 1985) and can be graphically presented in terms of the potential V in (13). In Fig. 1 (upper row) the potential V is plotted in dependence of R_+ and R_- as the control parameter Ω increases from left to right. Here the R_--axis points out of the page. The corresponding isoclines of V are plotted below in arbitrary units of R_+, R_-. Below the critical frequency bistability is present (left two pictures), i.e. either the symmetric or the antisymmetric mode is present. At the critical frequency Ω_c (third picture) the antisymmetric mode becomes unstable and only the symmetric mode remains for higher frequencies (right picture).

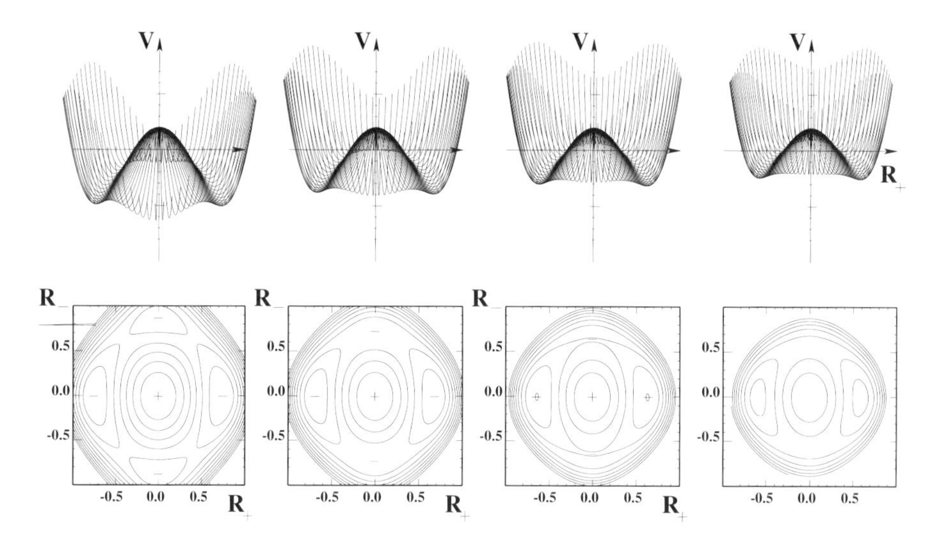

Fig. 1. The potential V is plotted (top row) in dependence of R_+ and R_- as the frequency increases from left to right. The axis of R_- points out of the page. The scale on the axes is in arbitrary units. The isoclines of the potential V are plotted in dependence of the oscillator amplitudes R_+, R_-. The plus sign marks a local maximum, the minus sign a local minimum

3 The Level of Brain

In the framework of this paper we will use the behavioral mode system (5) as a guideline to traverse scales of organization from the behavioral level to the brain level. First we briefly review a field theoretical description of neural activity recently developed by Jirsa & Haken (1996a, 1996b, 1997), then we specify the neural field equation with respect to the bimanual coordination experiment and discuss it in detail.

3.1 Field Theoretical Description of Neural Activity

Let us consider a $(n-1)$-dimensional closed surface Γ representing the neocortex in a n-dimensional space. This medium Γ shall consist of neural ensembles to which we assign two state variables describing their activity: Dendritic currents generated by active synapses cause *waves* of extracellular fields which can be measured by the EEG (Freeman 1992) and intracellular fields measurable by the MEG (Williamson and Kaufman 1987). Action potentials generated at the somas of neurons correspond to *pulses*. We call the magnitude of the neural ensemble average of the waves the wave activity $\psi_j(x,t)$ with $j = \mathrm{e}, \mathrm{i}$ where the indices distinguish excitatory and inhibitory activity and the magnitude of the neural ensemble average of the pulses the pulse activity $\psi_j(x,t)$ with $j = \mathrm{E}, \mathrm{I}$ distinguishing excitatory and inhibitory pulses. The location in Γ is denoted by x, the time point by t. These scalar quantities are related to each other via conversion operations (Freeman 1992) which we define as

$$\psi_j(x,t) = \int_\Gamma \mathrm{d}X\, f_j(x,X) H_j(x,X,t)\,, \tag{14}$$

where $j = \mathrm{e}, \mathrm{i}, \mathrm{E}, \mathrm{I}$. Here the function $H_j(x,X,t)$ represents the output of a conversion operation and $f_j(x,X)$ the corresponding distribution function depending on the spatial connectivity. From experimental results of Freeman (1992) it is known that the conversion from wave to pulse is sigmoidal within a neural ensemble, however the conversion from pulse to wave is also sigmoidal, but constrained to a linear small-signal range. Assuming that excitatory neurons only have excitatory synapses, inhibitory neurons only inhibitory synapses (which is generally true (Abeles 1991)) we obtain the following relations between conversion output and pulses

$$H_\mathrm{e}(x,X,t) = S\left[\psi_\mathrm{E}\left(X, t - \frac{|\, x - X \,|}{v}\right)\right] \approx a_\mathrm{e}\psi_\mathrm{E}\left(X, t - \frac{|\, x - X \,|}{v}\right) \tag{15}$$

$$H_\mathrm{i}(x,X,t) = S\left[\psi_\mathrm{I}\left(X, t - \frac{|\, x - X \,|}{v}\right)\right] \approx a_\mathrm{i}\psi_\mathrm{I}\left(X, t - \frac{|\, x - X \,|}{v}\right) \tag{16}$$

and between conversion output and waves

$$H_\mathrm{E}(x,X,t) = S_\mathrm{e}\left[\psi_\mathrm{e}\left(X,t-\frac{|x-X|}{v}\right)-\psi_\mathrm{i}\left(X,t-\frac{|x-X|}{v}\right)\right.$$
$$\left.+\,p_\mathrm{e}\left(X,t-\frac{|x-X|}{v}\right)\right], \qquad (17)$$

$$H_\mathrm{I}(x,X,t) = S_\mathrm{i}\left[\psi_\mathrm{e}\left(X,t-\frac{|x-X|}{v}\right)-\psi_\mathrm{i}\left(X,t-\frac{|x-X|}{v}\right)\right.$$
$$\left.+p_\mathrm{i}\left(X,t-\frac{|x-X|}{v}\right)\right] \qquad (18)$$

External input $p_j(x,t)$ is realized such that afferent fibers make synaptic connections and thus $p_j(x,t)$ appears only in (17), (18). Here a_e, a_i are constant parameters denoting synaptic weights, v the propagation velocity and S, S_j the sigmoid functions of a class j ensemble.

Let us now make the following considerations: We are interested in a spatial scale of several cm and temporal scale of 100 msec which is relevant in EEG and MEG. Intracortical fibers (excitatory and inhibitory) typically have a length of 0.1 cm, corticocortical (only excitatory) fiberlengths range from about 1 cm to 20 cm (Nunez 1995). Cortical propagation velocities have a wide range from 0.2 m/sec (Miller 1987) up to 6–9 m/sec (Nunez 1995). With an average velocity of 1 m/sec this yields propagation delays of 1 msec for the intracortical fibers and 10 msec to 200 msec for the corticocortical fibers. Synaptic delays and refractory times are of the order 1 msec, the neuronal membrane constant is in the range of several msec (Braitenberg and Schüz 1991). From this brief summary we see that the spatial and temporal scales vary considerably. The distribution of the intracortical fibers is very homogeneous (Braitenberg and Schüz 1991), but the distribution of the corticocortical fibers is not (estimates are that 40% of all possible corticortical connections are realized for the visual areas in the primate cerebral cortex (Felleman and Van Essen 1991)). We assume the corticocortical fiber distributions to be homogeneous as a first approximation. Using the discussed temporal and spatial hierarchies the dynamics of the system (14)–(18) can be systematically reduced (see (Jirsa and Haken 1996a,1996b,1997) for details): the fast dynamics ($<<$ 100 msec) becomes either instantaneous or can be eliminated and the spatial scales smaller than 1cm become point-like. Then the entire dynamics of the system can be described in terms of the slowest variable $\psi_\mathrm{e}(x,t)$ and a modified external input now denoted by $p(x,t)$. The dynamics of $\psi_\mathrm{e}(x,t)$ is given by

$$\psi_\mathrm{e}(x,t) = a_\mathrm{e}\int_\Gamma \mathrm{d}X\, f_\mathrm{e}(x-X) \qquad (19)$$
$$\times S_\mathrm{e}\left[\rho\,\psi_\mathrm{e}\left(X,t-\frac{|x-X|}{v}\right)+p\left(X,t-\frac{|x-X|}{v}\right)\right],$$

where ρ is a density of excitatory fibers, modified due to the elimination of the other variables. Note that from the equations (14)-(18) the models by Wilson & Cowan (1972,1973) in terms of pulse activities and by Nunez (1974,1995) in terms of wave activities can be derived and are connected by our approach.

Until now the dimension of the cortical surface has been kept general. Here we want to specify $n = 2$, meaning that Γ represents a closed 1-dimensional loop. Such a geometry has been reported by Nunez (Nunez 1995) to be a good approximation when macroscopic EEG dynamics is considered under more qualitative aspects of dynamics like changes of dispersion relations. In the following sections we will perform a low-dimensional mode decomposition in which case the chosen geometry suffices for a discussion of the temporal mode dynamics. Using the method of Green's functions (Jirsa and Haken 1997) the above integral equation (19) can be rewritten as a nonlinear partial differential equation

$$\ddot{\psi} + (\omega_0^2 - v^2\triangle)\,\psi + 2\omega_0\dot{\psi} = a_e\left(\omega_0^2 + \omega_0\frac{\partial}{\partial t}\right) S[\rho\,\psi(x,t) + p(x,t)] \qquad (20)$$

where $\omega_0 = v/\sigma$ and we dropped the indices e. Here we call $\psi(x,t)$ the neural field. The interaction of functional units with the cortical sheet Γ is represented by the external input signals $p_j(x,t)$, where $p(x,t) = \sum_j p_j(x,t)$ and the output signals $\bar{\psi}_j(t)$. A functional unit can include subcortical structures such as the projections of the cerebellum on the cortex or specific functional areas like the motor cortex. Anatomically these areas are obviously defined via their afferent and efferent fibers connecting to the cortical sheet. In the context of the present theory dealing with dynamics on a larger spatiotemporal scale, i.e. wavelengths in the regime of several cm, it is more appropriate to identify the spatial localizations of the functional input units with the spatial structures which are generated by the time dependent input signals $z_j(t)$, open to observation in the EEG/MEG. In the case of a finger movement this spatial structure $\beta_j(x)$ corresponds to the well-known dipolar mode in the EEG/MEG located over the contralateral motor areas. Thus such a functional input unit is defined as

$$p_j(t) = \beta_j(x)\, z_j(t)\ . \qquad (21)$$

Similarly, an output signal $\bar{\psi}_j(t)$ sent to non-cortical areas is picked up from the cortical sheet according to

$$\bar{\psi}_j(t) = \int_\Gamma \mathrm{d}x\ \beta_j(x)\psi(x,t)\ , \qquad (22)$$

where $\beta_j(x)$ defines the spatial localization of the jth functional output unit.

In summary, the field theoretical approach presented here aims at a description of the spatiotemporal brain dynamics on the scale of several cm and 100 msec. These scales emphasize the corticocortical connections and allow the derivation of (19) in one field variable $\psi(x,t)$ governing the spatiotemporal dynamics. Focussing on the dynamical aspects of the interaction of only a few modes in the following sections, a cortical representation of a closed strip is chosen.

3.2 Neural Field Theory of Bimanual Coordination

The neural areas subserving bimanual coordination are numerous and diverse. The cortex, through intracortical connections and long loop, reciprocal pathways to the basal ganglia and cerebellum, obviously plays a crucial role. Propriospinal and brainstem networks are also involved. Wiesendanger et al. (Wiesendanger et al. 1994) in a recent review of lesion studies in humans and non-human primates implicate lateral premotor cortex, supplementary motor area, parietal association cortex and the anterior corpus callosum (among others) in goal-directed bimanual coordination. Though many kinds of cortical lesions can affect bimanual movements, objective measures of spatiotemporal organization are rare in studies of patient populations. In the context of the present work, Tuller & Kelso (1989) showed that in-phase and anti-phase movements of the fingers were preserved in split-brain patients. Other phase relations were much more difficult for split brains to produce compared to normal subjects. Anatomical and physiological evidence for bilateral control of each cortical area may explain why callosal damage and unilateral cortical lesions tend to produce only transient disturbances of bimanual coordination (Wiesendanger et al. 1994).

For present purposes, we consider a simplified scheme in which cortical areas interact in a cooperative fashion to produce goal-directed bimanual coordination (see Fig. 2). Evidence for bilateral activation of primary motor cortices during a bimanual task in which both index fingers are simultaneously moved (see e.g. (Kristeva et al. 1991)), is consistent with our double representation of "motor areas" in Fig. 2. Likewise, the presence of movement evoked fields in both postcentral cortices corresponding to reafferent activity from the periphery during bimanual movements (Kristeva et al. 1991) justifies the two "sensorimotor areas" in our model. Thus, motor signals are conveyed from the motor areas in the cortical sheet to the individual fingers, sensorimotor signals carrying information about the finger movements are conveyed to the sensorimotor areas of the brain. Please note that we assign the same index l (r) to the left (right) finger and its contralateral hemisphere in order to keep the notation in the following mathematical treatment as simple as possible. Here we deal with the following situation as shown in Fig. 2: two input units localized at $\beta_{ls}(x)$ (left hemisphere, sensorimotor), $\beta_{rs}(x)$ (right hemisphere, sensorimotor) and two output units localized at $\beta_{lm}(x)$ (left hemisphere, motor), $\beta_{rm}(x)$ (right hemisphere, motor) are embedded into a one-dimensional closed neural strip. The origin of the underlying coordinate system is located between the two hemispheres where L is the length of the neural strip. Anatomical considerations imply as a first approximation

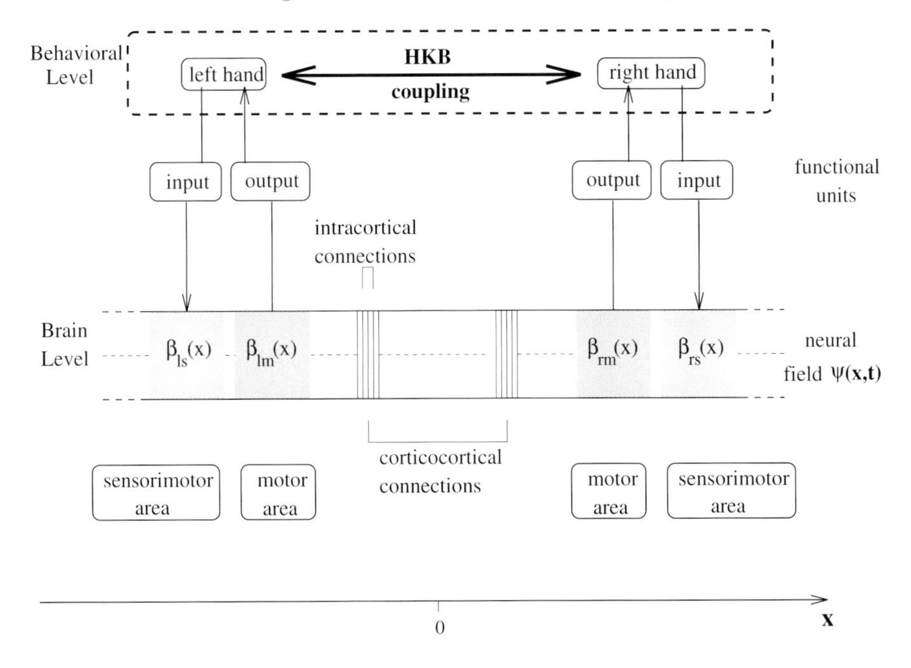

Fig. 2. Two input units localized at $\beta_{ls}(x)$ (sensorimotor, left hemisphere), $\beta_{rs}(x)$ (sensorimotor, right hemisphere) and two output units localized at $\beta_{lm}(x)$ (motor, left hemisphere), $\beta_{rm}(x)$ (motor, right hemisphere) are embedded into a one-dimensional closed neural strip whose activity is described by the field $\psi(x,t)$

the following symmetries:

$$\begin{aligned} \beta_{ls}(x) &= \beta_{rs}(-x) \\ \beta_{lm}(x) &= \beta_{rm}(-x) \end{aligned} \tag{23}$$

The output signal, here the motor field, is defined according to (22) and is conveyed to the corresponding finger where $z_l(t) = x_1(t)$, $z_r(t) = x_2(t)$ denote the extensions of the left and right finger movements, respectively. The motor movement $z_j(t)$ with $j = l, r$ shall be described phenomenologically as a function f of the motor field $\bar{\psi}_j(t)$ and, in order to take phase shifts into account, its derivative $\dot{\bar{\psi}}_j(t)$:

$$z_j(t) = f(\bar{\psi}_j(t), \dot{\bar{\psi}}_j(t)) \approx f_0 + f_1\, \bar{\psi}_j(t) + f_2\, \dot{\bar{\psi}}_j(t) + ... \tag{24}$$

where $f(\bar{\psi}_j(t), \dot{\bar{\psi}}_j(t))$ denotes a nonlinear function which we expanded into a Taylor function in terms of $\bar{\psi}_j(t)$ and $\dot{\bar{\psi}}_j(t)$ and truncated after the linear terms as an approximation. The constant f_0 describes a constant amplitude shift and can be set to zero for a rhythmic movement. See also (Jirsa and Haken 1997) for a treatment of the sensorimotor feedback loop in terms of driven oscillators. In the following discussion we will consider the limit case

$f_1 \neq 0, f_2 = 0$. A discussion of the other limit $f_1 = 0, f_2 \neq 0$ gives analogeous results. With (22),(24) and $f_1 \neq 0, f_2 = 0$ the sensorimotor feedback is now given by

$$p_j(x,t) = \beta_{js}(x) z_j(t) = c_0\, \beta_{js}(x) \int_{-L/2}^{L/2} \beta_{jm}(x) \psi(x,t)\, \mathrm{d}x \tag{25}$$

and the feedback loops of the motor and sensorimotor units are closed.

We are now in a position to specify the field equation (20) as follows

$$\ddot{\psi} + (\omega_0^2 - v^2 \triangle)\psi + 2\omega_0 \dot{\psi} = a_e \left(\omega_0^2 + \omega_0 \frac{\partial}{\partial t} \right) S[\rho\psi + p_l(x,t) + p_r(x,t)] \,. \tag{26}$$

The field $\psi(x,t)$ can also always be written in terms of symmetric and anti-symmetric contributions

$$\psi(x,t) = \underbrace{\frac{1}{2}(\psi(x,t) + \psi(-x,t))}_{\psi^+(x,t)} + \underbrace{\frac{1}{2}(\psi(x,t) - \psi(-x,t))}_{\psi^-(x,t)} \,. \tag{27}$$

Next, we make an assumption about the spatial pattern underlying the temporal dynamics of $\psi^+(x,t)$ and $\psi^-(x,t)$. In the bimanual coordination experiment a transition from one pattern to another is observed on the behavioral level. In this case the theory of dynamical systems, in particular synergetics (Haken (1983),1987), predicts low dimensional behavior of the system under consideration and we expect to observe low-dimensional transition phenomena on the brain level, too. Further, in previous analyses of brain-behavior experiments (Fuchs et al. 1992, Jirsa et al. 1995) involving behavioral transitions during coordination with external stimuli low-dimensional spatiotemporal brain dynamics was found and could be described in terms of two spatial modes. Here we deal with the different situation of the coordination of two limbs. But since a similar phase transition in behavior has also been observed, we assume to a first approximation that each contribution $\psi^+(x,t)$ and $\psi^-(x,t)$ is dominated by one spatial pattern and factorizes

$$\begin{aligned} \psi^+(x,t) &\approx g_+(x)\, \psi_+(t) \\ \psi^-(x,t) &\approx g_-(x)\, \psi_-(t) \end{aligned} \tag{28}$$

if only standing waves are present. The assumption of two dominating spatial modes is crucial for the following mathematical analysis and experimentally easy to test. If this assumption is confirmed, higher order structures of the dynamics can be included following the lines of (Jirsa et al. 1995) in which the reproduction of the experimental spatiotemporal signal could be improved by adding more spatial modes whose temporal dynamics depends only on the dynamics of the prior modes. The ansatz (28) can also be expanded for higher dimensions or traveling waves, but first, this will increase the complexity of the analytical treatment considerably and second, in most cases it will not lead to the same results as in the case of two dominating spatial modes. For these reasons the hypothesis (28) is the first to be tested experimentally.

A complex system whose dynamics is governed by a nonlinear evolution law may perform phase transitions from one stationary state to another when a control parameter is varied. Close to the transition point the dynamics of this system is governed by the first leading orders of the nonlinearities (Haken (1983),1987). Here we express the sigmoid function $S[n]$ by the logistic function

$$S[n] = \frac{1}{1+\exp(-4an)} - \frac{1}{1+\exp(4a)}\,, \tag{29}$$

where a denotes the sensitivity coefficient of response of the corresponding neural population. We expand (29) into a Taylor series around $n = 0$ up to third order in n and obtain

$$S[n] \approx an - \frac{4}{3}a^3 n^3 \tag{30}$$

which is a good approximation for values an smaller than 1. Projecting the neural field equation (26) onto $g_+(x)$ and $g_-(x)$ following the lines in (Jirsa and Haken 1996a,1997), we obtain

$$\ddot{\psi}_+ + 2\omega_0\dot{\psi}_+ + \Omega_+^2\psi_+ = a_e\left(\omega_0^2 + \omega_0\frac{\partial}{\partial t}\right)[af_{10}\psi_+ - \frac{4}{3}a^3(f_{30}\psi_+^3 + 3f_{12}\psi_-^2\psi_+] \tag{31}$$

$$\ddot{\psi}_- + 2\omega_0\dot{\psi}_- + \Omega_-^2\psi_- = a_e\left(\omega_0^2 + \omega_0\frac{\partial}{\partial t}\right)[af_{01}\psi_- - \frac{4}{3}a^3(f_{03}\psi_-^3 + 3f_{21}\psi_+^2\psi_-] \tag{32}$$

where

$$\Omega_i^2 = \omega_0^2 - v^2\int_{-L/2}^{L/2}\triangle g_i(x)\,\mathrm{d}x \qquad \text{with} \qquad i = +,- \tag{33}$$

and the terms f_{ij} with $i, j = 1, 2, 3$ are constant parameters which are given in the appendix.

So far we tackled the level of the brain, let us now traverse the scales of organization by referring back to the behavioral level where the equations governing the behavioral dynamics are known from (5). Here we take these equations as a guide to obtain conditions which restrict the solution space of (31), (32).

In (3) we expressed the behavioral modes in terms of finger displacements. With (24) we can express the behavioral modes in terms of the neural field

$$\tilde{\psi}_+(t) = z_l(t) + z_r(t) = c_0\int_{L/2}^{L/2}(\beta_{lm}(x) + \beta_{rm}(x))\,\psi(x,t)\,\mathrm{d}x \tag{34}$$

$$\tilde{\psi}_-(t) = z_l(t) - z_r(t) = c_0\int_{L/2}^{L/2}(\beta_{lm}(x) - \beta_{rm}(x))\,\psi(x,t)\,\mathrm{d}x \tag{35}$$

If we take our hypothesis (28) of two dominating spatial modes into account, the behavioral modes can be expressed as

$$\tilde{\psi}_+(t) = c_0\, \psi_+(t) \int_{L/2}^{L/2} (\beta_{lm}(x) + \beta_{rm}(x))\, g_+(x)\, \mathrm{d}x = c_+ \psi_+(t) \tag{36}$$

$$\tilde{\psi}_-(t) = c_0\, \psi_-(t) \int_{L/2}^{L/2} (\beta_{lm}(x) - \beta_{rm}(x))\, g_-(x)\, \mathrm{d}x = c_- \psi_-(t) \tag{37}$$

where c_+, c_- are constant. In the present framework, it turns out that in the bimanual coordination case the symmetric (antisymmetric) behavioral mode is proportional to the symmetric (antisymmetric) brain mode. As a consequence the dynamical system (5) of the behavioral modes and the system (31), (32) of the brain modes should be equivalent. This requires $\omega_0^2 \ll \omega_0 \partial/\partial t$ which implies on the considered time scales (see also (Jirsa and Haken 1997) where this limit was used) that the mean corticocortical fiber length is large. We rewrite (31), (32) as

$$\begin{aligned} &\ddot{\psi}_+ - \gamma\dot{\psi}_+ + \Omega_+^2 \psi_+ + b\tfrac{\partial}{\partial t}(f_{30}\psi_+^3 + 3f_{12}\psi_-^2 \psi_+) = 0 \\ &\ddot{\psi}_- - \gamma\dot{\psi}_- + \Omega_-^2 \psi_- + b\tfrac{\partial}{\partial t}(f_{30}\psi_-^3 + 3f_{21}\psi_+^2 \psi_-) = 2\dot{\psi}_-\,(\alpha + \beta\psi_-^2) \end{aligned} \tag{38}$$

where

$$\begin{aligned} b &= 4/3 a_e a^3 \omega_0 & \gamma &= a_e a \omega_0 f_{10} - 2\omega_0 \\ \alpha &= 1/2 a_e a \omega_0 (f_{01} - f_{10}) & \beta &= -2 a_e a^3 \omega_0 (f_{03} - f_{30}) \end{aligned} \tag{39}$$

The system (38) is structurally equivalent to the dynamical system (5) for the behavioral modes. Note that in the latter system the Rayleigh terms with the parameter B were introduced in order to obtain a frequency dependence of the oscillator amplitude (see (Haken et al. 1985)). An alternative way to achieve this dependence is to introduce frequency dependent parameters in (5), e.g. $A(\Omega) = A + 3B\Omega^2$ or $\gamma = \gamma(\Omega)$ which yields the same results as a Rayleigh term. If $f_{12} = f_{21}$ and $\Omega_+ = \Omega_-$, then the lhs of (38) represents a fully symmetric system with respect to the coupling terms. The rhs of the second equation in (38) is a consequence of the difference between the spatial overlap of *functional units – symmetric brain mode* and *functional units – antisymmetric brain mode*. From the behavioral mode system (5) we know the necessary condition $\alpha < 0, \beta > 0$ which leads to the nontrivial restriction $f_{01} < f_{10}$, $f_{03} < f_{30}$ and implies a greater spatial overlap of the considered functional units with the symmetric brain mode.

3.3 Numerical Treatment

In order to illustrate our more general results of the previous section we choose a simple example for a specific set of brain modes and localized functional units. The motor and sensorimotor areas on the right hemisphere are localized as follows

$$\beta_{rm}(x) = \beta_{rs}(x) = \begin{cases} \frac{2\pi}{L} & -\epsilon \leq x \leq \frac{L}{4} \\ 0 & \text{otherwise} \end{cases} \tag{40}$$

and on the left hemisphere

$$\beta_{lm}(x) = \beta_{ls}(x) = \begin{cases} \frac{2\pi}{L} & -\frac{L}{4} \leq x \leq \epsilon \\ 0 & \text{otherwise} \end{cases} \tag{41}$$

and satisfy the required anatomical symmetry (23). For reasons of simplicity we choose the motor and sensorimotor units on the same hemisphere to be identical in the numerical treatment. The sensorimotor feedback is specified for the limit, $f_1 \neq 0, f_2 = 0$, according to (24). We introduce a frequency dependent function $\gamma_0(\Omega)$ into the linear damping $G_0(\Omega) = 2\omega_0 + \gamma_0(\Omega)$ on the lhs of (26) which causes a frequency dependence of the wave amplitude and thus a frequency dependence of the finger movements as experimentally observed. For $\gamma_0(\Omega) = 0$ the original field equation is present. In our specific example (40)–(41) the mean field $\bar{\psi}(x,t)$, equivalent to the homogeneous mode, has to remain constant. To ensure this, we introduce a linear mean field damping $\dot{\bar{\psi}}(x,t)$ into (26). Using a semi-implicit Forward-Time-Central-Space procedure we integrate the field equation (26) numerically. The functional units are localized according to (40), (41) and the edges of the localization functions were smoothed for reasons of numerical stability. Periodic boundaries were chosen. The parameters used in the numerical simulations are: $\omega_0 = 2\pi\, 0.1$, $v = 0.152$, $a_e = 1$, $a = 0.4$, $\rho = 0.5$, $c_0 = 2$, extension of the neural strip $L = 1$, spatial overlap of the localization functions $\epsilon = 0.025$. Here the space unit corresponds to 1 m and the time unit to 100 msec. This parameter range is realistic: the corticocortical propagation velocity v is in the 1 m/sec range, the extension of the neural sheet L is in the 1 m range and the long range connectivity $\sigma = v/\omega_0$ is in the 10 cm range.

In the bottom left corner of Fig. 3 the localization of the functions $\beta_{li}(x)$, $\beta_{ri}(x)$ with $i = m, s$ is shown within the neural strip, above that the symmetric mode $g_+(x)$ and the antisymmetric mode $g_-(x)$ are shown. The color code used is given in the second row on the lhs. On the rhs of Fig. 3 four rows each consisting of a space-time plot and time series are given. In each space-time plot the color-coded field $\psi(x,t)$ is plotted where the spatial domain x is vertical and the temporal domain t horizontal as indicated by the arrows in the top left corner. In the graphs under the space-time plots the field $\psi(x,t)$ is projected onto the symmetric mode $g_+(x)$ (blue line) and the antisymmetric mode $g_-(x)$ (red line) plotted over time t (in sec). The first two rows describe the situation before the pretransition with $\gamma_0(\Omega) = 0$. Two possible states are shown: In the top row the antisymmetric mode $g_-(x)$ dominates; in the second row, it is the symmetric mode $g_+(x)$ that dominates. Increasing the damping to $\gamma_0(\Omega) = 0.3$, the antisymmetric mode becomes unstable and performs a transition to the symmetric state. This transition is shown in the third row. The symmetric mode remains stable in the posttransition regime as can be seen in the bottom row. Hence in the case $\epsilon > 0$, bistability is

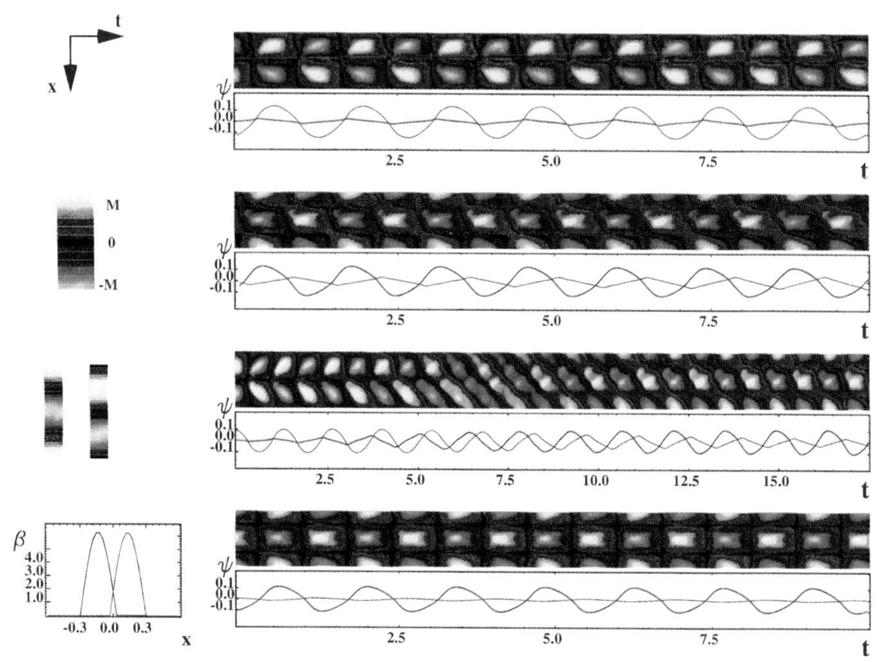

Fig. 3. The overlap of the localization functions $\beta_{ij}(x)$ is shown in the bottom left corner. The activity of the field $\psi(x,t)$ is plotted over the time t (horizontal) and the space x (vertical), together with the time series of the symmetric mode $g_+(x)$ (blue line) and the antisymmetric mode $g_-(x)$ (red line). The top two rows correspond to the pretransition region $\gamma_0(\Omega) = 0$, the bottom two rows to the posttransition region $\gamma_0(\Omega) = 0.3$

present in the pretransition regime and monostability in the posttransition regime. For $\epsilon = 0$ no transition is observed and bistability is preserved for the entire control parameter regime.

3.4 Preliminary Experimental Test of Theoretical Predictions

A brain-behavior-experiment has been performed by Kelso et al. (1994) in which subjects moved their left and right index fingers in time with an auditory metronome presented to both ears in ascending frequency plateaus of ten cycles each. The initial metronome frequency was 2.0 Hz and increased by 0.2 Hz for each plateau with a total number of eight plateaus. The subject was instructed to move the right finger anti-phase and the left finger in-phase with the metronome, and switched spontaneously to a movement pattern with both fingers in-phase as the metronome frequency increased to the fourth plateau (2.6 Hz). During the experiment magnetic field measurements of brain activity were obtained with a whole-head, 64-channel SQuID magnetometer at a sampling frequency of 250 Hz with a 40 Hz low-pass filter.

Each subject performed 5 blocks of 10 runs each. In the following we want to check these experimental data against the theoretical predictions above. Note that the experimental set-up (auditory metronome) and the subject's instructions differ somewhat from the experimental conditions of the present paper. Thus the following experimental results have to be considered a preliminary test of the presented theory.

In order to test our theoretical predictions of the previous sections we perform a Karhunen–Loève Decomposition (KL) (see e.g. Fuchs et al. (1992) and this volume) of the MEG data on each frequency plateau separately. A KL decomposition decomposes a spatiotemporal signal $\psi(x,t)$ into orthogonal spatial modes and corresponding time-dependent amplitudes such that a least-square error E is minimized and the KL modes have maximum variance. The normalized KL eigenvalue $\lambda = 1 - E$ is a measure for the contribution of a KL mode to the entire spatiotemporal signal. Figure 4 shows the first KL mode (top row) plotted over the movement frequency. The orientation of the modes is such that the nose is on the top indicated by the triangle and their color coding (after normalization) is given on the bottom. Before the transition at 2.6 Hz we observe a constant spatial pattern, after the behavioral transition (2.8–3.2 Hz) a different structure (increased activity on the right hand side of the mode contributing to a more antisymmetric shape) is observed in the first KL mode which is similar over the posttransition plateaus.

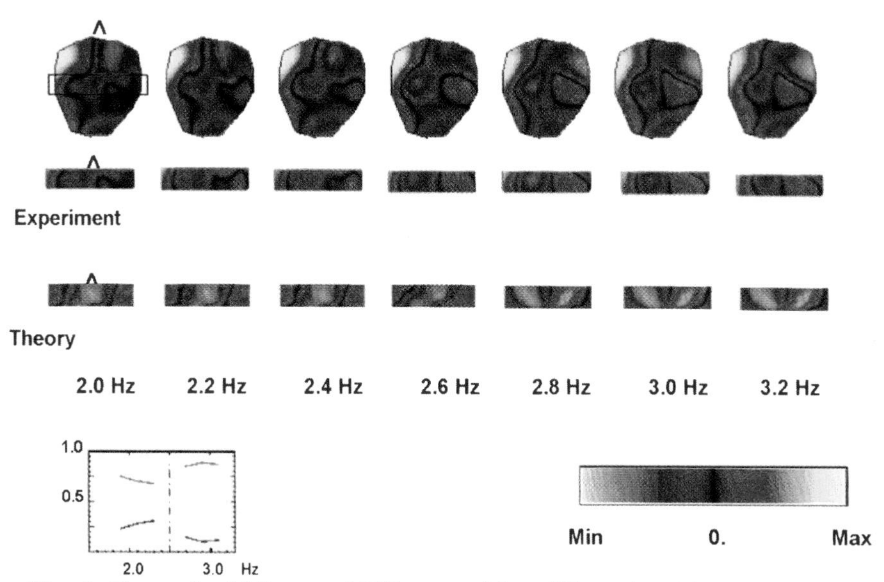

Fig. 4. Top and middle row: MEG map of first KL mode with extracted neural strip containing primary motor and sensorimotor areas plotted over movement frequency. Bottom row: Numerically simulated patterns. Bottom left: Quantification of symmetry of first KL mode (green: antisymmetric, blue: symmetric)

We extract a strip of activity from the spatial MEG patterns which is located mainly over the primary motor and sensorimotor areas (Fig. 4 middle row) and compare it with the activity of the numerically simulated data (bottom row). Note that in the numerical simulations the terms symmetric/antisymmetric apply to current distributions and hence to patterns to be observed in the EEG. Current flowing in apical dendrites of pyramidal cells generates a magnetic field in a plane orthogonal to the currents. Thus, spatial antisymmetric (symmetric) patterns in the MEG correspond to symmetric (antisymmetric) current distributions, and we reversed the symmetry of the numerically simulated patterns in Fig. 4 for better comparison.

The degree of symmetry s of the experimental first KL mode is quantified and plotted over the frequency in the bottom left corner. The transition plateau is indicated by the vertical dotted line. In the pretransition region the first KL mode is about 70% antisymmetric (green line) and 30% symmetric (blue line), whereas after the transition the antisymmetric contribution increases to 90% and the symmetric contribution decreases to 10%.

In the temporal domain (not shown in Fig. 4) almost the entire power of the first KL mode is in the Fourier component of the movement frequency for all plateaus. There is no transition in the relative phase between the amplitude of the first KL mode and the metronome which always remains in-phase with the non-switching finger movement as required by our theoretical predictions.

4 Summary

The main point of the present paper is to show how it is possible to derive the phenomenological nonlinear laws at the behavioral level from models describing brain activity. For the paradigmatic case of bimanual coordination we briefly reviewed the collective and component level of description of the behavioral dynamics. We proceeded by transforming the behavioral model on the component level onto a model describing the dynamics of the behavioral modes. The behavioral level was connected to the brain level by deriving the behavioral model on the mode level from a recently developed field theoretical description of brain activity. For the derivation the crucial points were the assumption of bimodal brain dynamics and the interplay between functional input and output units in the neural sheet. Here the comparison of behavioral and brain level serves as a guide to consistency of both descriptions. We made theoretical predictions about the global brain dynamics and presented a preliminary experimental test which gives strong indications that the predicted spatiotemporal dynamics is present during bimanual coordination tasks.

Acknowledgements

This research was supported by NIMH (Neursciences Research Branch) Grant MH42900, KO5 MH01386 and the Human Frontiers Science Program. Further, we wish to thank Hermann Haken for many interesting discussions. VKJ gratefully acknowledges a fellowship from the *Deutsche Forschungsgemeinschaft.*

A Explicit Forms of Coupling Integrals

We define

$$f_+(x) = \rho g_+(x) + c_0 \left(\beta_{ls}(x) + \beta_{rs}(x)\right) \int_{-L/2}^{L/2} \beta_{lm}(x) g_+(x)\,\mathrm{d}x \tag{42}$$

$$f_-(x) = \rho g_-(x) + c_0 \left(\beta_{ls}(x) - \beta_{rs}(x)\right) \int_{-L/2}^{L/2} \beta_{lm}(x) g_-(x)\,\mathrm{d}x \tag{43}$$

and give the explicit forms of the coupling integrals f_{ij} in (31),(32) as

$$f_{10} = \int_{-L/2}^{L/2} g_+(x)\, f_+(x)\,\mathrm{d}x \qquad f_{01} = \int_{-L/2}^{L/2} g_-(x)\, f_-(x)\,\mathrm{d}x \tag{44}$$

$$f_{30} = \int_{-L/2}^{L/2} g_+(x)\, f_+(x)^3\,\mathrm{d}x \qquad f_{03} = \int_{-L/2}^{L/2} g_-(x)\, f_-(x)^3\,\mathrm{d}x \tag{45}$$

$$f_{12} = \int_{-L/2}^{L/2} g_+(x) f_+(x) f_-(x)^2\,\mathrm{d}x \quad f_{21} = \int_{-L/2}^{L/2} g_-(x) f_+(x)^2 f_-(x)\,\mathrm{d}x \tag{46}$$

References

Abeles M. (1991), *Corticonics*, Cambridge University Press

Braitenberg V., Schüz A. (1991), *Anatomy of the cortex. Statistics and geometry*, Springer, Berlin

Buchanan J.J., Kelso J.A.S., de Guzman G.C. (1997), The selforganization of trajectory formation: I. Experimental evidence, Biol. Cybern. 76, 257-273

Carson R., Byblow W., Goodman D. (1994), The dynamical substructure of bimanual coordination, in: Swinnen S., Heuer H., Massion J., Casaer P., eds., *Interlimb coordination: Neural, Dynamical and Cognitive Constraints*, pp. 319-337, Academic Press, San Diego

Collins J.J., Stewart I.N. (1993), Coupled nonlinear oscillators and the symmetries of animal gaits, J. Nonlinear Sci. 3, 349-392

Cross M.C., Hohenberg P.C. (1993), Pattern formation outside of equilibrium, Rev. Mod. Phys. 65, 851

Felleman D.J., Van Essen D.C. (1991), Distributed hierarchical processing in the primate cerebral cortex, Cerebral Cortex 1, 1-47

Freeman W.J. (1992), Tutorial on neurobiology: From single neurons to brain chaos, Inter. J. Bif. Chaos 2, 451-482

Friedrich R., Fuchs A., Haken H. (1991), Spatiotemporal EEG patterns, in: Haken H., Koepchen H.P., eds., *Rhythms in Physiological Systems*, Springer, Berlin

Fuchs A., Haken H. (1988), Pattern recognition and associative memory as dynamical processes in a synergetic system I+II, Erratum, Biol. Cybern. 60, 17-22, 107-109, 476

Fuchs A., Kelso J.A.S., Haken H. (1992), Phase Transitions in the Human Brain: Spatial Mode Dynamics, Inter. J. Bif. Chaos 2, 917-939

Haken H., Kelso J.A.S., Bunz H. (1985), A Theoretical Model of Phase transitions in Human Hand Movements, Biol. Cybern. 51, 347 - 356

Haken H. (1983), *Synergetics. An Introduction*, 3rd ed., Springer, Berlin

Haken H. (1987), *Advanced Synergetics*, 2nd ed., Springer, Berlin

Haken H. (1991), *Synergetic Computers and Cognition, A Top-Down Approach to Neural Nets*, Springer, Berlin

Haken H. (1996), *Principles of brain functioning*, Springer, Berlin

Jirsa V.K., Friedrich R., Haken H., Kelso J.A.S. (1994), A theoretical model of phase transitions in the human brain, Biol. Cybern. 71, 27-35

Jirsa V.K., Friedrich R., Haken H. (1995), Reconstruction of the spatio-temporal dynamics of a human magnetoencephalogram, Physica D 89, 100-122

Jirsa V.K., Haken H. (1996), Field theory of electromagnetic brain activity, Phys. Rev. Let. 77, 960

Jirsa V.K., Haken H. (1996), Derivation of a field equation of brain activity, J. Biol. Phys. 22, 101-112

Jirsa V.K., Haken H. (1997), A derivation of a macroscopic field theory of the brain from the quasi-microscopic neural dynamics, Physica D 99, 503-526

Kelso J.A.S. (1981), On the oscillatory basis of movement, Bull. Psychon. Soc. 18, 63

Kelso J.A.S. (1984), Phase transitions and critical behavior in human bimanual coordination, Am. J. Physiol. 15, R1000-R1004

Kelso J.A.S., Scholz J.P., Schöner G. (1986), Nonequilibrium phase transitions in coordinated biological motion: critical fluctuations, Phys. Let. A 118, 279-284

Kelso J.A.S., Buchanan J.J., Wallace S.A. (1991), Order parameters for the neural organization of single, multijoint limb movement patterns, Exp. Brain Res. 85, 432-444

Kelso J.A.S., Bressler S.L., Buchanan S., DeGuzman G.C., Ding M., Fuchs A., Holroyd T. (1992), A Phase Transition in Human Brain and Behavior, Phys. Let. A 169, 134 - 144

Kelso J.A.S., Fuchs A., Holroyd T., Cheyne D., Weinberg H. (1994), Bifurcations In human brain and behavior, Society for Neuroscience , 20, 444

Kelso J.A.S. (1995), Dynamic Patterns. The Self-Organization of Brain and Behavior, The MIT Press, Cambridge, Massachusetts

Kristeva R., Cheyne D., Deecke L. (1991), Neuromagnetic fields accompanying unilateral and bilateral voluntary movements: Topography and analysis of cortical sources, Electroenceph. Clin. Neurophys. 81, 284-298

Miller R. (1987), Representation of brief temporal patterns, Hebbian synapses, and the left-hemisphere dominance for phoneme recognition, Psychobiology 15 , 241-247

Nunez P.L. (1974), The brain wave equation: A model for the EEG, Mathematical Biosciences 21, 279-297

Nunez P.L. (1995), Neocortical dynamics and human EEG rhythms, Oxford University Press

Scholz J.P., Kelso J.A.S., Schöner G. (1987), Nonequilibrium phase transitions in coordinated biological motion: critical slowing down and switching time, Phys. Let. A 123, 390-394

Schöner G., Jiang W.Y., Kelso J.A.S. (1990), A Synergetic Theory of Quadrupedal Gaits and Gait Transitions, J. theor. Biol. 142, 359-391

Tuller B., Kelso J.A.S. (1989), Environmentally-specified patterns of movement coordination in normal and split-brain subjects, Exp. Brain Res. 75, 306-316

Wallenstein G.V., Kelso J.A.S., Bressler S.L. (1995), Phase transitions in spatiotemporal patterns of brain activity and behavior, Physica D 84, 626-634

Wiesendanger M., Wicki U., Rouiller E. (1994), Are there unifying structures in the brain responsible for interlimb coordination?, in: Swinnen S., Heuer H., Massion J., Casaer P., eds., *Interlimb coordination: Neural, Dynamical and Cognitive Constraints*, pp. 179-207, Academic Press, San Diego

Williamson S.J., Kaufman L. (1987), Analysis of neuromagnetic signals, in: Gevins A.S., Remond A., eds., *Methods of analysis of brain electrical and magnetic signals. EEG Handbook*, Elsevier Science

Wilson H.R., Cowan J.D. (1972), Excitatory and inhibitory interactions in localized populations of model neurons, Biophysical Journal 12, pp. 1-24

Wilson H.R., Cowan J.D. (1973), A mathematical theory of the functional dynamics of cortical and thalamic nervous tissue, Kybernetik 13, 55-80

EEG-Detected Episodes of Low-Dimensional Self-Organized Cortical Activity and the Concept of a Brain Attractor

R. Cerf, E.H. El Ouasdad, and M. El Amri

Laboratoire de Dynamique des Fluides Complexes, U.M.R.n°7506,
Centre National de la Recherche Scientifique, Univesité Louis Pasteur,
4 rue Blaise Pascal, 67070 Strasbourg, France

1 In Quest for Low-Dimensional Self-Organized Cortical Dynamics

Our studies of EEG[1]-recorded spontaneous rhythms have shown that no low-dimensional cortical dynamics are observed as a general phenomenon, but that short-lived low-dimensional episodes are found, of varying life-times and probabilities, depending on the rhythm considered. On the other hand, as is known from the works of H. Haken [1], a self-organized complex system may be governed by a few unstable collective modes. Our low-dimensional episodes, then, are windows in which the self-organized cortical dynamics being characterized can be experimentally investigated in conditions where these dynamics reduce to a few collective modes.

A low dimension found in single-channel experiments need not bear unambiguous meaning. We should keep in mind that, in his model studies of coupled Lorenz attractors, E. Lorenz [2] has found both low and high dimensions, depending on how strongly the variable he analysed was coupled to other subsets of variables. Yet, a low dimension only showed up in this case in a system comprising low-dimensional sub-systems.

We shall assume here that for an EEG low-dimension to show up, collective dynamics of some cortical network must be low-dimensional. With this assumption, the very outcome of a low dimension is, in itself, meaningful.

Of course, low dimensionality renders deterministic chaos likely, because few non-chaotic modes would not produce as highly irregular EEG signals as is commonly observed. Chaos, in turn, is of major interest in studies of the central nervous system, since it is a long-lasting instability, which offers the system opportunities for switching between states.

The first publication by A. Babloyantz and her coworkers [3–7] announcing low-dimensional chaotic dynamics in the brain appeared twelve years ago. Their approach was based on the then newly applied magnificent mathematics allowing an attractor to be reconstructed from experimental data.

[1] EEG: electroecephalography

Single-channel analysis, based on time-delay reconstruction, first advocated by D. Ruelle (cited in [8]), and on the Grassberger-Procaccia algorithm [9] were extensively used. Low-dimensional attractor-ruled dynamics, thus high self-organization, were claimed for most EEG rhythms. For each of those rhythms, the effect was found for apparently all signal sections selected according to rules commonly used by electroencephalographers. The data included further a one minute deep-sleep δ-wave and a four minute α-wave [3,6]. Work on non-linear analysis of EEG signals had been started simultaneously, in particular by the group of P. Rapp [10].

The Grassberger–Procaccia algorithm produces an artefactual low dimension if short-time correlations are not eliminated. The algorithm, as is known, counts pairs of points on the phase trajectory, at a distance less than r from each other. It is indispensable not to count pairs that are close in space because they are close in time. We can avoid counting them, according to Theiler [11], by suppressing those pairs that are closer in time than w sampling units. We may call the artefact which occurs when the correction is not made the "single-strand artefact". It can, indeed, be viewed as being due, for each reference point on the trajectory (point P in Fig. 1), to the overwhelming weight at low r of one strand of the phase trajectory. But that is not all: after eliminating the single-strand artefact, we may encounter the "no-strand artefact", which will be described here for the first time.

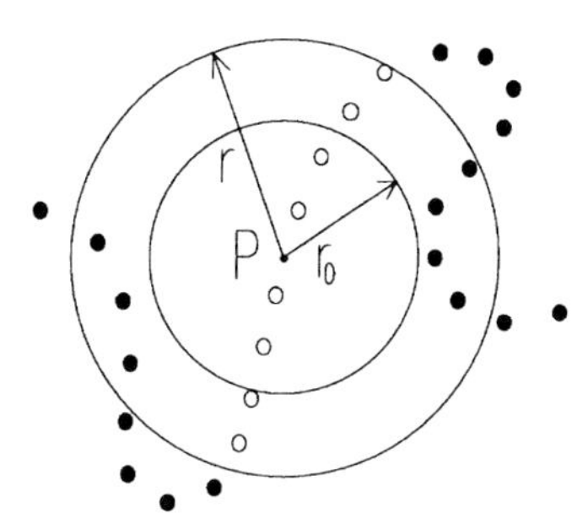

Fig. 1. Counting pairs on the phase trajectory. No point deleted: one strand has overwhelming weight, leading to single-strand artefact. Open points are deleted: single-strand artefact is corrected for, but when $r \leq r_0$ no-strand artefact comes into play (see Fig. 7).

The single-strand artefact was not taken into account at Brussels. The Theiler correction had, indeed, not been applied in the first place, and had then erroneously been stated to have negligible effect on the results (key to Table 1 in [6]). As a result, the dimension data did not characterize brain dynamics, and were a mere manifestation of the short-time correlations. This conclusion extends to Creutzfeldt-Jakob signals, as could be checked by using recordings kindly donated by Dr. Kees Stam. For the "petit-mal" absence seizure, however, it is merely the range of values of the correlation integrals in which scaling was observed that invalidates the Brussels data, as will be shown in Sect. 5 below. Our conclusions for petit-mal raise interesting questions regarding its theoretical analysis in terms of a small number of spatio-temporal collective modes.

The low-dimensional dynamics found by our group for some EEG rhythms are not of the form described at Brussels, and their observation requires use of the appropriate scaling properties of the correlation integrals (recalled in Sect. 2 below). We use single-channel time-reconstruction. So far, taking due care of the artefacts, we have not found any low dimension in a rhytm by using the multichannel space-reconstruction. The dimensions considered are lower than 7, the approximate upper limit at which our method of scaled structures works.

2 Scaled Structures Versus Scaled Correlation Integrals

First of all, no scaling of single correlation integrals exists with the EEG, other than as exceptions of no significance for Grassberger–Procaccia analysis [12–19]. When exceptionally observed after Theiler-correcting, such scaling is not robust to variation of the parameters: embedding dimension – which must be increased far beyond the strict mathematical requirements – time delay, origin and duration of the analysed time-series.

On the contrary, for certain EEG rhythms, and some of the signal sections short enough for the dynamics to be stationary, scaled structures may be observed that are robust to parameter variation. This is illustrated here by one of our recent examples, the syndrome of continuous spike-and-wave discharges during slow-sleep (CSWS).

This syndrome occurs exclusively in children between 3 and 10 years of age. It involves deterioration of cognitive functions that were previously normally acquired. The neuropsychological deficit may regress before the end of adolescence.

Figure 2 shows a 15-s signal section from an FP_2 derivation, of 21.6 pseudo-cycles. Figure 3a shows the result of single-channel time-reconstruction with a time delay of 32 ms. Figure 3a is a plot of the logarithmic derivative of the correlation integral $C(r)$ against the logarithm of $C(r)$, i.e. a Rapp-plot; 39 slope-curves of embedding dimensions 2 to 40, are represented. Further results on CSWS will be published subsequently (El Ouasdad, Cerf, Hirsch).

For classical scaling of single correlation integrals, a double logarithmic plot of C against r exhibits a scaling range of a slope equal to the correlation exponent ν. The slope curve then exhibits a horizontal segment of an ordinate equal to ν. The saturation of ν as a function of the embedding dimension, which is the signature of a low-dimensional attractor, results in several scaling

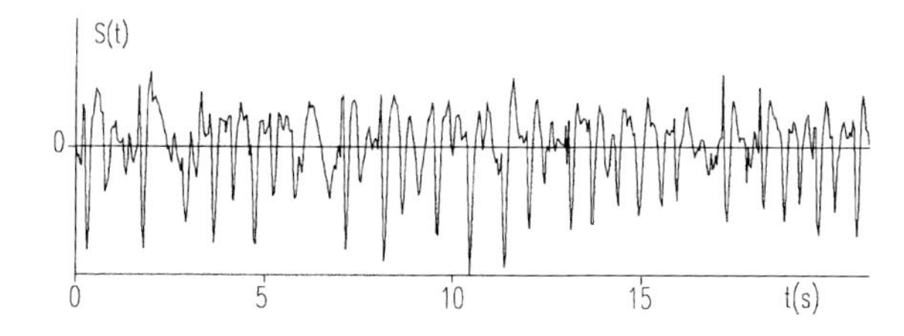

Fig. 2. Continuous spike-and-wave discharges during slow-sleep (CSWS); FP_2 derivation; recording frequency: 128 Hz.

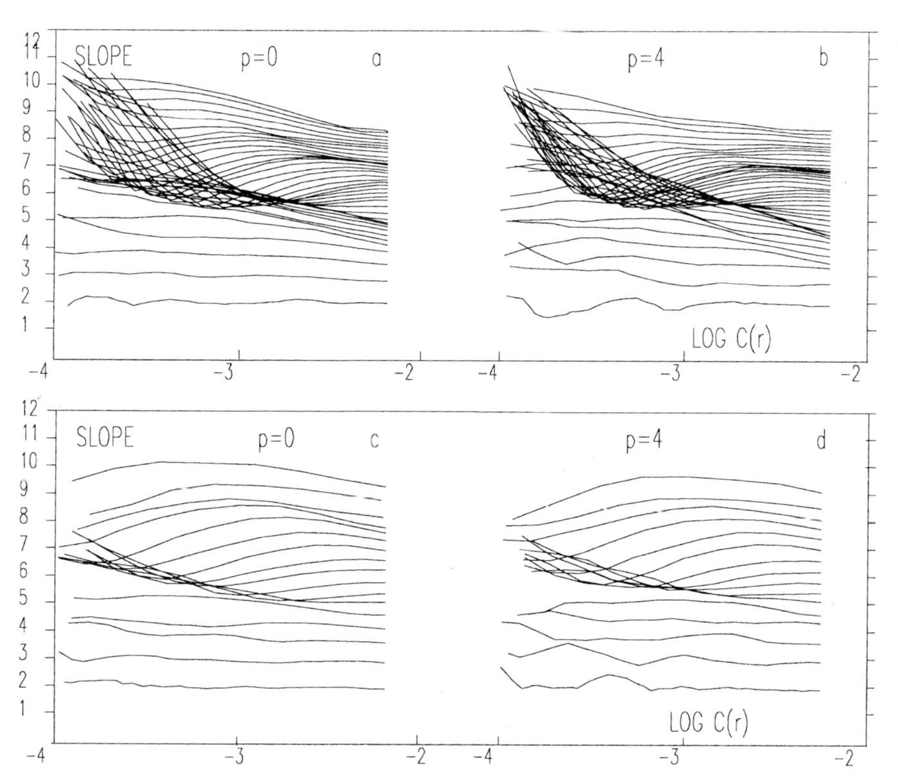

Fig. 3. CSWS scaled-structures for FP_2 15-s signal, $\tau = 32$ ms, $2 \leq n \leq 40$: a; time-reparametrized: b; $\tau = 88$ ms, $2 \leq n \leq 18$: c; $2 \leq n \leq 17$, time-reparametrized: d.

ranges being parallel to each other in the plot of $\log C$ against $\log r$. In the slope diagram, several horizontal segments are then concentrated in a narrow horizontal strip. No such behaviour is observed here.

However, there does appear a structure in the family of correlation-integral derivatives; its lower envelope is horizontal, and therefore it follows a scaling law. We call this a scaled structure. When does it occur?

Suppose the parallel scaling ranges in the plot of $\log C$ against $\log r$ shrink down to parallel inflection tangents. This leads to slope curves that show a common horizontal tangent, and we then observe a scaled structure.

The passage from the first regime to the second could be observed for a signal obeying the Mackey and Glass delay-differential equation. A long signal of a correlation dimension close to 5 was divided into short pieces. For the long signal the correlation dimension was obtained from classical scaling, and for the short signals from scaled structures. The measured correlation dimension remained constant when the duration was reduced from long signal values to those of the short signals, provided these were not shorter than 14 pseudo-cycles [17].

We call this property "tangentiality" to classical scaling. Often a few slope curves participating in a structure exhibit a short scaling range, thus confirming tangentiality.

In the case of the Mackey and Glass equation, as in the case of δ-sleep, two structures may occur in a slope diagram [14,17]. An example of a doublet is shown in Fig. 4 for a C_zO_1 15-s section of δ-sleep and a time delay of 32 ms. For the Mackey and Glass signal, the doublets have been obtained under the specifications of [17] using a TVP4 compiler. Depending on the compiler used, the signal may alter its tracing sufficiently to exhibit singlets instead of doublets. This is in line with the observation that a Mackey and Glass signal of even as long a duration as 40 pseudo-cycles, exhibiting almost classical scaling, may show variations in its correlation dimension up to 20%, depending on the number of iterations performed prior to start of the analysis.

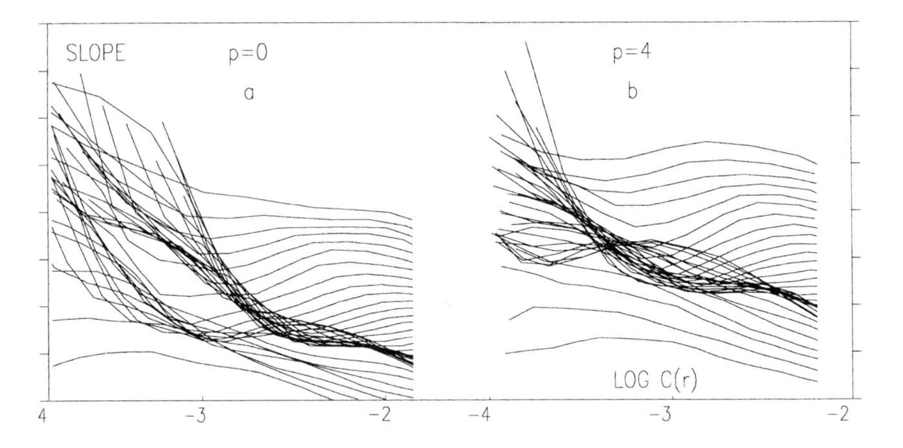

Fig. 4. Deep-sleep doublet for C_zO_1 15-s signal; recording frequency: 128 Hz; $\tau = 32$ ms, $5 \leq n \leq 40$: a; $6 \leq n \leq 40$, time-reparametrized: b.

A remarkable feature of scaled structures is their robustness to time delay τ, excepting occasional atypical values of τ. When τ is modified, the curves constituting a structure are replaced by new ones, of different embedding dimensions, at approximately constant embedding-window, i.e. at $(n-1)\tau \simeq$ const. The structure itself remains, however, conserving its shape and value of ν, over a range of values of τ, the τ-window. Figures 3a and 3c show the structure obtained for the preceding CSWS-signal for $\tau = 32$ ms and 88 ms. The lowest value of τ for which a structure is obtained is $\tau = 16$ ms; thus the τ-window reaches the value 5.5:1.

3 Episodes of Low-Dimensional Self-Organized Brain Dynamics

3.1 α and δ Episodes

Low-dimensional episodes are found e.g. in α and δ waves, of typically 5 s and 12 s, respectively [14, 15, 18–21]. Episodes may provide evidence of higher or lower quality for low-dimensional self-organized dynamics.

To assess the existence of an episode it is not sufficient to observe scaled structures in a τ-window for a given signal section. It is necessary to check the robustness of the data to the origin and duration of the analysed time-series. In other words, it is necessary to investigate an ensemble of signal sections of neighbouring origins and varying durations. We call such an ensemble of sections a "segment". When a segment provides evidence for self-organized dynamics, we have identified an episode.

The probability of δ-episodes is low, of a few percent, but could be shown to be modified by sleep deprivation [18]. Among α-signals of good quality, the probability of α-episodes may be of the order of 20% after signal processing.

Non-stationarity limits the number of pseudo-cycles available for non-linear studies less drastically in the case of α-waves than for the other cerebral rhythms, because α-waves have a relatively high frequency of about 10 Hz. A 5-s section then comprises about 50 pseudo-cycles, which provides an appropriate number of recurrences.

The two following examples [19] of α-episodes are exceptional, because they are observed with the unprocessed[2] signals, and are therefore free from the perturbations that pre-treatments may produce. Furthermore, these two episodes, which come from two different subjects, are of an exceptionally long duration of 10 s. Let us call the two subjects, and corresponding episodes, 1 and 2, respectively. All sections starting every second of a minimal duration of 4 s for episode 1 (P_1O_1 derivation), and of 5 s for episode 2 (P_3O_1 derivation), have been analysed, and graded for the evidence they provide for low-dimensional dynamics.

First, marks are given to each scaled structure, taking into account the shape of the structure and the number of embedding dimensions contributing to it. Then a mark equal to the sum of the structure marks is attributed to each section. This procedure accounts for the width of the τ-window.

Figure 5a shows the correlation exponent ν of the 5-s sections, and its variation during each episode. Each section is identified by its starting time,

[2] That is, without any processing after data recording at the clinic.

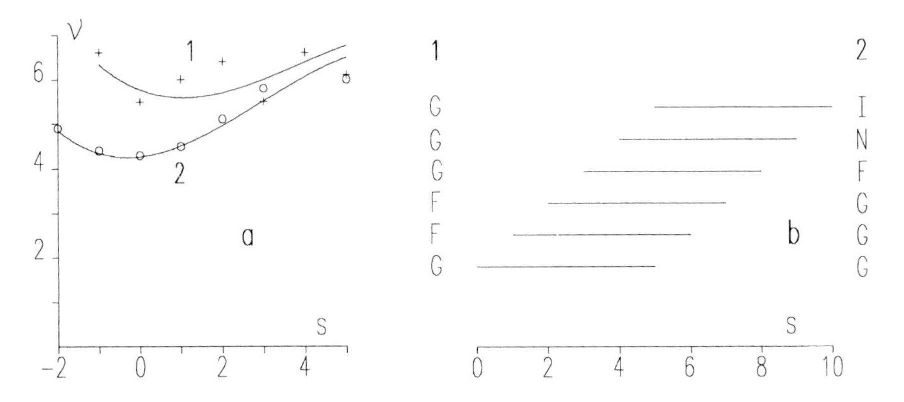

Fig. 5. 10-s α-episodes 1 and 2. Data for 5-s sections starting every second, in function of starting time; correlation exponent ν: a; quality (G, F, I, N = good, fair, indicative, nothing): b.

chosen to be zero for the section showing the lowest correlation exponent. The marks are shown in Fig. 5b for six 5-s sections of each episode. The marks are: good (G), fair (F), indicative (I), and nothing (N).

As a characteristic time for α-episodes, we may choose the signal duration required for the scaled-structure analysis of a section, i.e. 4 to 5s. On the other hand, over the time of 3 s separating the third and sixth 5-s sections of episode 2 (Fig. 5a), the correlation exponent ν varies by almost 2 units. The characteristic time of 5 s is by no means short compared to the time of 3 s over which the variation of ν takes place. Episode 2, therefore, cannot be described by an adiabatically varying parameter. Nor is it of the type, considered by Haken [22], when a system stays on an attractor for a fairly long time, and then jumps to another part of it, or to another attractor. For both an adiabatic parameter variation, and walks among attractors, we have at our disposal the quite appropriate term "quasi-attractor" introduced by Haken [22].

3.2 "Shadow-Attractor"

Our analyses of cortical dynamics suggest accepting attractors that can be of higher or lower "quality" (see Sect. 3.1). Often, the lower an episode's dimension, the better the quality. Fig. 5b illustrates the variation in quality for sections within episodes.

The variation in time of the section's correlation exponent shown in Fig. 5a is typical of EEG low-dimensional episodes: the exponent decreases first, after it becomes measurable, and it subsequently increases, until the effect vanishes. The data suggest that episodes start and finish smoothly.

In the order parameter picture, the number of sub-systems enslaved to each order parameter varies with time, and it does so smoothly. The term "enslaved" is used in the sense of Haken [1]. The variation in dimension shown in Fig. 5a reflects increased slaving, followed by escape from slaving. Non-stationarity may possibly include here the effect of incoming signals from other parts of the brain.

Our device for characterizing an attractor from scaled structures, which consists in using a series of controls and marking procedures for sections and episodes (Sects. 2, 3.1, 4.4, and [14,15,18–21]), just expresses the fact that attractor-ruled dynamics may be approximately valid under conditions where the mathematical definition does not hold. We say that a "shadow-attractor" rules these dynamics [19].

3.3 An Application: The Effect of Psychoactive Drugs on α-Waves

In general, no low-dimensional dynamics can be observed when α-signals are not processed for the elimination of other spectral components. Of course, care must be taken to avoid processing artefactually producing a low dimension.

In collaboration with the hospital at Rouffach, a study of pharmacological interest was carried out on the effect of psychotropics on α-waves [20]. In order to reduce the drawbacks of processing, mainly signals for which the α-band is well separated from other spectral components were studied. Two different approaches to processing were also used.

Dopamine is assumed to be involved in the pathogenesis of psychosis, in particular schizophrenia. It has, indeed, been observed that agents which cause excessive release of dopamine can provoke schizophrenia-like states.

To assess the dopaminergic activity of a putative antipsychotic, the pharmacological profiles both of apomorphine, an agonist of the dopaminergic receptors, and of the candidate substance, a sigma receptors ligand, were compared. The measured correlation dimension diminished after administration of apomorphine [20]. For the putative antipsychotic, the inverse effect anticipated was observed [21].

4 Further Artefacts and Further Tools

4.1 The Minima-Chart

Figure 6a shows the plot of the correlation exponent $\nu_{\min}$ at the minimum of each slope curve, in function of the embedding dimension n, for a 15-s δ-sleep signal of 21.6 pseudo-cycles exhibiting doublets. Each symbol corresponds to a value of τ in the range of 24 ms to 72 ms. This minima-chart illustrates the way in which robustness to time delay τ makes it possible to measure a correlation dimension.

Here we required, as we currently do (Cerf and Ben Maati [13]), that the normalized correlation integral $C(r)$ obey the inequality:

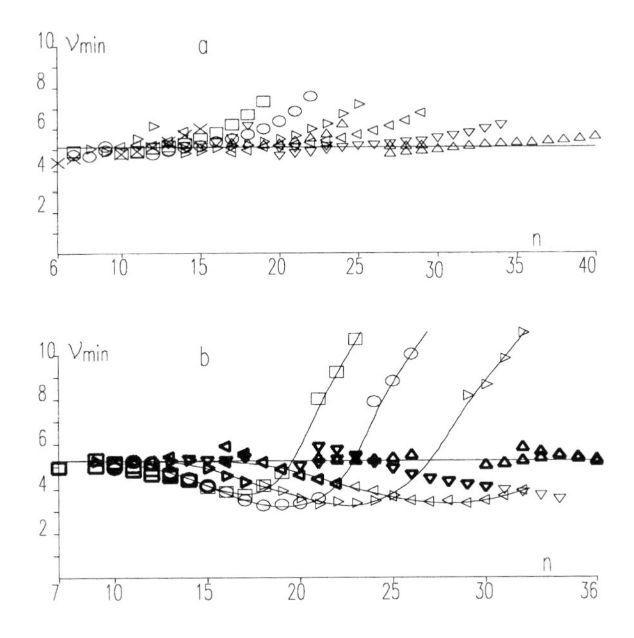

Fig. 6. Minima-charts for slow-sleep. Duration 15 s, $NC(r) \geq 1$: a. Duration 10 s: b. In b, heavy symbols: $NC(r) \geq 1$; light symbols: $NC(r) \geq 0.3$.

$$NC(r) \geq 1 \ . \tag{1}$$

Inequality (1) follows from the rule that noise comes into play for scaled structures when the average number of trajectory points per hypersphere, not counting the centre, i.e. $NC(r)$, is less than 1 [13] (N is the number of experimental data; $N^2C(r)$ is the number of trajectory pairs). Inequality (1) also gives a limiting value of $C(r)$, of fairly general validity, below which the effect of noise shows itself for the shortest time-series for which classical scaling still holds.

The argument which Eckmann and Ruelle (see [23]) proposed, in the case of classical scaling, for evaluating the number of experimental data required in order to measure a correlation dimension d, can be repeated taking (1) into account. The rule proposed earlier by Procaccia [24]:

$$N \geq 10^d \tag{2}$$

then follows [13], instead of Eckmann and Ruelle's condition $N \geq 10^{d/2}$. The latter condition assumes that $NC(r)$ can meaningfully reach its lowest values of order N^{-1}, which is ruled out by (1). However, because they concern classical scaling, neither Eckmann and Ruelle's condition, nor condition (2), are in fact relevant to EEG analysis. On the other hand, with scaled-structure analysis, conditions must be set on the signal's number of pseudo-cycles and on the number of recurrences (at $r = r_{\min}$, see Sect. 4.3 below); examples can be given where the value of N is not crucial (see Fig. 4 of [19]).

Fig. 6b shows the minima-chart for a 10-s section comprising 14.4 pseudo-cycles, cleaved from the 15-s δ-sleep signal in Fig. 6a. The heavy symbols represent results obtained under condition (1). The data are somewhat dispersed, but about the same dimension close to 5 seems to obtain, as for the 15-s section. The light symbols were obtained under the relaxed condition $NC(r) \geq 0.3$, and show downward deviations of the measured correlation exponents.

4.2 The "No-Strand Artefact"

These downward deviations are due to the "no-strand artefact". The first three slope-diagrams in Fig. 7 are for $\tau = 48$ ms and show the result of increasing the Theiler correction, from $w = 2$ to $w = 20$ (recording frequency

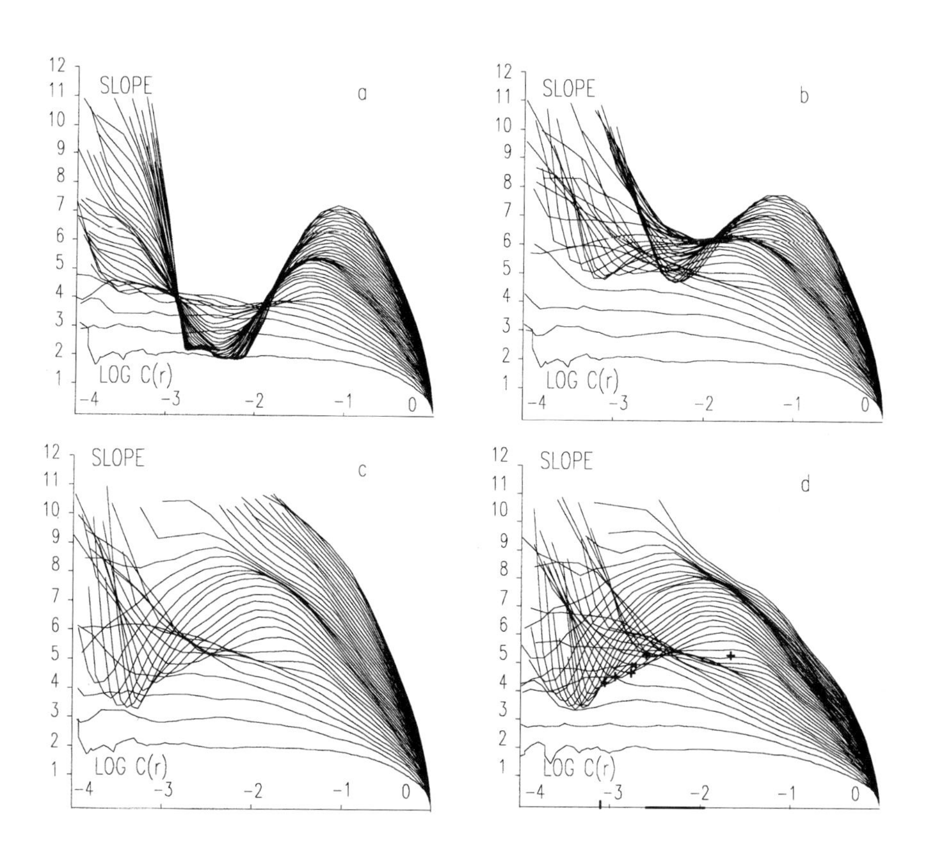

Fig. 7. Deep-sleep 10-s signal; recording frequency: 128 Hz; $2 \leq n \leq 40$. a-c: $\tau = 48$ ms; for increasing Theiler correction w the no-strand artefact gradually replaces the single-strand artefact; a-c: $w = 2, 5, 20$; $\tau = 32$ ms, $w = 20$: d. Vertical mark on abscissa in d: $NC(r) = 1$.

= 128 Hz). The artefactual low dimension due to the single-strand artefact, is gradually replaced by another artefact, in the form of pits in the slope curves.

For short time-series, hyperspheres may be empty of points along entire stretches of the phase trajectory, after Theiler correcting (see Fig. 1). The contribution of these stretches to the slope is zero, and the slope itself is low. However, some spheres still contain points. When r is lowered further, and the total number of remaining points becomes quite low, noise is important and the slope increases again.

4.3 Recurrence Counts

It is necessary to explore the phase trajectory for empty stretches. The quantity of interest is the number ξ_i of trajectory points in the hypersphere centred on point P_i, when radius r equals the value $r_{\min}$ at which the correlation-integral derivative reaches its minimum. We call "recurrence count" the linear array of dots showing the value of ξ_i all along the trajectory. A recurrence count differs from the recurrence plot of Eckmann, Kamphorst and Ruelle [25], which is mostly fitted to display time scales.

Figure 8a shows the recurrence count for a 5-s section of an unprocessed P_3O_1 α-signal, $\tau = 56$ ms, embedding dimension $n = 19$. The scale on the left fits the highest value reached by the number ξ_i of points per hypersphere. The more exacting case of a 10-s section of δ-sleep (see Sect. 4.1) is shown in Fig. 8b for $\tau = 32$ ms, $n = 16$, and in Fig. 8c for $n = 19$. Enormous inhomogeneity of the phase trajectory has come about as a result of a small change in the value of the embedding dimension. Recurrence counts make it possible to find out the range over which a parameter can be varied.

4.4 Time Reparametrization

The phase trajectories of the autonomous first-order differential equations:

$$\frac{\mathrm{d}\mathbf{X}}{\mathrm{d}t} = \mathbf{F}(\mathbf{X}) \tag{3}$$

modelling the system considered, obey invariance under time reparametrization [26]. Ould Hénoune and Cerf [27] use the following set of simple reparametrizations:

$$t_p = \int_0^t ||\mathbf{F}(\mathbf{X}(t'))||^p \mathrm{d}t' \tag{4}$$

with p integer, as a test for attractor-ruled dynamics.

The reparametrization test nicely detects an artefact observed when an attention process transiently desynchronizes an α-wave, as occurs when interfering with a state of relaxed wakefulness in a subject having his eyes closed [27]. Desynchronization lasted 3/4 of a second in one of our examples.

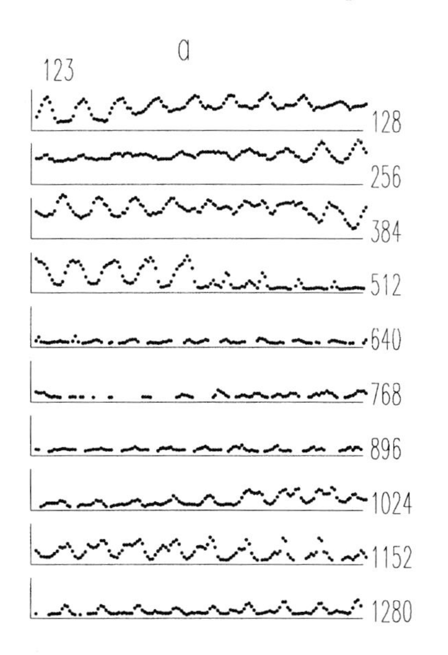

Fig. 8. Recurrence counts. 5-s α-wave, each line: 0.5 s; $\tau = 56$ ms, $n = 19$: a. 10-s deep-sleep δ-wave, each line: 1 s; $\tau = 32$ ms, $n = 16$: b; $n = 19$ c.

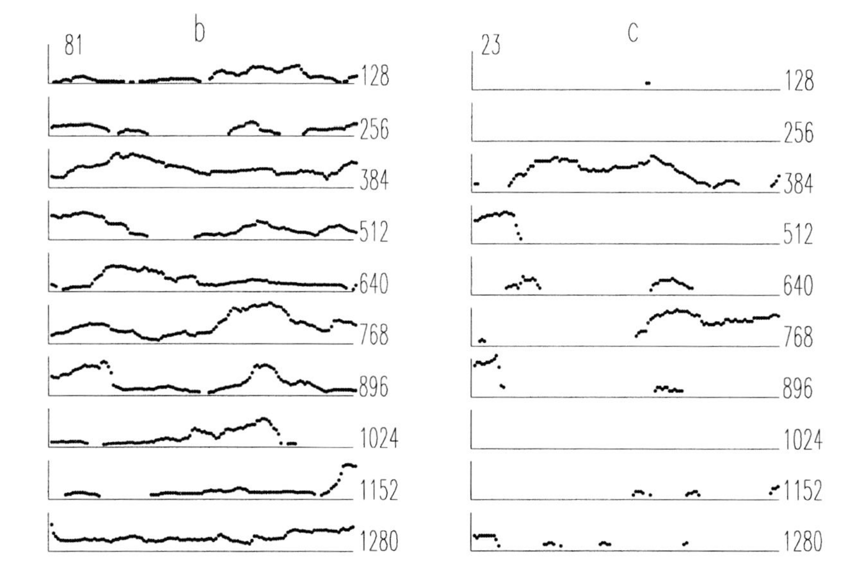

The test helps us to assess the quality of an attractor. α-attractors generally pass the test successfully, but attractors found in δ-sleep not quite so well, as shown in Fig. 4b for the 15-s section of δ-sleep considered in Sects. 2 and 4.1. As shown in Figs. 3b and 3d, the syndrome of continuous spike-and-wave discharges during slow-sleep passes the test very well. This is one of my reasons for believing that CSWS may well turn out to provide an interesting model for non-linear studies.

4.5 Randomization

The checking procedures described in previous sections, based on the determination of τ-windows for the scaled structures observed in a signal section, on identification of signal segments comprising sections of differing origins and durations, and on invariance of scaled structures to time reparametrization, each provide a necessary condition for the observed effects to characterize a low-dimensional episode.

We employed surrogate data [28] in order to check the extent to which the combined use of the preceding checking procedures can be fooled by signals that resemble those analysed, but in principle are not deterministic. Three 5-s α-signals, each taken from a low-dimensional 10-s episode, were phase-randomized, then wave-separated [16].

About one third of the resulting processed signals differed in the α-band from the original EEG signal by their Fourier spectrum, and were discarded.

Most randomized sections showed no scaled structure at all, but 25 to 42% of them, depending on the EEG section considered, still showed such structures. However, all randomized sections showing scaled structures were graded F (see Sect. 3.1), none of them G. When the randomized signals were time-reparametrized after randomization, the number of F sections dropped to 7.5 to 12.5%. The remaining F sections may possibly be spurious effects, that can either stem from the randomization procedure, or be rooted in general grounds, in particular in the fact that some non-deterministic short signals can mimic determinism to a certain approximation, or both.

The above data are, however, sufficient to demonstrate that our procedure has not been fooled in the identification of the 10-s α-episodes considered in Sect. 3.1. The preceding percentages, indeed, are values of the probability π that our analysis can spuriously identify a 5-s α-section as being F. The probability π^2 of spuriously identifying two consecutive 5-s sections both as F, is therefore in the range of 0.6 to 1.6%. In principle, no 5-s G section, or segment of any duration comprising such sections, might be identified spuriously. To reach this pay off, use had to be made of all our necessary conditions, recapitulated in Sects. 2, 3.1 and 4.4. The advantages that may be found in using minima-charts and recurrence counts have not yet been fully investigated.

Surrogate-data analysis always comes in complement to a more fundamental approach. Progress in the latter may make it possible to apply the bulk of the computations to the biological signal, rather than to surrogates.

5 Epilepsy

Potential application of non-linear dynamics to epilepsy has aroused a number of studies. Here we consider the "petit-mal" seizure. As an example of an accomplishment of our methodology, we demonstrate that the data of Babloyantz and Destexhe [4] on petit-mal are artefactual.

First of all, we were able to reproduce the data of their Fig. 4, however for only 3 out of the 39 signals we investigated for 2 patients and 6 seizures in all. The channels were selected according to the stationarity of their signals; the durations were in the range of about 5 to 8 s. Two of our signals are shown in Fig. 9.

The effect described in Fig. 4 of [4] consists in what may seem to be short Grassberger–Procaccia scaling ranges. It appears exclusively at fairly low values of the time delay τ, typically $\simeq$ 16 ms, and high values of $C(r)$, typically $\simeq 10^{-1}$. As a consequence of the latter value, the single-strand artefact is negligible, and overlooking it as in [4] is of no harm in this case.

On the other hand, however, the preceding values of τ and $C(r)$ are a contraindication – not a counter-proof – for interpreting the data in terms of an attractor dimension. Furthermore, using time reparametrization, we found an increase in the dimension from $d \simeq 2$ to $d \simeq 3$, which also contraindicates characterization of an attractor. At a dimension of 2 an attractor is, indeed, expected to be "of good quality", and d well invariant to reparametrization.

Our proof now follows from considering a signal for which we observed both the effect described in [4] and evidence from scaled structures for attractor behaviour at $d \simeq 6$ (see Fig. 10). The scaled structures were observed in the range of time delays $\tau = 12$ to 64 ms, at values of $C(r)$ of a few 10^{-3}, and were invariant to time reparametrization. The simultaneous observation of the two types of structures, the ones providing highly doubtful evidence for attractor behaviour, the others fulfilling the criteria recapitulated in previous sections, discards the former as measuring a correlation dimension.

To summarize data which will be published shortly (Cerf, El Amri, Hirsch): among the 39 single-channel signals analysed, 29 gave no evidence for a measurable dimension, 9 showed a dimension between about 5 and 7, and one signal showed a dimension as low as 4.7. Not unexpectedly, no measurable dimension was found with multichannel space-reconstructions. The dimension of 2 observed for $w = 1$ disappeared on Theiler-correcting,[3] and the remaining indication of a higher dimension disappeared with time reparametrization.

Our data need not contradict the results of the work by R. Friedrich and C. Uhl [29]. The theory, indeed, allows shorter – therefore more stationary – signals to be investigated than any experimental correlation-dimension analysis does. On the other hand, the signals used in the theoretical analysis had attenuated spike-and-wave characteristics in comparison to those shown by our signals, that were recorded at 256 Hz (see Fig. 9). This circumstance may, in fact, have been helpful in developing the first approach to the spatio-temporal reconstruction of petit-mal, but may, at the same time, result in affecting the significance of the computed dimension.

---

[3] The single-strand artefact must of course be eliminated in multichannel as in single-channel reconstruction.

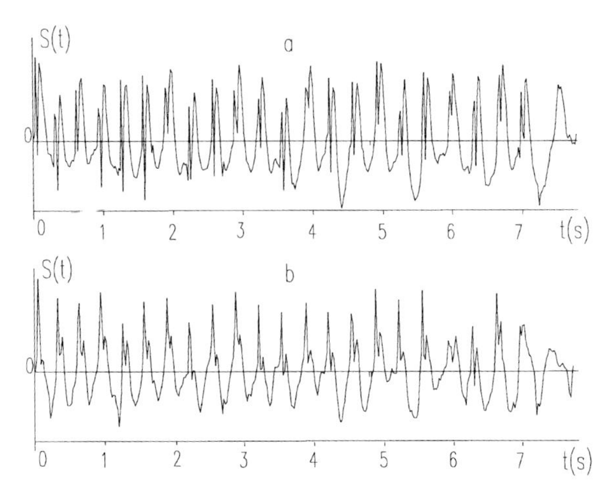

Fig. 9. Petit-mal seizure; recording frequency: 256 Hz; FP_1 derivation: a; C_4 derivation: b.

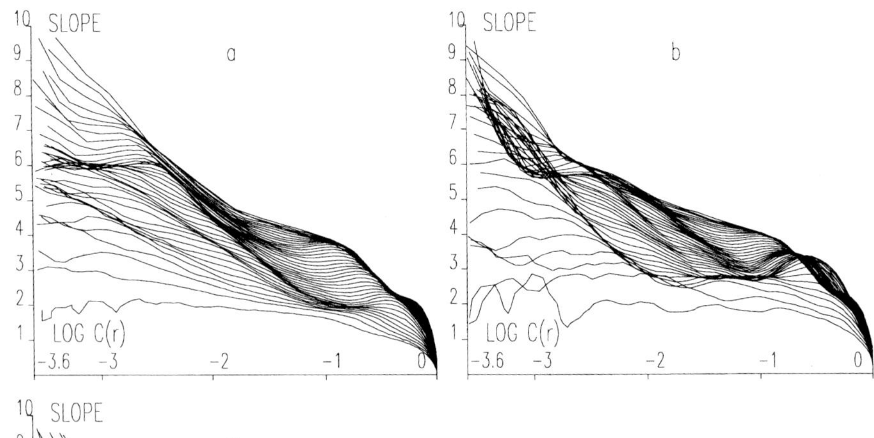

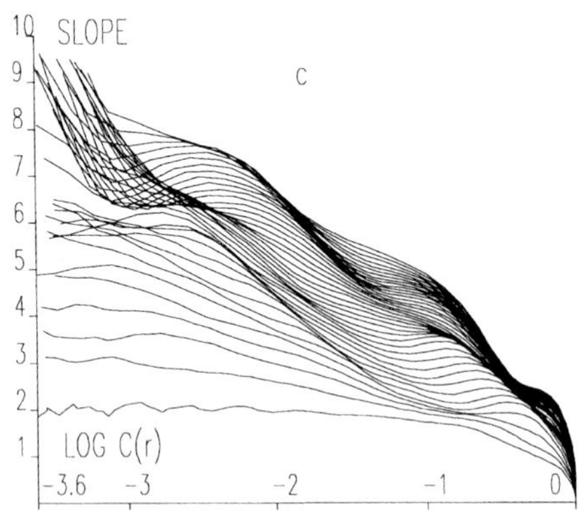

Fig. 10. Petit-mal seizure; $2 \leq n \leq$ C_4 derivation, $\tau = 12$ ms: a; tin reparametrized: b; $\tau = 24$ ms: c. In ea figure, bottom right: artefactual dim sion $d \simeq 2$; top left: dimension $d \simeq$ from scaled structures, robust to ti delay and time reparametrization.

6 The Largest Lyapunov Exponent

In determining the largest Lyapunov exponent λ_1, the exponential separation between neighbouring phase trajectories with increased evolving time must first of all be observed; the measurement must then be robust in ranges of values of time delay τ and embedding dimension n. If, for example, the separation is faster than exponential, it cannot be considered that the largest Lyapunov exponent has been measured, nor of course that the measurement establishes chaotic dynamics. Whenever in doubt about the significance of a measurement, the data must be checked against randomization.

Having identified by their scaled structures a number of low-dimensional episodes, for different cortical rhythms, we then measured λ_1 using the algorithm of Rosenstein et al. [30], suited to the study of short time-series, and found the following (El Ousdad and Cerf):

1. In general, the measurement of λ_1 was feasible for those signal sections that were part of an episode, but hardly for the others. This result confirms the significance of the low-dimensional episodes for brain dynamics, in particular in terms of deterministic chaos.
   The preceding rule was found valid for 31 signals among a total of 33 we investigated, including 10 α-waves, 14 δ-waves, and 9 CSWS-signals. Mostly, the measurements were either possible or not for both the dimension d and λ_1. However, in two cases did d perform better than λ_1, and in two cases did the opposite occur.
2. In two exceptional cases of interest, one α-wave, one CSWS-signal, the rule did not apply. In each case we could measure λ_1, whereas scaled structure analysis was no more than indicative of a dimension higher than accepted according to our criteria.
   However, for the α-wave, a dimension $d \simeq 5$ could be measured after wave-separation. For the CSWS-signal, good indication for a dimension between 7 and 8 was obtained on shifting the signal by 3 s in direction of time.
   Thus, in these exceptional cases, a success in measuring λ_1 allowed us, after complementary dimension analyses, to identify signal sections showing a low-dimensional behaviour. One of the dimensions found was a little above those complying with our criteria for the dimension. Seemingly, the combined measurements of d and λ_1 may help in extending the range of explorable EEG dimensions slightly above 7.

7 Conclusion

To conclude, we may ponder on the two main directions of research in non-linear analysis of spontaneous cortical activity, the search for a low correlation dimension, and the studies of the predictability of this activity, or the closely related measurement of the largest Lyapunov exponent.

Having recognized that EEG single correlation integrals exhibit no scaling of significance to Grassberger-Procaccia analysis, our group was able to show that low-dimensional dynamics do exist in various cortical rhythms.

We have established that a low dimension can be observed in single-channel analyses as short episodes, the practical significance of which shows up in the effects of sleep deprivation on the probability of δ-episodes [18], and of pharmaceuticals on the correlation dimension of α-episodes [20,21]. Low-dimensional episodes have the important potential of giving us access to the experimental investigation of unstable collective modes of spontaneous cortical activity. Interpretation of the measured dimension in terms of an attractor's dimension is warranted for the "good episodes"; otherwise the better the episode, the clearer the interpretation remains.

The significance for brain dynamics of the low-dimensional episodes identified by our methodology is exemplified by their correlation with those signal sections for which the measurement of the largest Lyapunov exponent λ_1 is either most reliable, or even feasible at all. On the other hand, the compatibility of our results for λ_1 with previously published data requires a detailed discussion, to be presented shortly.

Our results are of significance for further aspects in brain dynamics research. They suggest that comparative investigations of predictability for signal sections belonging to a low-dimensional episode or not should be of interest.

So far, our multichannel space-reconstructions showed no spontaneous spatially extended collective modes of low dimension. Spatially extended collective modes are known to exist from studies under appropriate stimulations [31]. Perhaps the brain has sufficient flexibility to exhibit spatially extended collective modes under certain stimulations, and yet to perform more locally in spontaneous collective modes.

Acknowledgment. Thanks are expressed to FORENAP (Rouffach, France, 68250) and to Doctor J.P. Macher for financial support.

References

[1] Haken H (1977) *Synergetics.* Springer, Berlin Heidelberg New York
[2] Lorenz E (1991) Dimension of weather and climate attractors. Nature 350:241–244
[3] Babloyantz A, Salazar M, Nicolis C (1985) Evidence for chaotic dynamics of brain activity during the sleep cycle. Phys Lett A 11:152–156
[4] Babloyantz A, Destexhe A (1986) Low-dimensional chaos in an instance of epilepsy. Proc Natl Acad Sci USA 88:3513–3517

[5] Babloyantz A, Destexhe A (1987) Strange attractors in the human cortex. In: Rensing L, and der Heiden U, Mackey MC (eds) *Temporal disorder in human oscillatory systems* (Springer series in synergetics, Vol 36). Springer, Berlin Heidelberg New York, pp 48–56
[6] Destexhe A, Sepulchre JA, Babloyantz A (1988) A comparative study of the experimental quantification of deterministic chaos. Phys Lett A 132:101–106
[7] Babloyantz A (1989) Estimation of correlation dimensions from single and multichannel recordings - A critical view. In: Basar E, Bullock TH (eds) *Brain dynamics* (Springer series in Brain Dynamics, Vol 2) Springer, Berlin Heidelberg New York, pp 122–130
[8] Packard NH, Crutchfield JP, Farmer JD, Shaw RS (1980) Geometry from a time series, Phys Rev Lett 45:712–716
[9] Grassberger P, Procaccia I (1983) Characterization of strange attractors. Physica D9:189–208
[10] Rapp PE, Zimmerman ID, Albano AM, de Guzman GC, Greenbaum NN, Bashore TR (1986) Experimental studies of chaotic neural behaviour: cellular activity and electroencephalographic signals. In: Othmer HG (ed) *Non-linear oscillations in biology and chemistry.* (Lecture notes in biomathematics, Vol 66) Springer, Berlin Heidelberg New-York, pp 175–205
[11] Theiler J (1986) Spurious dimension from correlation algorithms applied to limited time series data. Phys Rev A 34: 2427–2432
[12] Cerf R, Daoudi A, Khatory A, Oumarrakchi M, Khaider M, Trio JM, Kurtz D (1990) Dynamique cérébrale et chaos déterministe. CR Acad. Sci. Paris 311 II:1037–1044
[13] Cerf R, Ben Maati ML (1991) Trans-embedding-scaled dynamics. Phys Lett A 158:119–125
[14] Cerf R, Oumarrakchi M, Ben Maati ML, Sefrioui M (1992) Doublet-split-scaling of correlation integrals in non-linear dynamics and in neurobiology. Biol Cybern 68:115–124
[15] Cerf R (1993) Attractor-ruled dynamics in neurobiology: does it exist ? can it be measured ? In: Haken H, Mikhailov A (eds) *Interdisciplinary approaches in nonlinear complex systems.* (Springer series in synergetics, Vol 62) Springer, Berlin Heidelberg New-York, pp 201–214
[16] Cerf R, Ould Hénoune M, Ben Maati ML, El Ouasdad EH, Daoudi A (1994) Wave-separation in complex systems. Application to brain-signals. J Biol Phys 19: 223–233
[17] Cerf R, Ben Maati ML (1995) Attractor characterization from scaled doublet structures: simulations for small data sets. Biol Cybern 72: 357–363
[18] Cerf R, Sefrioui M, Toussaint M, Luthringer R, Macher JP (1996) Low-dimensional dynamic self-organization in δ-sleep:effect of partial sleep deprivation. Biol Cybern 74:395–403
[19] Cerf R, Daoudi A, Ould Hénoune M, El Ouasdad EH (1997) Episodes of low-dimensional self-organized dynamics from electroencephalographic α-signals. Biol Cybern 77:235–245
[20] Gasser B, Toussaint M, Luthringer R, Macher JP, Cerf R (1996) Wave-separation versus bandpass filtering: a comparative non-linear analysis of brain α EEG signals with and without psychotropic drug treatment. J Biol Phys 22:209–225

[21] Gasser B, Toussaint M, Luthringer R, Macher JP, Cerf R (1997) EEG-signal analysis with non-linear dynamic methods: a comparative study of brain α-EEG-signals with and without psychotropic drug treatment. In: Macher JP, Crocq MA, Nedelec JF (eds) *The bio-clinical interface, Neurosciences and Pharmacology, Psychiatry and Biology*, Rouffach, pp 217–224
[22] Haken H (1991) Synergetics - can it help physiology ? In: Haken H, Koepchen HP (eds) *Rhythms in physiological systems.* (Springer Series in Synergetics, Vol 55) Springer, Berlin Heidelberg New York, pp 21–31
[23] Ruelle D (1990) Deterministic chaos: the science and the fiction. Proc Roy Soc, London Ser A 427:241–248
[24] Procaccia I (1988) Weather systems: complex or just complicated? Nature 333:498–499
[25] Eckmann JP, Kamphorst S, Ruelle D (1987) Recurrence plots of dynamical systems. Europhys Lett 4:973–977
[26] Chechetkin VR, Ezhov AA, Knizhnikova LA, Kutvitskii VA (1992) Multiplicative group of invariance of phase space and natural time-scale for chaotic attractors. Phys Lett A 162:370–374
[27] Ould Hénoune M, Cerf R (1995) Time reparametrization of phase trajectories as a test for attractor-ruled dynamics: application to electroencephalographic α-waves. Biol Cybern 73, 235–243
[28] Theiler J, Eubank S, Longtin A, Galdrikian B, Farmer D (1992) Testing for non-linearity in time series. Physica D58:77–94
[29] Friedrich R, Uhl C (1996) Spatio-temporal analysis of human electroencephalograms: Petit-mal epilepsy. Physica D 98:171–182
[30] Rosenstein MT, Collins JJ, De Luca CJ (1993) A practical method for calculating largest Lyapunov exponents from small data sets. Physica D65:117–134
[31] Kelso JAS, Bressler SL, Buchanan S, de Guzman GC, Ding M, Fuchs A, Holroyd T (1992) A phase transition in human brain and behavior. Physics Letters A 169:134–144

Part II

Methods & Applications

Source Modeling

Sara Gonzalez Andino[1], Bob W. van Dijk[2], Jan C. De Munck[2],
Rolando Grave de Peralta Menendez[1], and Thomas Knösche[3]

[1] Functional Brain Mapping Lab., Dept. of Neurology, Geneva University Hospital, 1211 Geneva 14, Switzerland
[2] MEG Center KNAW, AZVU Rec. C, De Boelelaan 1118, 1081 HV Amsterdam, The Netherlands
[3] Max-Planck-Institute of Cognitive Neuroscience, Stephanstrasse 1a, 04103 Leipzig, Germany

Electroencephalography (EEG) is the measurement of the potential difference between two electrodes attached to the head, as a function of time. Usually, multiple channels are recorded with respect to a common reference electrode. Since the EEG is generated by electrical activity of the brain, a study of EEG signals may reveal the functioning of the brain. Similarly, the Magneto Encephalogram (MEG), which is the magnetic induction field corresponding to the electrical brain activity, can be recorded to study brain function. Although the generation mechanisms of EEG and MEG are very similar, the measurement of MEG is much more involved than EEG because the magnetic field of the brain is only very weak ($\mathcal{O}(10^{-13}\ \mathrm{T})$). The EEG can be recorded with electrodes and quite common electrical amplifiers, whereas for the MEG one needs SQUIDS (Super Conducting Quantum Interference Device) and usually a magnetically shielded room.

The human brain consists of about 10^{10} interacting neurons. Each neuron consists of a cell body or *soma*, about 10^2 to 10^5 *dentrites*, which act as electrical input, and an axon, which carries the electrical output signal of the cell to other neurons. When a neuron is at rest, the inside of the cell membrane has a negative potential with respect to the outside. This potential difference, called *polarization*, is caused by pumping mechanisms of the cell membrane. The polarization of the cell membrane can be locally reduced or inverted. Under certain conditions and with a rather complex process this disturbance of the potential, called *depolarization*, may travel along the axon to the end. The traveling depolarization, called *action potential*, coincides with electrical currents flowing inside and outside the axon. However, these are oppositely directed current dipoles (one at each side of the depolarization) and therefore the potential in tissues constituting the head (the *volume conductor*) falls off very quickly with distance. For this reason, the action potential itself does not contribute to the EEG and MEG.

When the action potential reaches the end of the axon, a complex process takes place which transfers the electric disturbance from the axon to the dendrites of the next neuron. This happens at the so-called *synapses*, the connections between different neurons. At the synapse, a post synaptic potential (PSP) builds up across the cell membrane of the post-synaptic dendrite. If PSPs are built up on different dendrites simultaneously and if a certain threshold is exceeded, a secondary action potential will be generated on that neuron. The electrical activity of simultaneously activated synapses acts as a current source, which brings the free charges in the volume conductor into movement. Contrary to the action potential the current source of the PSPs do not cancel and an electrical potential difference is generated on the skin, which can be measured as an EEG signal. The MEG is generated by this same current source and also by the resulting volume currents. Concluding from this rough sketch we may say that both the EEG and MEG originate from *interacting* neurons.

In order to localize the brain interactions from E/MEG signals, one must quantify the magnitude and distribution of the EEG and MEG signals as precisely as possible, using a mathematical model. The corresponding physical model consist of a current source reflecting the electrical activity at the neural interactions, which is embedded in a volume conductor which carries the secondary volume currents. The electrical and magnetic fields measured on, respectively outside the head, are determined by the geometry of the sources and conductivities of head tissues. The current sources are usually described with (a combination of) electrical current *dipoles*. The distribution of the conductivities in the head determines the volume currents and is usually referred to as the *volume conductor model*. Different models are in use for source localization, of different levels of complexity. The simplest model is the homogeneous sphere model. Here the head is described as a homogeneously conducting sphere. For this model a fast analytically closed formula can be derived. The effect of the low conducting skull is ignored. The next simplest model is the three concentric sphere model. This model takes into account that the conductivity of the skull is much smaller than the conductivity of the other tissues of the head. This model can be solved analytically using an expansion in spherical harmonics. The three sphere model can be generalized to an arbitrary numbers of concentric spherical or confocal spheroidal shells, which may all have an anisotropic conductivity. These models can still be solved analytically, using spherical harmonics expansions.

It is also possible to derive realistic volume conductor models, which take into account the true shape of the different tissues of the head. These shapes can be derived from MR images. These models have a large accuracy, but they can not be solved analytically. Instead, one has to use numerical methods like the boundary element or the finite element method.

The inverse problem of EEG and MEG is to estimate the locations of the current sources underlying the measurements. It can easily be shown that without a priori restrictions, there are infinitely many solutions for the inverse problem. In other words, there are infinitely many combinations of current sources that describe the EEG and MEG measurement without any error. The mathematical term to denote this type of problem is that the inverse problem is *ill posed*. Ill posed problems can be solved in many different ways. But a common feature in all approaches is to add a priori information, which restricts the solution space. One can e.g. restrict the number of dipoles that are fitted to the data or one can make restrictions in the temporal behavior of the sources. Furthermore, one can seek a solution which satisfies some well defined restrictions on expected features of the sources. Still another approach is the addition of a priori information in statistical terms. All these three approaches will be addressed in this chapter.

Spatio-Temporal Dipole Analysis

Thomas R. Knösche

Max-Planck-Institute of Cognitive Neuroscience, Stephanstrasse 1a,
04103 Leipzig, Germany

1 General Aspects of Dipole Analysis

In general, event-related potentials (ERP) and event-related fields (ERF) represent a mixture of components originating from functionally different brain processes. The task of disentangling these components requires firm assumptions about their properties. Often it is assumed that each *functional* component is characterised by one *temporal* component (i.e. a time series) and one *spatial* component (i.e. a scalp distribution). Although this view is somewhat simplistic, it seems likely that such *spatio-temporal* components are related to functional brain processes. Hence, the idea is to decompose the measurement into such spatio-temporal components:

$$\mathbf{f}(t) = \sum_{i=1}^{k} \mathbf{s}_i \cdot q_i(t) + \mathbf{n}(t). \tag{1}$$

The time-dependent scalp distribution $\mathbf{f}(t)$ is split into k spatio-temporal components, each consisting of a scalp distribution $\mathbf{s}_i$ and the corresponding time series $q_i(t)$, and noise $\mathbf{n}(t)$. If the time axis is sampled, (1) can be written as matrix equation:

$$F = S \cdot Q + N. \tag{2}$$

The measurement matrix F and the noise matrix N contain in each row the sampled time course of the respective channel and in each column a "snapshot" of the EEG/MEG at the respective time instant. The rows of S contain the spatial components $\mathbf{s}_i$ and the columns of Q the sampled versions of the temporal components $q_i(t)$.

This concept requires the completion of two tasks. First, we have to separate the signal part $S \cdot Q$ from the noise N. Second, we need to define criteria for selecting one out of the infinite number of possible decompositions of the signal part of the data. The separation of signal and noise will be addressed in the following section. The decomposition of the signal requires constraints that work on the spatial or temporal part of the signal components. They have to be chosen in such a way that the components represent functional entities. It is therefore plausible to look for such scalp distributions which can be elicited by a certain combination of intracranial generators. This directly leads to the *neuroelectromagnetic inverse problem*.

The classical way of approaching this problem can be described by three major components, as shown in Fig. 1.

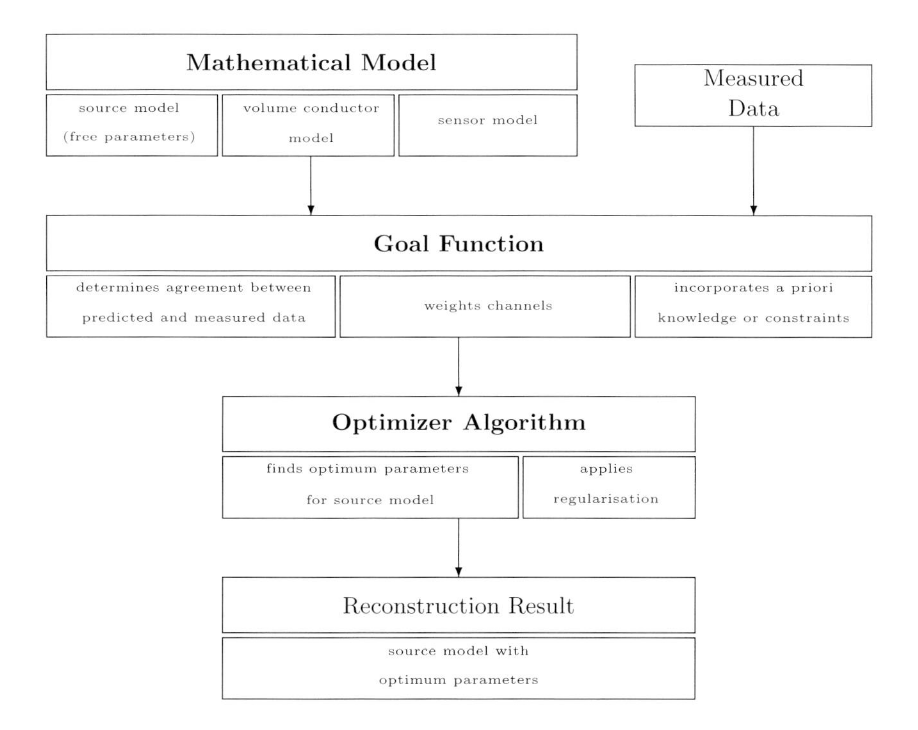

Fig. 1. Major components of approaches to the inverse problem.

First, it is necessary to construct a **mathematical model** of the entire situation. This mainly includes a *model of the sources* that incorporates certain (simplifying) assumptions on the nature of the activity. For example, if the extent of the active volume is small compared to the distance between source and sensors, it may be described by a point-like source model, the current dipole. The source model should contain a limited number of free parameters. The *volume conductor model*, together with a *model of the sensor configuration* and the Maxwell equations determines the *forward solution*. It allows us to predict the magnetic field or electric potential that would arise from a certain source. It is obvious that the accuracy of this prediction is crucial for the accuracy of the inverse reconstruction. Nevertheless, the forward solution is relatively detached from the inverse problem, because almost any forward solution can be combined with any inverse method.

The **goal function** provides a measure of agreement between predicted and measured data. Mean square differences or correlation are common choices.

Different weightings of the measuring channels and additional constraints on the source parameters can be applied here.

The **optimizer** procedure is needed to find the optimum of the goal function with respect to the source parameters. It may be a linear equation solver or a non-linear search routine, depending upon the nature of the goal function.

For the solutions to the inverse problem presented here, it is assumed that the intracranial activity can be described with sufficient accuracy by a *small number of focal sources*, so-called *current dipoles*. This model will be discussed in detail later. The choice of the source model has also implications on the goal function. Since the (generally unknown) positions of the current dipoles influence the predicted potential/field in a non-linear way, in most cases the goal function becomes non-linear with respect to at least part of the source parameters. This in turn requires non-linear search algorithms in order to find the optimum set of dipole parameters.

2 Separation of Noise Using Information Criteria

According to (1), each measurement can be split up into a signal and a noise part. By definition, only the signal part reflects relevant brain processes and should be further analysed, e.g. by source reconstruction. For the separation of these parts, the idea of signal and noise sub-spaces is useful. If a measurement is characterized by m registration channels (e.g. electrodes or MEG-sensors) and $n > m$ time samples, each recording ("snapshot") can be described as point in a m-dimensional *measurement space* (each axis represents a sensor). Now assume, a certain intracranial source, which only changes its strength, but not the spatial distribution or orientation of its currents, is the sole generator of the measurements. Then, the generated sensor outputs will exclusively lie on a line through the origin of the co-ordinate system. They form a spatio-temporal component with one spatial distribution (here defined by the angles of the line with the axes of the measurement space) and with one time series (defined by the distances of the sensor outputs from the co-ordinate origin). If two such components are simultaneously active and linearly independent, the two lines would span a plane in the measurement space, on which all possible sensor outputs would be located. This way, a sub-space of the measurement space is defined: the so-called *signal sub-space*. It must be of lower dimension than the measurement space. This means, the experiments should be designed in such a way that only a few relevant brain processes are present in the data.

The development of the signal in time is reflected in the measurement space by a trajectory in the signal sub-space. The consecutive time samples are points on this trajectory and the projection of these points onto the axes that define the spatial components yields the respective temporal components. However, since not only signal, but also noise is part of the measured data, the trajectory of the measurement does not lie exactly within the signal sub-space, but also has a (hopefully small) extent into the space orthogonal to it: the *noise sub-space*. Obviously, this concept of separating signal and noise assumes that the two sub-spaces are orthogonal, i.e. that noise and signal are linearly independent from each other. Any non-zero projection of the noise onto the signal sub-space will contaminate the signal.

Provided that the two sub-spaces are (nearly) orthogonal, the definition of the axes within each of them does not interfere any more with the task of separating signal and noise. We can choose them mutually orthogonal for simplicity. Hence, we have to rotate the original co-cordinate system, defined by the sensor outputs, in such a way that a sub-set of the axes span the signal sub-space. Since the signal sub-space is unknown in practice (in fact, it is what is searched for), this operation is not straightforward. Therefore, another assumption is made. In many cases, it is reasonable to state that the signal components have a much higher amplitude that the noise components. This assumption is backed by two facts. First, in many practical cases it is made sure that the signal-to-noise ratio is much bigger than one, e.g. the noise amplitude is small compared to the signal. Second, the number of active signal components is often small compared the the overall dimension of the measurement space, resulting in few dimensions in the signal and many in the noise sub-space. Because much of the noise is stochastic in nature, it is likely that the noise variance distributes more or less equally among all directions of the noise sub-space, diminishing the amplitude of each of the components even further.

Now, the following technique can be employed. First, we search the direction with the largest mean square amplitude in the measurement space and define it as first axis. Then, in the remaining space perpendicular to this axis, again the direction with the largest mean square amplitude is searched and defined as second axis. This procedure is repeated until the entire measurement space is included into the new co-ordinate system. This procedure is called *singular value decomposition* (SVD). Provided that (a) noise and signal are mutually independent, and (b) each of the signal components has a bigger variance than any noise component (see above), this decomposition allows a separation of noise and signal in such a way that the first k components (axes of the new co-ordinate system) span the signal sub-space, and the remaining $m - k$ components span the noise sub-space.

Mathematically, the SVD is carried out in the following way:

$$F = U \cdot W \cdot V^T \ . \tag{3}$$

U and V are orthogonal matrices and contain the so-called left and right eigenvectors. The columns of U represent the directions in the measurement space as defined above. The columns of V form the projections of the (discrete) trajectory onto each of the those directions. Each pair of them forms a spatio-temporal component.[1] A crucial role is played by the diagonal matrix W. It contains the so-called *singular values*. Each of these singular values represents the square root of the mean square amplitude of the respective component. According to the procedure described above, these values are arranged in descending order. These singular values play a crucial role in the separation of signal and noise components.

In practice, not the measurement matrix F, but the (estimated) spatial covariance matrix[2] of the measurements is decomposed:

$$D = F \cdot F^T = U \cdot W \cdot W \cdot U^T \quad . \tag{4}$$

This way, the temporal dimension is eliminated from the problem. The eigenvalues of this matrix are the squared singular values of the measurement matrix and represent the mean square amplitude of the respective components. In Fig. 2, typical eigenvalue spectra of data covariance matrices are plotted. If no noise is present, only the signal components have non-zero variances, as shown in the left panel. The identification of the signal sub-space is therefore trivial. Now assume that Gaussian, spatially uncorrelated noise is added to the data and there are enough time samples to ensure that the squared data matrix is a good approximation of the data covariance matrix D. Then, all noise eigenvalues would be about equal and form a horizontal "noise floor", as depicted in the middle panel of Fig. 2. Still, the separation of the two sub-spaces is straightforward. Unfortunately, the real situation is always similar to the one presented in the right panel. The noise is spatially correlated and the spatial covariance matrix is more or less inaccurate. The separation between signal and noise components by mere inspection becomes a very subjective matter. The identification of signals in noisy measurements is a problem not unique to biomedical recordings. It has been treated in general signal processing as well as biomedical literature (Akaike 1974, Bartlett 1954, Jagers et al. submitted, Knösche 1997, Knösche et al. in press, Lawley 1956, Rissanen 1978, Uijen 1991, Uijen and van Oosterom 1992, Wax and Kailath 1985, Wong et al. 1990, Zhao et al. 1986). From these studies, a number of so-called *information criteria* have emerged. For an overview, see Jagers et al. (submitted), Knösche (1997), or Knösche et al. (in press).

[1] Of course, one could also exchange the roles of spatial and temporal eigenvectors by spanning the measurement space in the time domain (each time sample one coordinate axis) and plotting every temporal component as direction in this system. If the channels would be given an order, one could even draw a trajectory. Signal and noise sub-spaces could be defined in this space, and the new orthogonal basis would be formed by the columns of V.

[2] The term *covariance matrix* is not entirely correct here, since this statistical concept (with the given definition) is only valid for zero-mean stochastic signals, which the rows of F are hopefully not. For simplicity, the term is used here for the matrix containing the expectation values of the scalar products between the time series of the different channels.

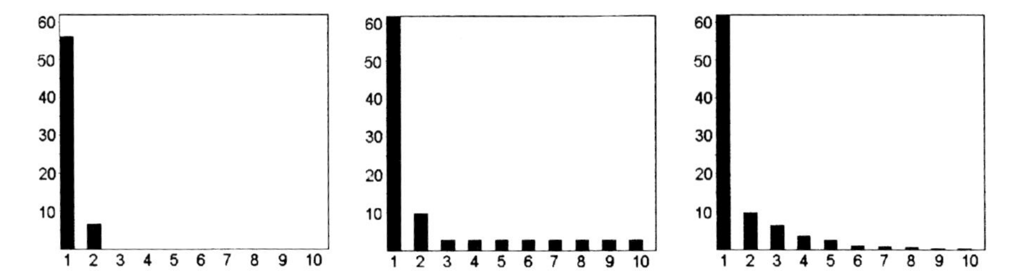

Fig. 2. Typical singular value spectra of measurement covariance matrices for the case of 10 channels and 2 independent source components. Units are arbitrary. (*left*) no noise at all. (*middle*) uniform Gaussian uncorrelated noise, covariance matrix exactly known. (*right*) correlated noise, only sample covariance matrix known. The eigenvalues are ordered according to size, the numbers at the horizontal axis represent their indices.

An information criterion is a function of the number of signal components (counted from the beginning of the descending spectrum), consisting of two additive parts: the log likelihood function and a correction term

$$\mathrm{IC}(k) = -2\log(g_{\hat{\theta}_k}(\mathbf{f})) + C(\nu_k, m). \tag{5}$$

The first term is formed by the natural logarithm of the likelihood g of the measurements $\mathbf{f}$ under the condition of a certain statistical model with an optimal parameter vector $\hat{\theta}$. The parameters of the model are optimized in such a way that the likelihood $g_{\hat{\theta}}(\mathbf{f})$ becomes maximum. The model g can be based on the *spatial eigenvectors and the eigenvalues of the measurements*, only the *eigenvalues of the measurements*, or on the *eigenvalues of the noise sub-space*. It also has to make certain assumptions on the noise, especially its spatial covariance matrix. In Table 1, and overview on different log likelihood functions is given. For more details, see e.g. Knösche (1997). It is obvious that the log likelihood function decreases with the number of signal components k,

Table 1. Summary of different log likelihood functions used in literature. Σ_N is the noise covariance matrix, $\mathbf{I}$ the identity matrix, σ the noise standard deviation, and $\mathbf{S}_N$ the estimated noise covariance matrix

assumptions on noise	based on the distribution of ...		
	the measurements	all eigenvalues	the noise eigenvalues
$\Sigma_N = \sigma^2\mathbf{I}$, σ unknown	Wax and Kailath (1985)	Wong et al. (1990)	Jagers et al. (submitted) Knösche (1997)
$\Sigma_N = \sigma^2\mathbf{I}$, σ known	Zhao et al. (1986)	Wong et al. (1990)	Jagers et al. (submitted) Knösche (1997)
Σ_N arbitrary, $\mathbf{S}_N$ known	Zhao et al. (1986)	Jagers et al. (submitted)	Jagers et al. (submitted) Knösche (1997)

i.e. the number of free parameters in θ. Hence, the model becomes the better, the more signal components are assumed. This would lead to excessively complicated models where noise components are tried to be explained by sources (e.g. dipoles). Therefore, a correction term C has been introduced into (5), depending upon the number of degrees of freedom ν (length of θ) and the number of channels m. In literature, two such correction terms are widely used: the *Akaike* correction term (Akaike 1974) and the *minimum description length* (MDL) correction term (Rissanen 1978). These terms form increasing functions of k, so that the sum of the two terms has a minimum at a certain $k = \kappa$, which is considered the optimum choice for the number of signal components.

The large number of information criteria, which are based on different models and assumptions, raises the question of their reliability and consistency. This question has been addressed by Knösche (1997) and Knösche et al. (in press) using computer simulations. These studies have revealed three factors that limit the performance of the information criteria. First, it is important that the model concerning the noise properties is consistent. If e.g. the noise is not spatially uncorrelated and equal in all channels (homoscedastic), the criteria of the first two rows in Table 1 can only be applied after a pre-whitening operation using the estimated noise covariance matrix. The second limiting factor is the noise level itself. The level from which on the estimation of the number of signal components fails, differs strongly among the criteria. It also increases with the numbers of channels and time samples. Thirdly, the relative inaccuracy of the estimated noise covariance matrix proved to be a crucial factor. Here, the increase in numbers of channels and time samples has a negative effect. Again, different criteria perform differently, and in general, the criteria that are tolerant against high noise levels require low inaccuracies in the covariance matrix and vice versa.

Hence, if the noise is strong and the available estimate of the noise covariance matrix is fairly accurate, high numbers of channels and time samples are advantageous. A recommended criterion for this case involves a log likelihood function based on the noise eigenvalues and correlated noise, and the Akaike correction term. On the other hand, if the noise is not too strong, but the accuracy of the noise covariance matrix is poor (e.g. because the number of averaged epochs is small), the number of channels and time samples should be reduced, e.g. by using a subset. The best criterion for such cases is based on all eigenvalues, assumes correlated noise, and also employs the Akaike correction term.

Information criteria not only promise a more accurate separation of noise and signal in the data, they also introduce a degree of objectivity into the choice of the number of signal components which is vital for standard analysis procedures of scientific or medical data.

3 Modeling Focal Brain Activity with Current Dipoles

As already pointed out in the introduction, the chief constraint of the spatial components of the signal should consist in the fact that they have to be explainable by intracranial sources. For this purpose, a model of these sources is needed. One of the most widely used models for representing focal generators is the *current dipole model.* A current dipole can be imagined as a current flow of neglible spatial extent, characterized by position, orientation and strength. In the following paragraphs, the theoretical justification of this model will be briefly sketched.

For simplicity, it is assumed that the material properties σ (conductivity) and μ (magnetic permeability) are constant and isotropic everywhere, sensors, i.e. the so-called volume conductor in infinite, homogeneous, and isotropic. For different volume conductor models, the formulae are different, but the principal approach remains the same. Forward solutions using numerical volume conductor models based on the boundary element method are described by e.g. Zanow and Peters (1995) and those utilizing the finite element method by van den Broek et al. (1997). For analytical solutions for spherical volume conductor models, see e.g. Sarvas (1987), Frank (1956), and de Munck (1989).

If we look at electrical activity of the brain in a macroscopic way, it can be described as a continuous distribution of the current density vector $\mathbf{j}(\mathbf{r})$. We start with the formulation of the *scalar electric potential* φ and the *magnetic vector potential* $\mathbf{a}$ at the observation point $\mathbf{r}_p$ for an infinitely extended homogeneous medium.

$$\varphi(\mathbf{r}_p) = \frac{1}{4\pi\sigma} \int_{\Omega} \frac{\mathbf{j}^T(\mathbf{r}) \cdot (\mathbf{r}_p - \mathbf{r})}{|\mathbf{r}_p - \mathbf{r}|^3} \, \mathrm{d}\mathbf{r}, \tag{6}$$

$$\mathbf{a}(\mathbf{r}_p) = \frac{\mu}{4\pi} \int_{\Omega} \frac{\mathbf{j}(\mathbf{r})}{|\mathbf{r}_p - \mathbf{r}|} \, \mathrm{d}\mathbf{r}. \tag{7}$$

The integration is carried out over the three-dimensional source region Ω, σ is the constant and isotropic conductivity in this region, and μ the magnetic permeability. The relationship between the magnetic vector potential and the magnetic induction $\mathbf{b}$ is $\mathrm{curl}(\mathbf{a}) = \mathbf{b}$.

The goal of source modeling is to describe the continuous current distribution by a finite number of parameters. One way to achieve this is the representation of (6) and (7) by a truncated series of orthogonal functions. This approach is referred to as current multipole expansion (see e.g. Katila, 1983; Haberkorn et al., 1992; Nolte and Curio, 1997). If the multipole expansion is truncated after the first term, the dipole solution is obtained. For the infinite homogeneous medium, it consists of the following approximations for the scalar electric potential $\varphi(\mathbf{r}_p)$, the magnetic vector potential $\mathbf{a}(\mathbf{r}_p)$, and the magnetic induction $\mathbf{b}(\mathbf{r}_p)$:

$$\varphi(\mathbf{r}_p) = \frac{1}{4\pi\sigma} \cdot \frac{\mathbf{d}(\mathbf{r}_0) \cdot (\mathbf{r}_p - \mathbf{r}_0)}{|\mathbf{r}_p - \mathbf{r}_0|^3}, \tag{8}$$

$$\mathbf{a}(\mathbf{r}_p) = \frac{\mu_0}{4\pi} \cdot \frac{\mathbf{d}(\mathbf{r}_0))}{|\mathbf{r}_p - \mathbf{r}_0|}, \tag{9}$$

$$\mathbf{j}(\mathbf{r}_p) = \frac{\mu_0}{4\pi} \cdot \frac{\mathbf{d}(\mathbf{r}_0) \times (\mathbf{r}_p - \mathbf{r}_0)}{|\mathbf{r}_p - \mathbf{r}_0|^3}, \tag{10}$$

with the dipole moment

$$\mathbf{d}(\mathbf{r}_0) = \int_{\Omega} \mathbf{j}(\mathbf{r}) \cdot d\mathbf{r}. \tag{11}$$

The electric potential or magnetic field caused by an arbitrary current distribution in Ω is now approximated by the potential or field due to a point-like current source at the position $\mathbf{r}_0$, the so-called *current dipole.* The source is characterized by a position, a direction or orientation, and a strength. This model is the basis for most EEG/MEG source reconstruction approaches. It is, however, only reasonably precise, if the extent of the source area Ω is small compared to the distance to the sensors. If this is not the case, or if Ω consists of several detached sub-areas (different brain regions), the source area may be divided into regions that can be described by a current dipole each. The electric potential or magnetic field is the described by a superposition of all the contributions of the separate dipoles. This approach is referred to as *multiple dipole model.* It is suitable for describing relevant neuronal activity, which is concentrated in a few small areas of the brain.

If active brain areas become bigger, higher order multipole terms, like quadrupoles and octupoles might be used (see e.g. Nolte and Curio, 1997). But their importance has remained limited so far. One reason for this may be the comparatively difficult interpretation of the parameters of such source models.

Having agreed to represent the intracranial activity by current dipoles only, each of the spatial components in (1) is chosen in a way that it can be explained by one such dipole. The strength of the dipole is variable over time and forms the respective temporal component. This approach is referred to as *multiple fixed dipole model,*[3] because the position and orientation of the sources remain fixed over the entire analysed time interval (Scherg and von Cramon, 1985, 1986; Scherg et al, 1989; Scherg and Berg, 1991; Scherg, 1992; Syder, 1991, Knösche 1997).

4 Spatial and Temporal Constraints

As pointed out earlier in this chapter, the decomposition of ERP or ERF data into functional entities and the assignment of these entities with intracranial generators is a non-unique problem and therefore needs additional

[3] This concept is also called spatio-temporal dipole modeling in literature.

constraints. Using (1) as the basic model, these constraints can act on the dimension of the signal space k, the spatial components $\mathbf{s}_i$, or the temporal components $q_i(t)$. The determination of the number of independent components k can be carried out with the help of information criteria, presented in Sect. 2. The chief constraint on the spatial components is the fixed dipole model, as described in Sect. 3. Hence, any further spatial constraints will work on the parameters of the dipole model, i.e. the position and orientation of the focal generators. Temporal constraints can be used to restrict the variety of possible time courses of the dipolar components. They are far more difficult to formulate than spatial ones. While the spatial distribution of EEG or MEG recordings is governed by the *physical* principles of the projection of currents in a certain brain region onto the sensors, the temporal activation of these regions is determined by complicated and by no way fully understood *physiological* principles.

Another way of classifying constraints is the division into *hard* and *soft* constraints. Hard constraint are rigid restrictions in the solution space, i.e. solutions that do not meet these constraints are not possible. Such constraints are often realized by a reduction of the source parameter set. Soft constraints only favour certain solutions and suppress others. They are often implemented as penalty terms in the goal function.

4.1 Spatial Constraints

Spatial constraints of the inverse solution can be based on *anatomical* information, on *physiological* knowledge or assumptions, or on the results of *other imaging methods*.

Anatomical constraints represent knowledge about the position and orientation of such neuronal populations that are capable of generating external signals. The most trivial case of an anatomical constraint is the restriction of the solution to the brain volume. Every dipole localisation procedure has to take care of this hard constraint. The brain area may be approximated by e.g. the inner compartment of the employed volume conductor model (if spherical or BEM models are employed). But, if single subject data (no grand averages across subjects) are considered and individual MRIs are available, more sophisticated anatomical constraints become feasible. It is generally assumed that the measurable external electric and magnetic fields are generated by currents that flow along the apical dendrites of cortical pyramidal cells due to post-synaptic potentials (Niedermeyer and Lopes da Silva, 1995). Hence, it would be sensible to allow only dipoles, which are located on the cortical sheet and which are oriented perpendicularly to this surface. These assumptions have frequently been used to restrict linear distributed solutions (e.g. Dale and Sereno, 1993). For the localization of focal generators, such constraint has been realized by Buchner et al. (1997). They used a triangulated cortical surfaces on the nodes of which the dipoles had to be located. However, because their reconstruction surface represented the outer hull of the brain, rather than the cortical sheet, the orientation constraint was not applied. Wolters et al. (1996) demonstrated the use of constrained positions and orientations on a simplified model of a cortical sheet.

Physiological constraints are used to select from the manifold of anatomically possible solutions only those that are physiologically plausible. Such restrictions are often highly theory-driven and therefore dangerous. One example is the symmetry constraint (Scherg and von Cramon, 1989). Since the brain is a highly symmetrical organ, in many cases bilateral symmetrical activity can be expected. In terms of constraints, this means e.g. that position and orientation of one dipole is computed by mirroring the same parameters of another one at the sagittal plane. The joint number of free (spatial) parameters of the two sources is cut in half. Similar constraints on the parameters of a multiple dipole model, like fixed distances between two sources, are also possible. Such possibilities have been thoroughly discussed by Scherg and Berg (1991). Quite often, certain hypotheses on the location of the generators of certain signals exist. In such cases, the number of free spatial dipole parameters (position, orientation) can even be reduced to zero and the solution of the inverse problem reduces to the testing of a hypothesis. Such an approach is the *signal space projection* (SSP), combined with the use of current dipoles for the specification of the signal space components, as presented by Tesche et al. (1995, 1996). They analysed MEG recordings of an auditory oddball experiment. Dipolar sources in the hippocampus and in the auditory as well as somatosensory primary cortices were postulated and the magnetic fields computed from them formed the spatial components according to (1). The temporal activation of the generators was then computed by solving (2) for the temporal components.

Instead of physiological hypotheses, also the results of other imaging or localisation methods may be used to restrict the dipole localization procedure. For example, the results of brain imaging methods, like PET or functional MRI studies (Menon et al. 1997) may be employed. This way, the milisecond-by-millisecond temporal resolution of EEG and MEG can be combined with the higher spatial resolution power of the other techniques. However, it is not guaranteed that the hemodynamic responses seen in PET or fMRI scans are always associated with a measurable electric or magnetic response and vice versa. Since PET and fMRI yield a temporal integral of the activity, prominent but short-lasting activity may be missed. On the other hand, much of the neuronal activity may not be seen in EEG or MEG, because the generators are not aligned or do not fire in a synchronous fashion. Nevertheless, the combined use of different techniques is a very promising way to overcome the intrinsic shortcomings of the separate methods.

If generators with different temporal activations and different orientations are located closely together, they may be represented by a single dipole that rotates in time. Such a *rotating dipole* may be modeled by three fixed ones with one shared position and fixed orientations pointing into the axes of the cartesian co-ordinate system (e.g. Knösche, 1997). This way, the $5 \cdot k$ free parameters of the k generators have been reduced to a mere 3, representing the position of the activity, which is quite often the most interesting information. However, the orthogonal orientations, and consequently also the time courses, of the three generators lost their connection to functional entities. In order to cure this problem at least partially, Scherg (1992) proposed the *regional source* concept. A rotation dipole is used to explain the activity of an entire brain region. Then the directions are fitted separately to e.g. certain prominent peaks in the data, which are likely to mainly originate from one functional entity, always keeping the directions mutually orthogonal. Of course, the orthogonality constraint is artificial and allows in most cases only one direction to be reconstructed correctly, but this is the maximum extractable information about a cluster of generators, the positions of which are not distinguishable by the available data.

4.2 Temporal Constraints

As already mentioned, the formulation of restrictions on the temporal components is somewhat more difficult. Several ideas how plausible constraints on the source waveforms can be used will be sketched in the following paragraphs.

It is generally assumed that the main generators of EEG and MEG are postsynaptic potentials (Niedermeyer and Lopes da Silva, 1995). Action potentials are likely to make only small contributions, because with a signal propagation speed of about 1 m/s in unmyelinated cortical axons and a duration of approximately 1 ms, the depolarization and repolarization fronts are only 1 mm apart, yielding quadrupolar contributions to EEG and MEG. Furthermore, the summation of the short-lasting action potentials is far less effective than the one of the postsynaptic potentials, which last 10 to 20 ms. However, weak high frequency activity (around 600 Hz) has been reported recently riding on top of somatosensory evoked fields and potentials, which might be produced directly or indirectly by action potentials (Curio et al., 1994, 1995). Nevertheless, in most cases, it may be assumed that the temporal components of the EEG and MEG exhibit a certain smoothness. Scherg and Berg (1993) use this *smoothness criterion* to constrain the time courses of spatio-temporal dipole models.

If there is some reason to believe that a certain generator is only or mainly active within a certain time interval, one may constrain the solution in such a way that outside this time interval the variance of the respective temporal component is minimum. This concept is called *variance criterion* by Scherg and Berg (1993).

Another hard physiological fact on the temporal components is the finite amplitude. About $4\,\mathrm{cm}^2$ of cortex cause a generator strength of $100\,\mathrm{nAm}$ (see Freeman, 1975; Stok, 1986; Wieringa, 1993). Therefore, extremely large amplitudes are implausible. Scherg and Berg (1993) use the so-called *energy criterion* in order to minimize the overall variance of a temporal component.

Both smoothness, variance, and energy constraint are typical examples of soft constraints. A more rigid way to apply temporal constraints to a dipole solution is the construction of a model for the waveforms, similar to the dipole model for the spatial components. Scherg and von Cramon (1985) used a biphasic activation pattern to model the time courses of the generators of late auditory evoked potentials (AEP). The pattern was defined by 6 parameters, defining the amplitudes and the latencies of the onset, the peaks, and the offset. Between the defining latencies, spline interpolation was used to produce a smooth curve.

5 Optimization Strategies

In this section, a number of approaches to the spatio-temporal dipole model are discussed and examples for their application are given. They differ not only in their way of constraining the spatial and temporal components, as discussed in Sect. 4, but also in the method to find the optimal parameter set of the constrained model. For the latter purpose, a *goal function* (also called cost function or objective function) has to be defined, which quantifies the degree of agreement of (1) the predicted with the measured data, and (2) the source model with certain (soft) constraints. It may e.g. assume the following form:

$$f(\mathbf{p}) = \frac{(w_{\mathrm{LS}} f_{\mathrm{LS}}(\mathbf{p}) + w_{P_1} f_{P_1}(\mathbf{p}) + w_{P_2} f_{P_2}(\mathbf{p}) + \ldots)}{w_{\mathrm{LS}} + w_{P_1} + w_{P_2} + \ldots} \,. \tag{12}$$

The subscript LS refers to the least square error function:

$$f_{\mathrm{LS}}(\mathbf{p}) = \frac{||F - \tilde{F}(\mathbf{p})||^2}{||F||^2} , \tag{13}$$

where F denotes the measured and $\tilde{F}(\mathbf{p})$ the predicted EEG or MEG data. P_1, P_2, etc. to the penalty functions. The argument $\mathbf{p}$ represents the parameter vector of the problem (e.g. dipole parameters). An adjustable weight is also assigned to every term. A possible penalty term would involve e.g. the energy criterion (Scherg and Berg, 1993). Since dipole parameters such as positions and orientations generally have a non-linear impact on the predicted signal, a non-linear optimization problem has to be solved.

5.1 Goal Function Scan

The simplest way of finding the optimum of a non-linear function is just scanning the entire argument space with a suitable step length. Such an approach has been used by e.g. Fuchs et al. (1996)[4] and Knösche (1997). The

[4] Called here: *spatial deviation scan*.

method is only computationally feasible, if the number of free parameters is small. If only one dipole is assumed as source model, 5 spatial parameters have to be determined (3 for position and 2 for orientation). If a rotating dipole model (3 fixed dipoles with one shared position and orthogonal directions) is used, only 3 position parameters remain. Now, the volume of interest (in the simplest case the entire brain) can be divided into small voxels. Then each voxel is tested for its capability of generating the measured signal. For this purpose, a rotating dipole is placed at the centre of the voxel. The time courses are computed linearly by solving (2) for the temporal components[5]. The result of this operation is called *locally optimal dipole*. The goal function value is computed, e.g. according to (12). The voxel with the smallest goal function value defines the position of the solution dipole.

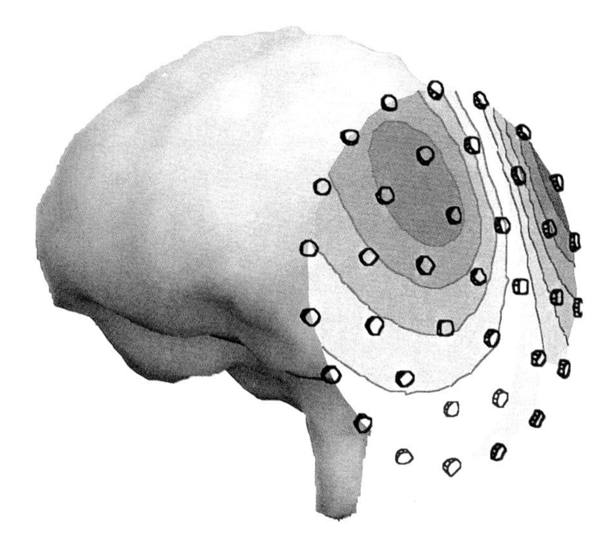

Fig. 3. Left frontal view of the volume conductor model (consisting of 935 triangles), the positions of 37 magnetometers, and the spline interpolated map for the primary auditory response due to the presentation of spoken sentences 138 ms after stimulus onset. The scale goes from −300 fT (black) to +300 fT (white).

This strategy has two important advantages. First, it is very easy to constrain both position and orientation of the dipole. One may, for example, select certain brain regions or restrict the solution to cortical generators as described in Sect. 4. Second, besides the position of the optimum dipole, also the sharpness and uniqueness of the goal function minimum can be scrutinized. Under certain circumstances, the surfaces of equal root mean

[5] Of course, either the noise term N has to be ignored or the separation of signal and noise has to be carried out first, e.g. by use of information criteria. Then we solve $F_s = S \cdot Q$ with $F_s = F - N$

square error can be roughly interpreted as bounds of confidence regions, as has been demonstrated by Knösche (1997).

One severe drawback of the goal function scan method is its virtual limitation to the single dipole case. Already taking two dipoles into account would require scanning a six-dimensional parameter space, resulting in millions of forward solutions. The computational effort becomes considerable. Theißen et al. (1997) proposed a more sophisticated approach called *hierarchical multidipole deviation scan*, which reduced the calculation time by factor 100 for 2 dipoles and factor 1000 for 3 dipoles.

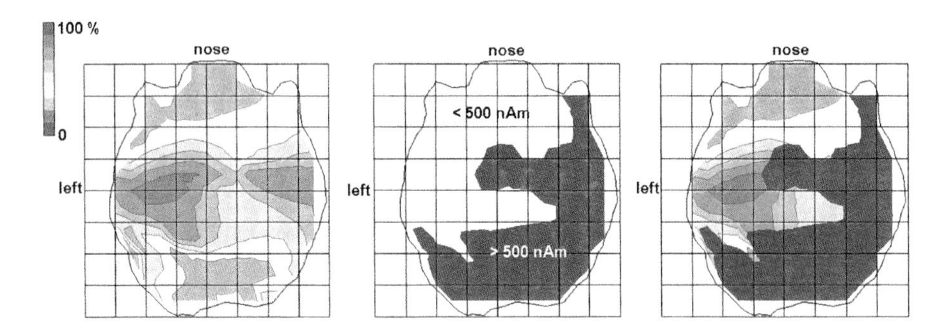

Fig. 4. Result of Goal Function Scan on primary auditory response to the presentation of spoken sentences. A slice through the primary auditory cortex is depicted. The left panel shows the result of the pure goal function scan. On the middle panel, the areas, where the dipole proposed by the scan exceeded 500 nAm are marked black. The right panel shows the overlay of the two others, indicating the regions, were plausible dipoles could explain the data in a satisfactory way.

Figures 3 and 4 show an example for the application of the Goal Function Scan method. In a neuropsychological MEG experiment, spoken sentences were presented to both ears of the subject. For test purposes, the primary auditory responses between 100 and 150 milliseconds after the beginnings of the 520 presented sentences were averaged. The signals were recorded with 2×37 magnetometers, arranged in two circular clusters over the fronto-temporal regions of both hemispheres. In order to apply the GFS method for the localisation of the primary auditory response, only the left sensor cluster was used for analysis, which should mainly reflect left hemisphere activity. A boundary-element model of the brain, containing 935 triangular elements, was used for the forward calculation (Fig. 3). The scan was carried out on a grid with 10 mm distance between points. In Fig. 4, a slice approximately through the auditory cortex is depicted. On the left panel, the result of the goal function scan is plotted. Besides the expected maximum near the left auditory cortex, another deflection in the opposite hemisphere is apparent, where also high explained variance values are present. In order to find out whether this second maximum is real or spurious, the strength of the proposed dipoles at every scan point was checked. If we assume that the activated cortex area does not exceed 2000 mm^2, then the dipole moment should be smaller than 500 nAm (see Freeman, 1975; Stok, 1986; Wieringa, 1993). The middle panel of Fig. 4 shows the areas that are excluded by the chosen criterion. This pattern is relatively insensitive to the exact value of the dipole moment limit. The right panel shows the overlay of the two plots, revealing that only the source in the left temporal cortex is real.

5.2 Multiple Signal Classification (MUSIC)

The MUSIC algorithm originates in information theory. Its application to MEG was first presented by Lewis et al. (1992). What follows, is a brief summary of the basic concepts used by Lewis et al. (1992) and by Mosher et al. (1992).

A single probe dipole is scanned through the region of interest in the same way as with the goal function scan method. The MUSIC metric or cost function is now simply the squared normalized projection of the spatial component defined by the probe dipole onto the signal sub-space. It yields peaks at all true dipole locations. One can also project onto the noise sub-space which gives much sharper peaks. The separation of the noise and signal sub-spaces is crucial with this method. The concepts presented in Sect. 2 may be useful for this task. It can happen that peaks appear in the metric that do not correspond to actual sources. That means some linear combination of true sources produces the same output as a single source at the location where the false peak occurs. This is called an array ambiguity by Lewis et al. (1992). It can be detected by observing that more sources are found than there are dimensions of the signal sub-space. The only way to resolve it is to add additional measurements. One important limitation of the MUSIC metric is that linearly dependent (correlated) sources cannot be identified. This effect is called temporal ambiguity. If synchronous sources are expected to appear, special source models become necessary. A detailed description of the MUSIC algorithm has been given by Mosher et al. (1992). There, one also finds a comparison to other dipole fitting methods. Also results from simulations as well as from the analysis of experimental data are presented. More background, however not in connection with MEG or EEG, may be obtained from Cadzow (1990).

Here, we present a brief example on the application of the MUSIC method to the same sort of data as already described in Sect. 5.1, however, this time measured with a 148-channel whole-head magnetometer system. Since for this subject no MRI data was available, a spherical head model was employed for the forward calculation. Figure 5 shows sensor configuration, volume conductor model, and spline interpolated map. Figure 6 shows the computed MUSIC metric in a slice approximately through the primary auditory areas. Two maxima in the left and right temporal cortices are clearly visible. This result was obtained, if the signal sub-space was assumed to have 2 or 3 dimensions.

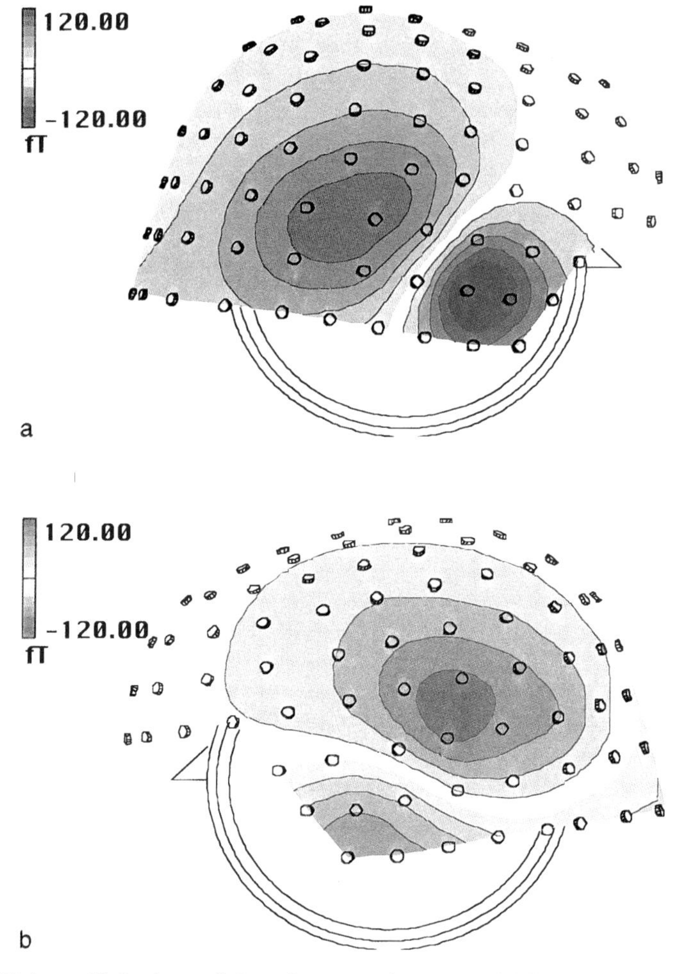

Fig. 5. Right and left views of the volume conductor model (concentric spheres), the positions of 148 magnetometers, and the spline interpolated map for the primary auditory response due to the presentation of spoken sentences 105 ms after stimulus onset.

With only 1 dimension, exclusively the right hemisphere source appears, while with more than 3 dimensions spurious sources in central regions emerge. The question may arise, why the relatively synchronous primary activity in both auditory cortices due to biauricular stimulation does not cause the MUSIC algorithm to fail. However, it has been shown that even for correlations as high as 96 % between two generator waveforms, they can still be resolved by the MUSIC algorithm (Knösche et al., 1993).

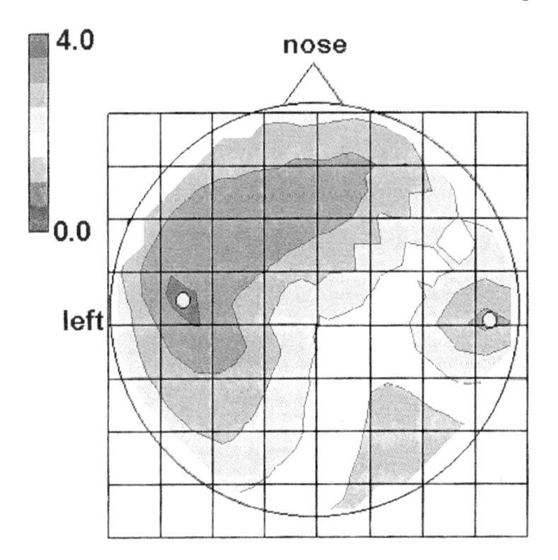

Fig. 6. Result of MUSIC scan on MEG data of the primary auditory response to the presentation of spoken sentences. A slice approximately through the primary auditory cortices is depicted. Two maxima are visible (*black dots*), indicating probable source positions.

5.3 Non-linear Dipole Fit

There is a large body of search strategies that can be used to optimize a nonlinear function like (12). Such non-linear algorithms have been widely used to find the parameters of current dipoles, e.g. the Gauß-Newton algorithm (Deuflhard and Apostulescu, 1980; Scherg and von Cramon, 1985), a quasi-Newton method (Ishiyama and Kamai, 1992), the Levenberg-Marquardt algorithm (Marquardt, 1963; Knösche, 1997), the simulated annealing algorithm (Buchner et al. 1997), simplex algorithms (Scherg, 1992). It is relatively easy to apply spatial and temporal constraints, as discussed e.g. by Scherg and Berg (1991), Scherg and von Cramon (1985), Buchner et al. (1997). The major drawback of many of these algorithms is the possibility to get stuck in local minima of the goal function. The results may be dependent upon the starting values of the parameter vector. Methods which avoid this problem, like Monte Carlo schemes or the simulated annealing procedure, may require a considerably higher computational effort.

Dipole fitting is frequently used for the localisation of the sources of evoked potentials and fields. Less common are attempts to use this method to characterise the generators of signals related to higher cognitive functions. In the following paragraphs, the localisation of the sources of the Early Left Anterior Negativity (ELAN) is presented very briefly. This component is associated with syntactic violations (in this case word category errors) in auditorily presented connected speech and appears about 150 to 250 ms after the onset of the critical (violation carrying) word. For more details see e.g. Friederici (1995) or Friederici et al. (1993). In our example, 128 electrodes were used to record the event-related potentials. Both correct and incorrect sentences were presented to the subject and the respective event-related potentials were averaged separately. For the solution of the forward problem, a 3-shell boundary element model was created from the subject's individual MRI. Figure 7 shows the sensors, volume conductor model and spline interpolated map of the potentials for the syntactically incorrect condition at the peak latency of the early syntactic effect. The following dipole fitting strategy was employed.

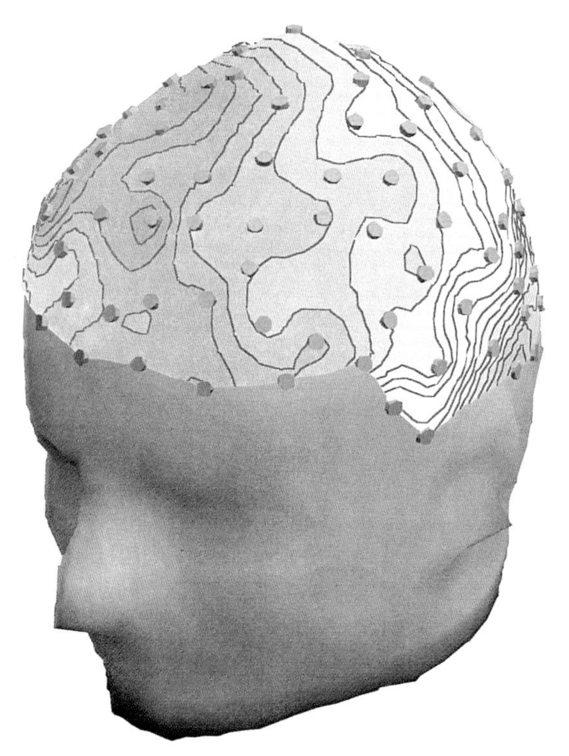

Fig. 7. Left frontal view of the volume conductor model, the positions of 128 electrodes, and the spline intepolated map for the early response due to a syntactic violation in spoken sentences 160 ms after stimulus onset. The scale goes from $-2\,\mu V$ (blue) to $+2\,\mu V$ (red).

First, the latency of the primary auditory N100 component was determined. Then, two temporally fixed dipoles were fitted to a time window of about 30 ms around the N100 peak. The fitted dipoles were interpreted to represent activity in the primary auditory cortex. Then, a time window of about 50 ms length around the maximum difference between the potentials related to the syntactic violation and the corresponding correct sentences was defined. Dipole analysis was carried out simultaneously on both the incorrect and correct data sets. That means, positions and orientations had to be the same, while the magnitudes were allowed to be different between the two conditions. The positions and orientations of the auditory cortex dipoles were kept constant and additional dipoles were added one by one. After each addition, the dipole parameters were optimized (except for the auditory cortex dipoles). In order to prevent the solution to settle in local minima, 4 different initial guesses of the position were used for each newly added dipole, being approximately left anterior, right anterior, left posterior, and right posterior. The solution with the smallest residual variance was accepted. After a model

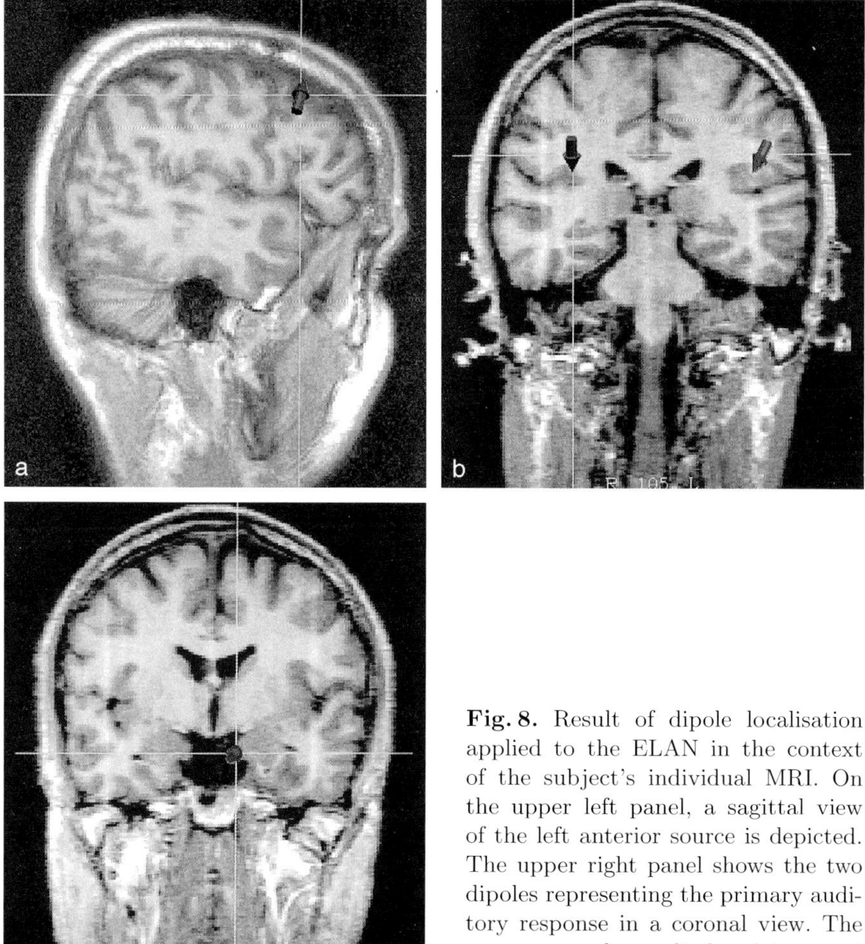

Fig. 8. Result of dipole localisation applied to the ELAN in the context of the subject's individual MRI. On the upper left panel, a sagittal view of the left anterior source is depicted. The upper right panel shows the two dipoles representing the primary auditory response in a coronal view. The source near the medio-basal temporal lobe is shown on the lower panel.

had been found that explained the combined data set with sufficient accuracy (residual variance less than 5 %) or the addition of further dipoles did not yield substantial improvement, these models were considered to represent all activity occurring in either condition. In Fig. 8, the localised dipoles are shown in MRI context. Besides the auditory cortex, a left anterior and a left medio-basal temporal source seem to be active.

5.4 Hypothesis Testing Using the Dipole Model

In clinical or psychological research it often occurs that one or more hypotheses on the exact location of the generators can be formulated. This is even more likely, if results of other functional imaging techniques, like PET or functional MRI are available. If such an hypothesis consists of a small number of active areas, the spatial extent of which is smaller than their distance to the sensors, dipoles may be placed at their centres. Their orientation can be left free or fixed, e.g. perpendicular to the cortical surface. The free parameters of the dipoles are then fitted to the measured data using linear techniques. The goal function value, e.g. the correlation between measured and predicted data or the square error according to (13), can be used to assess, whether the hypothesis can be maintained or not. The exact level of acceptance depends on the signal-to-noise ratio of the data. The testing of alternative hypotheses gives additional confidence.

In the following paragraph, a brief example is given, how a hypothesis on the generators of slow parietal positivity during a delayed matching-to-sample (DMS) task was confirmed by dipole analysis. These slow potentials occurred while subjects actively maintained object information in their visual working memory. The data have been recorded at 74 electrodes in 24 subjects. After averaging over subjects and subtraction of a neutral control condition, a prominent parietal positivity accompanied by some left anterior

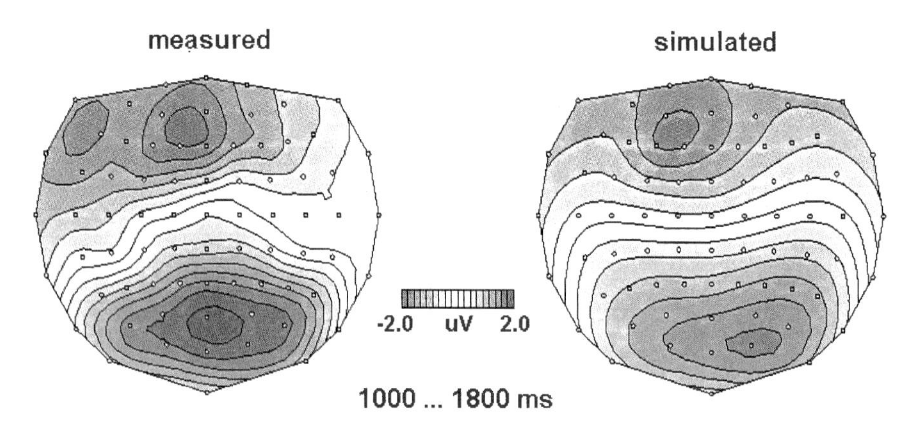

Fig. 9. Spline interpolated maps of measured and simulated slow potentials during DMS task (electrodes projected to plane, nose pointing upwards, integrated over time).

negativity became apparent (see Fig. 9, left panel). It was hypothesized that these potentials at least partially originate from generators in both fusiform gyri. This hypothesis was based on evidence from functional MRI in literature (e.g. Courtney et al., 1997). In order to evaluate this hypothesis, dipoles were placed in both fusiform gyri of a typical brain[6] using the MR image. Their directions were chosen perpendicular to the average brain surface in that region. A 3-shell boundary element model of the person's head was employed to estimate the potentials at the electrode positions. Two more dipoles were

[6] The person did not participate in the experiment.

used to explain the negativity at the frontal electrodes (see Fig. 10). The model could explain about 90 % of the variance of the measured data (Fig. 9, right panel). The strength of the dipoles did not exceed 50 nAm, corresponding to about 2 cm^2 of cortex (Freeman, 1975; Stok, 1986; Wieringa, 1993).

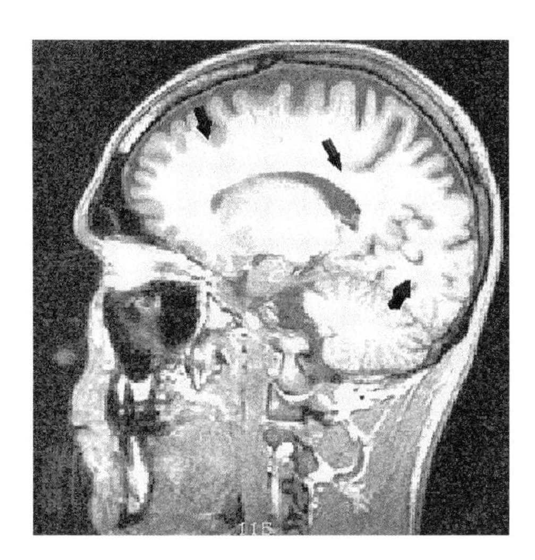

Fig. 10. Sagittal MRI scan with (projected) positions of the dipolar source used for the hypothesis-driven explanation of slow potentials associated with the presence of object information in the visual working memory. The two dipoles in the fusiform gyri are plotted on top of each other.

Although the usage of non-matching MRI and volume conductor model in connection with grand average ERP data certainly introduces considerable errors, this analysis impressively shows that the broad parietal positivity related to the maintaining of object information in the visual working memory can be explained by two symmetric generators of plausible strength in the fusiform gyri. The other two dipoles play a lesser role and only show that also the frontal potentials can be explained by dipolar generators. Note, however, that this is strong evidence, only because the hypothesis has been formed independently, based on evidence from other imaging methods.

6 Interpretation of the Results

If conclusions are drawn from the results of dipole localization procedures, one has to bear in mind the condition these algorithms are subject to. These condition have been tried to clarify in this paper and will be briefly summarized in this paragraph.

- Only the signal part of the data has to be explained by the dipolar sources. If the noise is not sufficiently small, the separation of noise and signal in (1) has to be carried out properly. In section 2, this problem is discussed.

- The dipole model only describes the centre-of-mass of a certain active area. It depends upon the shape and extent of such a generator, how complete this description is.
- If more than one dipole is used, the uniqueness of the solution may be lost. Alternative source configurations can yield (nearly) equal goal function values. Especially, noise in the data can increase this problem (confidence regions).
- In order to cope with the above-mentioned uniqueness problems, it is important to use as much external information as possible to constrain the solution.

If these problems are taken care of, dipole modeling is a powerful tool for characterizing the sources of electric or magnetic signals of the brain. The combination of this method with other brain imaging methods, like PET or functional MRI seems to be the most promising direction of development at present.

References

Akaike, H. A new look at the statistical model identification. IEEE Trans. Automat. Contr., 1974, 19:716–723.

Bartlett, M.S. A note on multiplying factors for various χ^2 approximations. Journal of the Royal Statistical Society, Series B, 1954, 16:296–298.

Buchner, H., Knoll, G., Fuchs, M., Rienäcker, A., Beckmann, R., Wagner, M., Silny, J., and Pesch, J. Inverse localization of electric dipole current sources in finite element models of the human head. Electroenceph. clin. Neurophysiol., 1997, 102:267–278.

Cadzow, J. A. Multiple source location – The signal subspace approach. IEEE Transactions on Acoustics, Speech, and Signal Processing, 1990, 38(7):1010–1025.

Courtney, S.M., Ungerleider, L.G., Keil, K., and Haxby, J.V. Transient and sustained activity in a distributed neural system for human working memory. Nature, 1997, 386:608–611.

Curio, G., Mackert, B.-M., Burghoff, M., Koetitz, R., Abraham-Fuchs, K., and Härer, W. Localisation of evoked neuromagnetic 600hz activity in the cerebral somatosensory system. Electroenceph. clin. Neurophysiol., 1994, 91:483–487.

Curio, G., Mackert, B.-M., Burghoff, M., Müller, W., and Marx, P. Somatotopic ordering of evoked high-frequency (600hz) neuromagnetic fields recorded non-invasively from the human primary somatosensory cortex. Eur. J. Neurosci., 1995, suppl. 8:111.

Dale, A.M. and Sereno, M.I. Improving localisation of cortical activity by combining EEG and MEG with MRI cortical surface reconstruction: A linear approach. J. Cog. Neurosci., 1993, 5:162–176.

de Munck, J. A mathematical and physical interpretation of the electromagnetic field of the brain. PhD thesis, University of Amsterdam, The Netherlands, 1989.

Deuflhard, P. and Apostolescu, V. A study of the Gauß-Newton algorithm for the solution of non-linear least squares problems. In: Frehse, J., Pallaschke, D., and

Trottenberg, U., (Eds.), Special topics on applied mathematics, North-Holland Publ., Amsterdam, 1980, 129–150.

Frank, E. An accurate clinically practical system for spatial electrocardiography. Circulation, 1956, 13:737.

Freeman, W.J. Action in the Nervous System. Academic Press, 1975.

Friederici, A.D. The time course of syntactic activation during language processing: A model based on neuropsychological and neurophysiological data. Brain and Language, 1995, 50:259–281.

Friederici, A.D., Hahne, A., and Mecklinger, A. The temporal structure of syntactic parsing: Early and late effects elicited by syntactic anomalies. Journal of Experimental Psychology: Learning, Memory and Cognition, 1996, 5:1–31.

Fuchs, M., Wischmann, H.-A., Wagner, M., Drenckhahn, R., and Köhler, T. Source reconstruction by spatial deviation scans. In: Proceeding of the BIOMAG, 1996.

Haberkorn, W., Burghoff, M., and Trahms, L. Multipolquellenanalyse biomagnetischer felder. Biomedizinische Technik, 1992, 39 (suppl.):156–157.

Ishiyama, A. and Kamai, I. Source estimation by a method that combines MEG and EEG. In: Satellite Symposium on Neuroscience and Technology, 14th Ann. Int. Conf. of the IEEE Eng. Med. Biol. Soc., Lyon – Book of Abstracts, 1992.

Jagers, H.R.A., Knösche, T.R., and Peters, M.J. Estimation of the number of independent sources: Information criteria can help to solve inverse problems. submitted to IEEE Transactions of Signal Processing.

Katila, T. E. On the current multipole presentation of the primary current distribution. Il Nuovo Cimento, 1983, 2D:660–664.

Knösche, T.R. Solutions of the Neuroelectromagnetic Inverse Problem – An Evaluation Study. PhD thesis, University of Twente, The Netherlands, 1997.

Knösche, T.R., Berends, E.M., Jagers, H.R.A., and Peters, M.J. Determining the number of independent sources of the eeg – a simulation study on information criteria. Brain Topography (in press).

Knösche, T.R., Zanow, F., Peters, M.J., Gunter, T.C., and Brauer, H. The MUSIC approach in EEG source localisation. In: Eiselt, M., U.Zwiener, and Witte, H., (Eds.), Qualitative and topological EEG and MEG analysis, 51–58. Universitätsverlag Druckhaus-Mayer GmbH Jena, 1993.

Lawley, D.N. Test of sigificance of the latent roots of covariance and correlation matrices. Biometrika, 1956, 43:128–136.

Lewis, P. S., Mosher, J. C., and Leahy, R. M. A new approach to neuro-magnetic source localization. In: Satellite Symposium on Neuroscience and Technology, 14th Annual International Conference of the IEEE Engineering in Medicine and Biology Society, Lyon – Book of Abstracts, 104–111, 1992.

Marquardt, D. An algorithm for least squares estimation of nonlinear parameters. SIAM J. Appl. Math., 1963, 11:431–441.

Menon, V., Ford, J.M., Lim, K.O., Glover, G.H., and Pfefferbaum, A. Combined event-related fmri and eeg evidence for temporal-parietal cortex activation during target detection. NeuroReport, 1997, 8:3029–3037.

Mosher, J. C., Lewis, P. S., and Leahy, R. M. Multiple dipole modeling and localization from spatio-temporal MEG data. IEEE Transactions on Biomedical Engineering, 1992, 39(6):541–557.

Niedermeyer, E. and Lopes da Silva, Fernando . Electroencephalography – Basic Principles, Clinical Applications, and Related Fields. Williams and Wilkins, 1995.

Nolte, G. and Curio, G. On the calculation of magnetic fields based on multipole modeling of focal biological current sources. Biophysical Journal, 1997, 73:1253–1262.

Rissanen, J. Modeling by shortes data description. Automatika, 1978, 14:465–471.

Sarvas, J. Basic mathematical and electromagnetic concepts of the biomagnetic inverse problem. Phys. Med. Biol., 1987, 32(1):11–22.

Scherg, M. Functional imaging and localisation of electromagnetic brain activity. Brain Topography, 1992, 5:103–111.

Scherg, M. and Berg, P. Use of prior knowledge in brain electromagnetic source analysis. Brain Topography, 1991, 4:143–150.

Scherg, M. and Berg, P. BESA Brain Electric Source Analysis, Version 2.0, 1993.

Scherg, M., Vajsar, J., and Picton, T.W. A source analysis of the late human auditory evoked potentials. J. Cog. Neurosci., 1989, 1(4):336–355.

Scherg, M. and von Cramon, D. Two bilateral sources of the late AEPs as identified by a spatio-temporal dipole model. Electroenceph. clin. Neurophysiol., 1985, 62:32–44.

Scherg, M. and von Cramon, D. Evoked dipole source potentials of the human auditory cortex. Electroenceph. clin. Neurophysiol., 1986, 65:344–360.

Snyder, A. Dipole source localisation in the study of EP generators: A critique. Electroenceph. clin. Neurophysiol., 1991, 80:321–325.

Stok, C.J. The Influence of Model Parameters on EEG/MEG Single Dipole Estimation. PhD thesis, University of Twente, The Netherlands, 1986.

Tesche, C.D., Karhu, J., and Tissari, S.O. Non-invasive detection of neuronal population activity in human hippocampus. Cognitive Brain Research, 1996, 4:39–47.

Tesche, C.D., Uusitalo, M.A., Ilmoniemi, R.J., Huotitainen, M., Kajola, M. , and Salonen, O. Signal-space projections of meg data characterise both distributed and well-localised neuronal sources. Electroenceph. clin. Neurophysiol., 1995, 95:189–200.

Theißen, A., Fuchs, M., Wischmann, H.-A., Wagner, M., Köhler, T., and Drenckhahn, R. Hierarchical multidipole deviation scans. Biomedizinische Technik, 1997, 42 supplement 1:239–242.

Uijen, G.J.H. The information content of the electrocardigram. PhD thesis, Katholieke Universiteit Nijmegen, The Netherlands, 1991.

Uijen, G.J.H. and van Oosterom, A. The performance of information-theoretic critera in detecting the number of signal in multilead ECG's. Meth. Inform. Med., 1992, 31.

van den Broek, S.P. Volume Conduction Effects in EEG and MEG. PhD thesis, University of Twente, 1997.

Wax, M. and Kailath, T. Detection of signals by information theoretic criteria. IEEE Trans. Acoust., Speech, and Signal Processing, 1985, 33:387–392.

Wieringa, H.J. EEG, MEG and the Integration with Magnetic Reonance Images. PhD thesis, University of Twente, The Netherlands, 1993.

Wolters, C., Rienäcker, A., Beckmann, R., Jarausch, H., Buchner, H., Grebe, R., and Louis, A. K. Stable inverse current reconstruction in real anatomy using combinatorial optimization techniques combined with regularization methods.

In: Proceeding of the Third International Hans Berger Congres. Friedrich Schiller University, Jena, Germany, 1996.

Wong, K.M., Zhang, Q.T., Reilly, J.P., and Yip, P.C. On information theoretic criteria for determining the number of signals in high resolution array processing. IEEE Trans. Acoust., Speech, and Signal Processing, 1990, 38:1959–1971.

Zanow, F. and Peters, M.J. Individually shaped volume conductor models of the head in EEG source localisation. Med. Biol. Eng. Comp., 1995, 7:151–161.

Zhao, L.C., Krishnaia, P.R., and Bai, Z.D. On detection of the number of signals in presence of white noise. Journal of Multivariate Analysis, 1986, 20:1–25.

Zhao, L.C., Krishnaia, P.R., and Bai, Z.D. On detection of the number of signals when noise covariance matrix is arbitrary. Journal of Multivariate Analysis, 1986, 20:26–49.

Distributed Source Models: Standard Solutions and New Developments

Rolando Grave de Peralta Menendez, and Sara Gonzalez Andino

Functional Brain Mapping Lab., Dept. of Neurology, Geneva University Hospital, 1211 Geneva 14, Switzerland

1 General

When solving the electro-magnetic inverse problem we face the solution of a linear inverse problem with discrete measurements (Bertero et al. 1985, 1988) that can be written as:

$$d_i = \int_R L_i(x) \circ j(x) \, \mathrm{d}R + n_i \qquad i = 1..N_s, \tag{1}$$

where $L_i(x)$ stands for the vector lead field associated to the ith sensor, that is operated on (scalar or vector product $\circ$) with the unknown current density vector $j(x)$. The integration of this product over the whole region R results in the ith measurement. The n_i stands for the noise contribution.

Without attempting a formal definition, we term in what follows a distributed solution any solution that assumes a distributed source model when solving (1) or its discrete version. In contrast to the so called discrete solutions, associated for example with spatio-temporal dipolar models (Scherg and von Cramon 1990), distributed source models lead to underdetermined inverse problems, i.e., the number of unknowns exceeds the number of measured data. This property is a consequence of the impossibility, in many practical situations, of conceiving a more restrictive source model, that means, no a priori information is available about the number or positioning of the sources. On the other hand due to the generality of their assumptions, distributed source model overcome most of the problems that appear in spatio-temporal models, related with the non linear optimization process and the determination of the number of sources (Achim et al. 1991; Grave de Peralta Menendez and Gonzalez Andino 1994; Cabrera Fernandez et al. 1995).

In this chapter we will adopt the following notation. Matrices and vectors will be denoted by uppercase italics and lowercase bold letters, respectively. For a matrix, X^{-1}, X^{+}, X^{-}, represent the inverse, the Moore Penrose inverse and the generalized inverse respectively. The superscript t will be used to denote transposition of matrix and vectors. The $X_{i\bullet}$ will be used to denote the ith row of X and $\tilde{X}_{i\bullet}$ denotes the diagonal matrix constructed with it.

There are always two alternatives for solving (1): (a) Continuous solution in an infinite-dimensional space that may be discretized at a last stage to produce a graph or a numerical result and (b) discretization of the integrals in

(1) over a finite set of solution points in order to arrive to an underdetermined algebraic system of equations. While the discretization of the data (measurements) is the result of the present technical limitations, the discretization of the integrals in (1) is not so obvious. Even though we prefer the most powerful former approach, for the sake of simplicity we will here consider the later one, which reduces the analysis to simple geometrical considerations expressed in terms of the standard matrices and vector formalism. Let's remark that this discretization makes sense only when the discrete version of the operator is, in a well defined mathematical sense (Linz 1979, Chap. 4), close to the continuous operator.

By discretizing the system (1) for a set of points in R we obtain the algebraic linear system

$$d_i = \langle L_{i\bullet}, \mathbf{j} \rangle + n_i \qquad i = 1, \ldots, N_s, \tag{2}$$

which can be rewritten in more compact form as

$$\mathbf{d} = L\mathbf{j} + \mathbf{n}. \tag{3}$$

Where $L_{i\bullet}$ are the rows of matrix L determined by the discretization of the lead field functions over the set of solution points and the N_s vectors $\mathbf{d}$ and $\mathbf{n}$ encompass all the measured data and the noise contribution respectively. The vector $\mathbf{j}$ contains the discrete values of the unknown function that we want to determine, and $\langle \mathbf{x}, \mathbf{y} \rangle$ stands for the scalar product between vectors. With this notation most of the following results can be extended almost straightforwardly to the continuous case represented in (1). Future references to problem (2, 3) should be understood as references to any of the two equivalent problems (2) or (3).

1.1 Spaces Associated to a System of Linear Equations

Three basic spaces with a clear geometrical interpretation can be derived from problem (2, 3). The first one, frequently called the *space of the visible objects*, is composed by the vectors $\mathbf{j}$ that can be written as a linear combinations of the rows of the lead field matrix, i.e., the space spanned by the set $\{L_{i\bullet}\}_{i=1}^{N_s}$, usually represented as $\mathrm{M}(L^t)$. This space will be represented hereafter by $\mathbf{V}$. The second space that can be naturally defined is the set of vectors $\mathbf{j}$ that are orthogonal to $\mathbf{V}$, i.e., that are orthogonal to the set $\{L_{i\bullet}\}_{i=1}^{N_s}$. This set, frequently denoted by $\mathrm{Ker}(L)$, for kernel (or null space) of L, is also called the *space of the invisible objects*. In the following we will represent this vector space as $\mathbf{O}$.

The denomination in terms of visible or invisible objects arises from the capability of vector $\mathbf{j}$ to produce a non-null data measurement vector. From the previous definition all the visible objects, i.e., all the elements of $\mathbf{V}$ will produce a non-null data measurement. In contrast, all the elements of the invisible space have no contribution to the measured data.

Since **V** and **O** are orthogonal, any vector **j** that can be decomposed as $\mathbf{j} = \mathbf{j}^v + \mathbf{j}^0$ where $\mathbf{j}^v$ is an element of **V** and $\mathbf{j}^0$ belongs to **O**. In particular, in absence of noise, i.e., when $\mathbf{n} = 0$, the solution set (**J**) of (2, 3) can be represented by all the **j** vectors that can be written as $\mathbf{j} = \mathbf{j}^+ + \mathbf{j}^0$ where $\mathbf{j}^+$ is the unique solution of (2, 3) that belongs to **V**, i.e., the minimum norm solution. Note that the solution set **J** is not a vector space but a translation of the vector space **O**.

The third space is $\mathbf{R}=\mathrm{M}(L)$, i.e., the space spanned by the columns of the lead field matrix. The relevance of this space arises when one considers electric measurements, a case in which the effect of the reference has to be included in the model. This effect can be taken into account either by adding a constant column of ones to the lead field matrix L or by applying the centering matrix (that transforms a vector to average reference) to both member of (2, 3). In the latter case the rank of L is reduced by one, i.e., the space spanned by its columns is of dimension $N_s - 1$ and the matrix L does not have to have full row rank. For this reason we will in the following consider the general case where the lead field matrix has no full row rank.

Even though the case without noise hardly arises in practical applications, its study has been the basis of a strong mathematical development corresponding to the theory of generalized inverse of matrices and operators. Useful characterization of the inverse matrices in terms of the previously defined spaces for the noise free case can be found in Sect. 2.1.

1.2 Principles of Solution Construction

In practice, the system (2, 3) is purely underdetermined, i.e. there are infinitely many solutions even in the case of perfectly accurate data ($\mathbf{n} = 0$). In principle, inverse solutions can be classified into linear and non-linear. Linear solutions have been extensively investigated theoretically. Therefore, although both, linear and non linear solutions are considered here, we will mainly be dealing with linear solutions, i.e. those in which the inversion procedure can be represented as a matrix that, when applied to the data, produces the estimated solution:

$$\hat{\mathbf{j}} = G\mathbf{d}. \tag{4}$$

A general frame to derive a solution to problem (2, 3) is to consider the optimization of a functional that depends on all the available information:

$$\text{optimize } F(\mathbf{d}, L, \Theta). \tag{5}$$

where Θ is s set of parameters containing among others:

- information about the noise statistics, e.g., the covariance matrix.
- a priori information about the unknown vector **j** or its module, e.g. information extracted from other functional brain images.
- the metric associated with the measurement and/or solution space, e.g. constraints such as smoothness, interpolation law, etc.

- the matrix representing a linear inverse procedure, e.g., G in (4).
- the parameters that define a non-linear solution.

There are no limitations to the definition of functional F. It should simply reflect any features we expect in the solution. The next sections consider two different alternatives to define F based on properties of the solution vector $\mathbf{j}$ and the inverse matrix, respectively.

Note that we define a solution as linear independently of the functional F. That means, even if the associated optimization problem is non-linear, the solution is defined as linear when it can be written as in (4). An example of non-linear F that results in non-linear solution is presented in Sect. 5.2.

2 Methodology for Solution Construction

The next sections consider two different approaches that can be used to solve the system of (2, 3). The first approach, the variational approach, is based on the idea of selecting metrics in the solution space and the measurement space, respectively. These two metrics are characterized by symmetric matrices and express our idea of closeness in these spaces. With this approach the functional F is composed of two terms (see (6)), one that evaluates how well the solution explains the data, and the other that measures the closeness of the solution to an a priori selected $\mathbf{j}_p$. We include one subsection that discusses the selection of the metrics, presenting some general strategies to construct these matrices and some examples already described in the literature.

In the second approach the functional F is a measure of the closeness between the resolution matrix associated with the inverse G, and the ideal resolution matrix (see 17). By optimizing this functional we try to improve the resolution properties of the inverse procedure.

2.1 Variational Approach

Several of the approaches that can be used to yield a unique solution to the problem stated in (2, 3) converge to the following variational problem for $\mathbf{j}$ (Menke 1989, Sect. 5.8; Grave de Peralta Menendez and Gonzalez Andino 1998):

$$\min\ (L\mathbf{j} - \mathbf{d})^t W_d (L\mathbf{j} - \mathbf{d}) + \lambda^2 (\mathbf{j} - \mathbf{j}_p)^t W_j (\mathbf{j} - \mathbf{j}_p), \tag{6}$$

where W_d and W_j are symmetric positive (semi) definite matrices representing the (pseudo) metrics associated with the measurement space and the source space respectively. Vector $\mathbf{j}_p$ denotes any a priori value of the unknown current density available, e.g., from other modalities of brain functional images. The regularization parameter is denoted by λ. Independently of the rank of L, the solution to (6) is unique if and only if the null spaces of W_j and $L^t W_d L$ intersect trivially, i.e. $\mathrm{Ker}(W_j) \cap \mathrm{Ker}(L^t W_d L) = \{0\}$. In this case the estimated solution vector $\hat{\mathbf{j}}$ can be obtained using the variable change $\mathbf{j} = \mathbf{j}_p + \mathbf{h}$ and solving the resulting problem for $\mathbf{h}$, i.e.:

$$\hat{\mathbf{j}} = \mathbf{j}_p + [L^t W_d L + \lambda^2 W_j]^{-1} L^t W_d [\mathbf{d} - L\mathbf{j}_p]. \tag{7}$$

If and only if matrices W_j and W_d are positive definite (7) is equivalent to:

$$\hat{\mathbf{j}} = \mathbf{j}_p + W_j^{-1} L^t [L W_j^{-1} L^t + \lambda^2 W_d^{-1}]^{-1} [\mathbf{d} - L\mathbf{j}_p]. \tag{8}$$

In the case of null a priori estimates of the current distribution ($\mathbf{j}_p = 0$) and perfectly accurate data, i.e. λ approaching zero, the solution can be written as (Rao and Mitra 1971, Sect. 3.3.3):

$$\hat{\mathbf{j}} = G\mathbf{d} \text{ with } G = W_j^{-1} L^t W_d L (L^t W_d L W_j^{-1} L^t W_d L)^{-} L^t W_d. \tag{9}$$

When the metrics are selected as the identity matrix in both the source and the measurement spaces, the inverse G reduces to the well-known case of Moore–Penrose pseudoinverse (Penrose 1955; Rao and Mitra 1971, Sect. 3.3.1; Hämäläinen and Ilmoniemi 1984) (to prove it consider the singular value decomposition of L^t). An important characterization of the inverse in (9) in terms of the spaces defined in 1.1 is that this is the unique generalized inverse (Ben Israel and Greville 1974, Chap. 2, Sect. 5) with kernel and range as complementary spaces of $\mathbf{R}$ and $\mathbf{O}$ and defined by kernel$(G) = W_d^{-1}\mathbf{R}^{\perp}$ and range$(G) = W_j^{-1}\mathbf{O}^{\perp}$. Note that $\mathbf{O}^{\perp} = \mathbf{V}$. This relationship sheds light on the dependence that exists between the selected metrics and the kind of sources that can be perfectly retrieved (those that belong to range(G)). Again, these results show that the minimum norm solution is obtained when both metrics are the identity matrix.

Another useful characterization of the inverse (9) is given by the following four equations (Rao and Mitra 1971, Sect. 3.3.2) that can be used to assess the algorithm that computes the generalized inverse:

$$LGL = L, \quad (LG)^t W_d = W_d LG, \quad GLG = G, \quad (GL)^t W_j = W_j GL. \tag{10}$$

Two crucial points in the approach presented above are the selection of the regularization parameter and the metrics. For the former we refer the reader to Whaba (1990) and references therein. The latter aspect will be the subject of the following section.

2.2 Constructing the Metrics

For the selection of the metrics to be used in equations (6–9) we will distinguish between the data and the solution space.

(a) Metric for the data or measurement space. Since changing this metric changes the norm by which the size of the residual vector ($\mathbf{n}$) is assessed, it is common to use some information about the noise statistics. A typical selection in discrete problems is the inverse of the covariance matrix of the noise. This is the so called Mahalanobis metric.

(b) Metric for the solution space. Considering first the case without noise, note that the use of a metric different from the identity will generally (except for equivalent metrics, Rao and Mitra 1971, Sect. 3.3.4) produce solutions with non-null projections over the invisible space. For this reason the selection of a metric should be suggested by sound physical expectations. This means, since we will incorporate in the estimated solution a part that bears no relationship with the data, it is expected that this invisible part is selected on a reasonable basis. Keeping this aspect in mind we can continue the discussion of the selection of the metric for the noisy case. The simplest possible selection for the metric matrix (besides the identity matrix) is to choose a diagonal matrix that corresponds to a column scaling (Lawson and Hanson 1974, Chap. 25, Sect. 3). Possible alternatives are:

- Based on the norm of the columns of the lead field matrix (Lawson and Hanson 1974, Chap. 25)
- Based on the covariance data matrix. See for example PROMS solution in Greenblatt (1993).
- Imposing natural physical constraints , e.g., that the currents are bounded to the brain volume and thus the components should go to zero when approaching the surface of the brain. This results in the Radially Weighted Minimum Norm (RWMN) (Grave de Peralta Menendez and Gonzalez Andino 1998).
- Regularized location-wise normalization (Fuchs et al. 1994).
- Others heuristic reasons (Pascual Marqui 1995; Knösche 1997).

Other approaches that do not necessarily imply the use of diagonal matrices are:

- The covariance matrix of the sources (Sekihara and Scholz 1995, 1996; Dale and Sereno 1993).
- The product between the discrete laplacian operator and a diagonal weighting matrix (Pascual Marqui et al. 1995).
- Solutions with variable resolution based on averages that approximate a blurred version of the delta function (Grave de Peralta Menendez and Gonzalez Andino 1998).

Let's remark, before considering a general strategy for the selection of the metric, that those metrics that depend upon the columns of the lead field matrix, i.e depend on the electrode position, are quite sensitive to asymmetric sensor configurations (Spinelli et al. 1997).

If we consider the solution of the purely undetermined problem (2, 3) it is clear that the number of equations is not sufficient to determine uniquely the solution. Thus we need to incorporate some a priori information that allows to restrict the feasible solution set. One flexible alternative can be obtained if we try to propose some structure for the solution that we search for. That means, to impose some relationships between all the unknown values of the function on a local or a global scale. This can be interpreted as creating new equations to be fulfilled by the unknowns, corresponding with some *large scale evidence*. With this philosophy, the solution of the inverse problem is reduced to find such binding equations (the relationships) on the basis of common sense. For this we will consider the following mathematical tautology:

$$j(x) = \int \delta(x-y) j(y) \, \mathrm{d}y. \tag{11}$$

Applying the standard formalism that relates a partial differential operator D, its Green function $G(*,*)$ and the delta function (Roach 1970, Sect 9.1) the previous expression can be rewritten as:

$$j(x) = D_x \int G(x,y) j(y) \, \mathrm{d}y. \tag{12}$$

This expression clearly suggests that the function $j(x)$ has a local structure determined by the differential operator D and a global one determined by the integration with the Green function. This means, the differential operator imposes a relationship between x and the points in the vicinity of x, while the integration procedure, extends this relationship all over the solution space. Now, when solving for a solution of (2, 3) we can try to find a function with such structure, i.e to minimize $||j(x) - D_x \int G(x,y) j(y) \, \mathrm{d}y||$ or $||D_x \int G(x,y) j(y) \, \mathrm{d}y||$ for an a priori selected differential operator. This is a clear mathematical justification for the use of differential operators combined with weighting functions. Other approximations to a delta function are also possible and has already been used (Grave de Peralta Menendez and Gonzalez Andino 1998).

The previous strategy can be applied to the discrete case using the discrete interpolation formulas to create new equations that when imposed to the solution, define it uniquely. One example is the generalization of Shepard's interpolant (Allasia et al. 1995):

$$j(x) = \sum_{k=1}^{n} T_k^p(x) w_k(x) \quad \text{with} \quad w_k(x) \geq 0, \tag{13}$$

where $T_k^p(x)$ is the truncated Taylor expansion up to order p evaluated at the point x_k and referred to the displacement $x - x_k$. Note that for $p = 0$, $T_k^p(x) = j(x_k)$. We can use (13) to create new constraints for the unknown function. For this purpose consider the interpolation at point x_i on the basis of the other solution points:

$$j(x_i) = \sum_{\substack{k=1 \\ k \neq i}}^{n} T_k^p(x_i) w_k(x_i). \tag{14}$$

Adding the difference between both sides over all solution points we obtain the following functional:

$$\sum_i \left[j(x_i) - \sum_{\substack{k=1 \\ k \neq i}}^{n} T_k^p(x_i) w_k(x_i) \right]^2 . \tag{15}$$

Minimization of (15) forces the unknown function to have some structure that usually determines uniquely the solution. Note that from this general expression some well-know particular cases can be derived, e.g. the laplacian minimization, that corresponds to a linear estimation from the neighbors. Note also that this additional information arises from elementary reasoning rather from any assumption about the generators and that we can consider (14) like a new set of binding equations to be solved together with the system (2, 3) in a (weighted) least square sense.

The main limitation of the formulas presented hitherto is that they force an identical structure everywhere in the solution space, e.g. a non singular laplacian, forces the solution to be zero all over the surface that bounds the source space. Even though this condition seems to be reasonable for some regions, such solution exhibits a vicious tendency to penalize sources in the lowest plane of the solution space. For this reason we introduce a Homogenizer Operator (HO) that imposes a different structure on different regions allowing to selectively penalize the regions. To illustrate it let's consider a solution space formed by 4 points regularly distributed over the segment (0,1), e.g. $(0, 1/3, 2/3, 1)$. Denoting the function values by $f = (f_1, f_2, f_3, f_4)$ and considering the 4 solution points in the order described, we obtain for the HO and the non singular laplacian (B):

$$H = \begin{bmatrix} 1 & -1 & 0 & 0 \\ \alpha & 0 & 0 & 0 \\ 0 & 1 & -1 & 0 \\ -1 & 1 & 0 & 0 \\ 0 & 0 & 1 & -1 \\ 0 & -1 & 1 & 0 \\ 0 & 0 & 0 & \beta \\ 0 & 0 & -1 & 1 \end{bmatrix} \qquad B = \begin{bmatrix} 2 & -1 & 0 & 0 \\ -1 & 2 & -1 & 0 \\ 0 & -1 & 2 & -1 \\ 0 & 0 & -1 & 2 \end{bmatrix}$$

$$\begin{aligned} ||H\mathbf{f}|| = & \quad (f_1 - f_2)^2 + (\alpha f_1)^2 + (f_2 - f_3)^2 + (f_2 - f_1)^2 + (f_3 - f_4)^2 \\ & + (f_3 - f_2)^2 + (\beta f_4)^2 + (f_4 - f_3)^2. \end{aligned}$$

$$||B\mathbf{f}|| = (2f_1 - f_2)^2 + (-f_1 + 2f_2 - f_2)^2 + (-f_2 + 2f_3 - f_4)^2 + (-f_3 + 2f_4)^2$$

Since the maximum number of neighbours possible is two, we associate to each solution point two rows of H, corresponding with the two possible neighbours, independently of the real number of neighbors. This results to be relevant for the points on the border that will have a number of neighbors lower than interior points. Note that if we decrease α (second row) we do not penalize the first point for having only one neighbor. On the other hand, for a high value of β (seventh row), the last point is penalized. Accordingly, the solution computed with the metric $W_j = H^t H$ is compelled to have lower values at $x = 1$ but not at $x = 0$. In contrast the metric induced by the laplacian penalizes the values at both ends. The same happens with any other differential operator or interpolation rule that apply a fixed rule for all the solution points. This example illustrates the flexibility of the Homogenizer Operator. In our practical implementation we use weights different from 1 or -1 to increase the versatility, penalizing in different ways the different regions of the brain. In absence of any a priori information we just penalize the borders of the cortex without additional penalties to the lowest planes.

2.3 Optimizing the Resolution Matrix

In contrast to the previous approach where the solution is constructed on the basis of expected features of the estimated source $\mathbf{j}$, in this section we consider an alternative that imposes the constraints directly to the inverse matrix. If we substitute the data according to (2, 3) in (4) we obtain a fundamental equation for underdetermined linear systems:

$$\hat{\mathbf{j}} = GL\mathbf{j} + G\mathbf{n} = R\mathbf{j} + \mathbf{n}'. \tag{16}$$

The product $R = GL$ is termed the resolution matrix. In absence of noise this matrix reveals the relationship between the estimated solution $\hat{\mathbf{j}}$ and any solution to (2, 3), i.e. any $\mathbf{j}$ yielding the same data. If R equals the identity matrix, the estimates are exactly the original parameters that generated the data. Thus it is reasonable to try to construct an inverse matrix G such that the corresponding resolution matrix is as close as possible to the ideal one (the identity matrix). This is the philosophy behind the method of Backus and Gilbert applied originally for the continuous case (Backus and Gilbert 1967, 1968, 1970). For the discrete case we can consider the discrete Backus and Gilbert spread (Menke 1989 Sect. 4.6; Grave de Peralta Menendez et al. 1977a), that measures the distance between two matrices :

$$s(R, I) = \sum_i \sum_j W_{ij}(R_{ij} - I_{ij})^2, \tag{17}$$

where the W_{ij} are positive otherwise arbitrary weighting factors. After some algebraic transformations and considering that $R = GL$, (17) can be rewritten as:

$$s(GL, I) = \sum_i (G^t_{i\bullet} L \tilde{W}_{i\bullet} L^t G_{i\bullet} - 2W_{ii} G^t_{i\bullet} L_{\bullet i} + W_{ii}) \tag{18}$$

$$= \sum_i (G^t_{i\bullet} S_i G_{i\bullet} - 2W_{ii} G^t_{i\bullet} L_{\bullet i} + W_{ii}), \tag{19}$$

where $S_i = L\tilde{W}_{i\bullet} L^t$ is called the spread matrix. The summation index runs over the resolution kernels, i.e. the rows of the resolution matrix. Note that the ith term in the summation represents the distance from the ith resolution kernel to the ideal one and depends only upon the ith row of the inverse matrix G. This property facilitates the computation of the solution, reckoned by optimizing the spread $s(GL, I)$ with respect to the elements of G and using additional constraints when needed to ensure unique estimators.

From this formalism several solutions can be obtained ranging from the simple minimum norm to the generalized Wiener estimator when considering noise (Grave de Peralta Menendez et al. 1997a). A new alternative obtained from this framework was the Weighted Resolution Optimization (WROP) method which showed good performances in simulations as well as in the analysis of experimental data (Grave de Peralta Menendez et al. 1997b). In this case the weights are chosen to avoid the need of auxiliary constraints while trying to extend to the case of vector fields the philosophy of Backus Gilbert. This leads for a vector field of dimension N:

$$W_{ij} = [f(d_{ij}) + \beta] + (1 - \delta_{rs})\alpha, \tag{20}$$

where $i = (p-1)N + r$, $j = (q-1)N + s$, $1 \leq r, s \leq N$ represent the rth and sth components on points p and q respectively, and δ_{rs} stands for the discrete delta function, with value one for $r = s$ and zero otherwise. Function $f(x)$ is an increasing monotone function such that $f(0) = 0$. See Grave de Peralta Menendez et al. (1997a) for more details.

3 Evaluation and Comparison of Linear Inverse Solutions

Once we have designed a distributed solution we need: (a) to evaluate it's performance, (b) to compare it with other existing solutions and (c) to identify its basic limitations. This analysis can be done either by using theoretically designed figures of merit and/or synthetic data or by directly evaluating the solution for experimentally data sets, i.e., human intracranial data or scalp recorded data where the position of the generators is known from some independent information. The former theoretical approach is more adequate for a direct comparison among solutions or for the identification of common or particular drawbacks. However, the experimental approach is the only possible one to evaluate the plausibility of a given source model or that of some of the constraints selected to insure a unique solution. In this section we describe some of the theoretical methods that can be used for this purpose, exposing the main results that has been obtained in comparisons based on both the theoretical and the experimental approach. The section ends with a description of five basic limitations inherent to any linear distributed solutions and which have been identified from the computer simulations.

3.1 Theoretically Based Analysis of Solutions

The analysis of linear distributed solutions was, up to recently, exclusively based on measures formerly designed to evaluate the performance of single or multiple dipole fitting algorithms, i.e. the dipole localization error (DLE). There are several drawbacks in using such restrictive analysis, namely:

- The DLE is a non linear magnitude and thus the superposition principle does not hold for the DLE. This implies that by using this magnitude we can only evaluate the error committed in the estimation of the position of the maximum of the modulus of the currents in the reconstructed maps when assuming single dipoles. The DLE provides absolutely no information about the performance of the solution in the presence of simultaneously active sources which is precisely the situation for which distributed solutions are designed.
- The position of the maximum of the modulus bears no information about the quality of the reconstruction over the whole solution space and thus neither evaluates the existence of spurious sources at other solution points nor the level of blurring associated with the reconstruction at the different points.
- Usually, this oversimplified analysis does not even include a report of the errors committed in the estimation of the dipole moment or dipole strength. Since a dipole is defined by its position and its moment, it is impossible from a comparison based on the DLE to infer the actual performance of different solutions in the reconstruction of single dipolar sources. An adequate estimation of the source strength is a basic requirement for the adequate reconstruction of simultaneously active sources.

A more appropriate tool to analyze, compare and evaluate distributed linear inverse solutions is the model resolution matrix R defined in (15) (Menke 1989, Sect. 4.3; Grave de Peralta Menendez et al. 1996). The columns of the resolution matrix, impulse responses, contain all the required information about the reconstruction of single point sources at all the solution points and thus about the DLE as a particular case. Nonetheless, the main information resides in the rows of R, the resolution kernels, since they allow for a global analysis of linear solutions for the whole solution space, an analysis which is not restricted to single or multiple dipoles at previously selected knots. The reason for this is that each resolution kernel tells us how all other possible active sources influence the reconstruction at the point associated to that row. Thus resolution kernels make possible an exhaustive analysis of the reconstruction everywhere in the brain that includes all possible combinations of active sources.

Obviously, such an enormous analysis is impractical, especially for a fine discretization of the solution space, and this information needs to be summarized using figures of merit which emphasize particular aspects of the reconstruction procedures. Examples of figures of merit, already proposed and applied in the field are (Grave de Peralta Menendez et al. 1996, 1997a):

- Bias in Dipole Localization: Given by the plot of the distance between the grid points where the actual maximum of each impulse response is obtained and the expected ideal one. It can be interpreted as the localization error expectable with an inverse solution for each Cartesian component of unitary dipoles all over the solution space.
- Bias in Feature Position: Given by the plot of the distance between the grid points, where the actual maximum of each resolution kernel is obtained and the expected one. It can be interpreted as the grid points which are favoured by the considered solution when several generators are simultaneously active.
- Source identifiability: This measure is defined for each component k of the estimated $\hat{\mathbf{j}}$, as:

$$\mathrm{I}_k = \frac{(\mathrm{D} - \mathrm{d}_{ki})R_{kk}}{\mathrm{D}}, \tag{21}$$

  where D represents the maximum distance between grid points, R_{kk} denotes the absolute value of the k-th element of the k-th row of R (the diagonal element) and d_{ki} is the Euclidean distance between the grid point associated with this row and the point (r_i) where the maximum value of the k-th row was found. According to this definition, a high value of I indicates that the activity at this point can be correctly retrieved. A value near zero expresses either that the main peak is far away from the target point or that the amplitude of the resolution kernel at the point is small. Both are undesirable features resulting in poor identifiability of the current distribution.
- Source visibility: This figure of merit is not directly derived from the model resolution matrix but instead depends only on the lead field. It measures, for a given sensor configuration, a selected volume conductor model and a discretization of the solution space, the part of the total current that belongs to the visible space. For single point sources, it is given by $\mathrm{V} = ||L^+L_{\bullet i}||$, where $L_{\bullet i}$ is the ith column of the lead field matrix.

3.2 Results on the Theoretical Comparison and Analysis of Linear Distributed Solutions

In this section we start by describing some of the results obtained by authors who have carried out theoretical comparisons among different distributed solutions. The section finishes with a discussion that explains why some of these results contradict each other.

Pascual-Marqui (1995): Presents a comparison between the minimum norm, a lead field column weighted minimum norm and LORETA. Considers both electric and magnetic measurements (128 sensors). The basis for the comparison are the DLE and the level of blurring. There is no report about the error in the estimation of the dipole moment. The main conclusions reached by the author are:

- No basic differences are found between the compared solutions in terms of the level of blurring.
- LORETA leads in general to smaller DLE than minimum norm and weighted minimum norm. The DLE is for LORETA independent of source eccentricity.
- LORETA is able to accurately localize dipolar sources everywhere in the brain with a certain level of smearing. If the true current distribution is smooth, LORETA will retrieve it exactly.

Leahy et al. (1996): Presents a comparison among minimum norm, a weighted minimum norm, LORETA, a reweighted minimum norm and a Bayesian solution. Considers magnetic measurements (122 sensors). The basis for the comparison are maps obtained in the simulated reconstruction of multiple dipoles (also include the percent of residual error) and experimental phantom data. The main conclusions reached are:

- All linear methods produce a relatively large degree of smoothness.
- There is a wide variation in the characteristics of the solution obtained using different weighting functions.
- If the sources exhibit sparse focal characteristics, the Bayesian method of Phillips et al. (1997) is generally superior to those obtained using minimum norm methods.

Hauk et al. (1996): Presents a comparison between the Backus and Gilbert for scalar fields and the minimum norm solution in a simple model of a cortex fold. Consider two configurations with 37 and 148 magnetic sensors. The conclusions are:

- The resolution kernels of the minimum norm solution have more sidelobes but with narrower main peaks. It makes it difficult to decide which method should be preferred.
- Resolution significantly decreases with depth for both configurations and both inversion methods.
- The spread of the maxima, interpretable as a rough estimate of the resolving length, is at best in the range of centimeters.

Grave de Peralta et al. 1996: Presents a comparison among minimum norm and Backus-Gilbert solution. Considers only electric measurements (19 sensors placed according to 10/20 system). The basis for the comparison are figures of merit derived from the model resolution matrix: the bias in feature position, the bias in dipole localization and the dipole localization error. The main conclusions are:

- The original implementation of the Backus Gilbert method, intended to obtain the best achievable resolution from a given set of data, does not exhibit any appreciable differences over the classical minimum norm in terms of localizing single dipoles.
- In terms of the resolution kernels and the bias in dipole position the classical Backus and Gilbert method excels the classical minimum norm.

Yvert (1997): Evaluates the sensitivity of the minimum norm, weighted minimum norm and LORETA to the following factors: a) uncertainties in the electrode positions, b) variations of the volume conductor mode (realistic or spherical) c) Electrodes positioning. They consider electric measurements and use synthetic data associated with one or two dipoles. The basis of the comparison are the reconstructed maps of the currents. The main conclusions reached are:

- LORETA generally localizes one or two dipoles better than minimum norm or weighted minimum norm, even if the maxima may be displaced from their actual position by one or two centimeters.
- LORETA is sensitive to the symmetry and positioning of the electrodes.
- Uncertainties on the volume conductor model and the electrodes positioning provoke ghost maxima in the reconstructions for the cases of minimum norm and LORETA.

Grave de Peralta et al. (1997a): Includes a comparison among the minimum norm, the Backus-Gilbert, the Corrected Backus-Gilbert and the WROP method. Considers magnetic measurements (148 Magnetic sensors, BTI system). The basis for the comparisons are the 3D resolution kernels at different eccentricities. The main conclusions of this comparison are:

- The quality of the 3D resolution kernels for superficial sources (near the sensors) is quite adequate for all the considered solutions. The WROP method is optimal in the sense of minimizing the influence of farther away sources on the estimates at the superficial sites.
- The quality of the 3D resolution kernels decreases with the decrease of the source eccentricity for all the considered solutions. Below a certain eccentricity value none of the considered solutions is able to provide adequately centered 3D resolution kernels.

Grave de Peralta and Gonzalez (1998): Includes a comparison among an averaged solution, the minimum norm and LORETA. Considers electric measurements (41 sensors). The basis for the comparison are the impulse responses, the resolution kernels and two figures of merit: a) the source visibility and b) the source identifiability. The main conclusions are:

- All the analyzed solutions systematically underestimate the dipole moment or strength. The error committed in the estimation of the dipole strength depends for all the solutions on the source eccentricity, i.e. the error increases with the distance from the sources to the sensors.
- Some differences are found between the compared solutions in terms of their impulse responses, i.e. the considered solutions mainly differ in their capabilities to localize the position of single point sources. In this aspect the Averaged solution and LORETA are advantageous over the minimum norm.
- No differences were found among the solutions in terms of their resolution kernels, i.e. the three compared solutions behave similarly in the presence of simultaneously active sources.

The main reason for the contradictory results of these comparisons is the incorrect analysis of distributed linear solutions using either synthetic data associated to a few dipoles or measures designed to evaluate single or multiple dipole fitting algorithms with all their aforementioned drawbacks. Some theoretical aspects of linear inverse solutions, proved or stated in Grave de Peralta and Gonzalez (1997) might be helpful to shed light on what can be and cannot be expected from a linear solution. Worthy to mention are:

- No linear solution is able to accurately retrieve arbitrary source distributions.
- No linear solution, except for the minimum norm, is able to accurately retrieve even all the source distributions that fulfill the main property that this solution is designed to deal with, e.g., LORETA which searches for a smooth source configuration is not able to retrieve even all source configurations that fulfill this property.
- The use of specific a priori information mainly influence the impulse responses. The resolution kernels depend basically on the physical properties of the problem (the lead fields). Thus while the impulse responses might (to some extent) be manipulated at will, this is not possible with the resolution kernels.

We also have to be aware of some improper theoretical comparisons that have appeared in the literature (Riera et al. 1997a, Riera et al 1997b). These authors tried to compare the Backus and Gilbert, LORETA and the minimum norm solution. The results presented are invalid since in the former paper, they derived the Backus and Gilbert solution from a least square criterion and without the auxiliary unimodularity constraint which actually leads to the minimum norm solution (Tarantola 1987, Sect. 7.4; Grave de Peralta Menendez et al. 1997a). On the latter paper they misconstrued LORETA solution when considering the minimization of the laplacian without including the weighting function and which leads to a solution with features very similar to the minimum norm. On this basis it appears natural that they do not found any differences between the three methods compared.

We have so far identified six practical implications of these theoretical facts that manifest in the reconstruction provided by any linear solution in the presence of experimental or simulated data. These *main limitations of linear distributed solutions* are:

- Incorrect estimation of the position of the maximum of the modulus in the reconstructed maps.
- Incorrect estimation of the source strength, especially for sources farther away from the sensors.
- High level of spatial blurring.
- Existence of spurious sources even in the reconstruction of a single source and absence of noise.
- Deeper sources are frequently missed in the reconstruction when more external sources are simultaneously active.
- When, in computer simulations, we assumed as the source distribution a large number (near the number of electrodes) of randomly distributed dipoles of unitary moment we observed that each solution tends to reconstruct a fixed particular pattern independently of the position of the actual sources. So far we have found no theoretical explanation for this finding.

3.3 Experimentally Based Analysis of Solutions

The analysis of linear solutions on the basis of experimental data have been disdained by several theoretical researchers. However, when using the theoretical alternatives we are implicitly assuming that the selected source model is adequate and we evaluate where and to what extent the constraints used to guarantee a unique solution insure the adequate estimation of the sources. Still, the theoretical analysis does not allow to test the neurophysiological plausibility of some possible constraints or that of the selected source model. Thus the experimental analysis is a required step in the validation of any solution.

Earlier components of visual and auditory evoked responses have been already used to asses the consistency of particular linear inverse solutions (Pascual Marqui et al. 1995, Grave de Peralta Menendez et al. 1997b). However, these studies as well as those that use intracranial recordings (Lanz et al. 1997) are restricted to compare the map of the modulus obtained as the outcome of one particular inverse solution with either the neurophysiologically expectable localization or the intracranial recording. This procedure neither allows to asses whether or not different linear inverse solutions lead to the same results for the same data set nor to prove the advantages of one solution over the others. Furthermore, we have to be aware of several facts when interpreting these studies: (1) Intracranial recordings provide information about potential differences within the brain while the maps of the modulus obtained for the available inverse solutions reflect the estimated maximum of the current distribution. Although currents and potentials are physically related, the maxima for both magnitudes do not necessarily have to coincide. Section 5.1 exposes a solution which leads to potential maps within the brain, which facilitates a direct experimental verification. (2) Maps constructed on the basis of linear inverse solutions are highly blurred. Thus only a rough validation based on grossly defined regions of interest is possible. (3) Intracranial recordings generally cover limited brain areas which impedes a more complete evaluation of the solutions in terms of ghost sources, missed sources or excessive blurring.

Recently we carried out in our lab an evaluation of different linear inverse solutions: (a) Two Radially weighted minimum norm solutions (RW1, RW2), (b) the WROP method, (c) LORETA, using the data recorded from eighteen epileptic patients being evaluated for surgery. On these patients the position of the epileptic focus was known from independent information and all the patients were seizure free after resection of the presumed area. In agreement with our computer simulation results, no substantial differences in the localization of the focus were found over this set of patients for the compared solutions. Slight differences were found on the level where the solution reached the maximum. In contrast to the RW1 and WROP, LORETA and RW2 never showed a maximum at the lowest plane, where for some patients the focus was known to be localized from intracranial recordings and the neuropsychological evaluation. The RW2 is a weighted minimum norm designed to mimics LORETA solution. The behavior of these two solutions is completely similar, demonstrating that the same results can be obtained using a simple weighted minimum norm (with diagonal metric) or a more complex solution with a non diagonal metric. Even if the non diagonal metric can be written as a Kronecker product, the computational load considerably exceeds that of a diagonal metric.

4 Non-Linear Solutions

For the sake of brevity we will in this section only mention some of the non-linear solutions that have been applied to the solution of the neuroelectromagnetic inverse problem. This section does not pretend to be a review and leaves out interesting approaches which do not assume distributed source models like the MUSIC algorithm (Mosher et al. 1992) or it's improvements (Sekihara et al. 1997a, 1997b) among others. In brief we have the following solutions:

- Magnetic Field Tomography (Clarke et al. 1989; Ioannides et al. 1989): Produces a three-dimensional estimation of the primary current density using an iterative probabilistic approach.
- Hierarchical Minimum Norm Solution (Okada et al. 1992): Attempts to minimize the spatial blurring inherent to the minimum norm. It is an

heuristically based algorithm that uses constraints for the maximum current dipole moment.
- Electromagnetic Temporal Tomography (Grave de Peralta et al. 1993a, 1993b): A non linear solution is constructed by minimizing a functional that includes constraints on the spatio-temporal behavior of the sources as well as bounds to the maximum value of the source strength. The authors propose the use of Markov random fields to model the spatio-temporal constraints. More details about this approach are given in Sect. 5.2.
- FOCUSS (Goronidsky and Rao 1997): Use a recursive iterative weighted minimum norm algorithm which updates at each step the metric on the source space based on the solution obtained in the previous step.
- In Srebro (1996) an iterative procedure is proposed which at each step uses a regularized inverse which limits the amount of brain (size of the region) that need be searched in the next step.
- Bayesian approach: This approach has been considered among others by Phillips et al. (1997) where they use a Gibb's prior to reflect the sparse focal nature of the sources. Also Baillet and Garnero (1997) use a Bayesian approach to introduce anatomo-functional priors in the neuroelectromagnetic inverse problem. In this later approach some energy functionals that comprise spatial and temporal constraints are considered.
- L1 Norm minimization: This approach has been considered to increase the resolution and to diminish the blurring inherent to minimum norm solutions by Matsuura and Okabe (1995) and Wagner et. al. (1996).

5 New Trends

As described before, there are limits in the reconstruction of arbitrary current distributions that no inverse solution described so far seems to be able to surpass. The reason for that is that the measurements do not contain enough information about the sources. The only possible solution is then to include independent a priori information that can be incorporated in the estimation procedure in several different ways, examples are:

- Confining the solution points to specific brain regions, e.g. the gray matter, the cortex (Dale and Sereno 1993).
- Selecting specific basis functions to expand the unknown, (Gonzalez et al. 1989).
- Defining metrics that constrain the unknown or its derivatives (see Sect. 2.2).
- Selecting specific source models, e.g., Grave de Peralta et. al.(1997c).
- Using complementary information extracted from other functional brain images or related experiments.

Note that when using (9) we are implicitly assuming an a priori information about the sources since we are forcing all components of vector $\mathbf{j}$ to be close to zero. When the available a priori information concerns the whole unknown vector $\mathbf{j}$ it is possible to incorporate it in (6–8). However, existing complementary experimental techniques (e.g., intracranial recordings or functional brain images) are far from providing a complete information about a current density vector, except perhaps by its strength. Available information about the modulus of the vector can be incorporated in the metric for the solution space. Nonetheless, the relationship existing between the information provided by the fMRI, PET or SPECT and the electrical activity is not yet clear.

The next two sections discuss plausible a priori information that can be used in the solution of the neuroelectromagnetic inverse problem. While the first approach changes the source model, in the second one a group of constraints about the spatio-temporal properties of the unknown or its strength is considered. A natural third approach, the combination of both, remains to be evaluated

5.1 Constraining the Source Model for Electric Measurements: ELECTRA

One possibility to circumvent the lack of information is to change the biophysical model that underlies the statement of the inverse problem, i.e. searching for magnitudes which can be better determined from the available data, and which are neurophysiologically interpretable. This section describes an alternative to do so, which is applicable to the case of electric measurements and which has been termed ELECTRA (Grave de Peralta et al. 1997c).

According to Helmholtz theorem, any arbitrary current distribution $\mathbf{j}$ can be decomposed into an irrotational part $\mathbf{j}^i$ and a solenoidal part $\mathbf{j}^s$. Although, it is relatively well known that solenoidal currents cannot produce electrical measurements such constraint have neither been incorporated into the solution of the inverse problem nor used to reformulate the problem in more restrictive terms.

Considering electric measurements V and irrotational currents, i.e. $j^i(x) = \nabla_x \varphi(x)$, (1) can be written as:

$$V(x) = \int_R \nabla\varphi(y) \circ \nabla\psi(x,y)\,\mathrm{d}R = \int_R \nabla \circ j(y)\psi(x,y)\,\mathrm{d}R, \qquad (22)$$

where the vector lead field $L(x,y) = \nabla_y \psi(x,y)$, and ψ denotes the Green function.

The approach used in ELECTRA is based on the fact that if one assume irrotational currents as the generators, then (1) can be equivalently solved for three different physical magnitudes all consistent with the assumed source model: 1) The estimation of an irrotational current density vector $j^i = \nabla\varphi$

with the vector lead field $\nabla\psi$. 2) The estimation of a scalar field, the current source density (CSD), $\nabla \circ j^i(y) = I(y)$ with the scalar lead field ψ. 3) The estimation of a scalar field, the potential distribution in depth φ with a transformed scalar lead field $\nabla\psi(x,y) \circ \nabla$.

The use of this approach yields some advantages over the classical formulation in terms of the current density vector, namely: 1) The number of unknowns can be reduced which is equivalent to increase the number of independent measurements. 2) The constraints used to reformulate the problem are undeniable since they do not imply any hypothesis about brain function but are instead based on the character of the measurements. 3) Existing experimental evidence (Wikswo et al. 1979; Plonsey 1982) suggests that the proposed source model characterizes the type of currents that arise in excitable tissues. If the latter fact proves to be true for brain tissues then no additional information is added to the inverse problem by using magnetic measurements or a more general source model than the one proposed here. Images obtained using this method for the case of early and middle components of human visual evoked responses to checkerboard stimuli are presented in Grave de Peralta Menendez et al. (1997c). Such images show results consistent with available neurophysiological evidences as well as higher levels of spatial details than the images obtained for the modulus of the currents when using the same data.

5.2 Electromagnetic Temporal Tomography (EMTT)

On this section we want to present a non linear inverse solution that considered constraints on both the spatial and temporal behavior of the sources as presented in Grave de Peralta Menendez et al. (1993a, 1993b), and briefly described in Valdes et al. (1994). A solution is proposed trying to fulfill some expected properties in the ideal solution, namely:

- Source Resolution: The solution should be able to retrieve concentrated as well as distributed sources.
- Temporal Continuity: The current flow has to be *continuous and smooth in time*, i.e., the solution should satisfy physiological constraints regarding the temporal sequence of activation of generators. This means, solutions obtained for consecutive time instants should be related.
- Spatial Continuity: The current flow has to be *spatially continuous* with smooth changes of direction. Closely spaced points cannot have opposite flow directions.
- The solution should be concordant with all the available information. That means, all the additional information should be incorporated in the estimation procedure.
- The solution must be recovered for the whole volume conductor in a spatial representation similar to that provided by other tomographies.

- The current density strength should not exceed a certain physiologically defined value.
- The solution should be unique.

Then the solution is obtained minimizing the functional:

$$\min \; F(j) = R(j) + T(j) + S(j) + E(j), \tag{23}$$

Where $R(j)$ is the residual term, relating the data and the unknowns. $T(j)$ and $S(j)$ express the temporal and spatial continuity, respectively. The latter is defined as in Shimogawara and Higuchi (1992) in terms of the angle between the current in neighbor points. Note that this condition controls the direction of the flow and not the intensity of the current source density. Thus this is not a smooth constraint as imposed by differential operators. This point was misconstrued in Pascual Marqui and Michel (1994). The term $E(j)$ is selected to favor concentrated or distributed sources as well as to include a priori information about the strength of the currents.

Problem (23) is a highly non-linear problem that can be solved with the standard non linear optimization techniques. For more details about the terms and a possible minimization procedure see Grave de Peralta Menendez et al. (1993a). A very efficient way to consider the constraint on the strength is the method of projections over convex sets (Youla and Webb 1982; Sezan and Stark 1982; Simard and Mailloux 1990), that results in quite easy to implement routines. This method also allows the inclusion of pre-conditioners to speed out the convergence procedure.

A description of the results obtained with this solution is presented elsewhere. Here we would only like to mention that from the set of properties described above, the one that seems to be more difficult to fulfill is the source resolution. The only way to retrieve concentrated as well as distributed sources we have found so far is by changing the last term in (23) according to some a prior knowledge, e.g. for distributed sources we use the entropy for vectors or a weighted L2 norm while for concentrated sources we use the spatial autocorrelation function.

6 Conclusions

The mathematical problem stated in (2, 3), or its continuous version (1), is purely underdetermined and thus has infinite solutions. That means that if we have M unknowns and N independent measurements ($M \gg N$), the values at $M - N$ sites can be arbitrarily assigned. This implies that we can construct a solution for alpha rhythm data with activity at the frontal lobes or wherever we want. This uncertainty is precisely the "curse of the non uniqueness". As seen in this chapter the way of dealing with this curse is to include a priori information in order to obtain a unique solution. Still not every a priori information added to (2, 3) is reasonable or appropriate nor does it necessarily lead to a unique solution. For instance, although the constraint used in ELECTRA (irrotational sources) is completely reasonable from a physical point of view, it is not enough to define a unique solution. An example of unreasonable constraint is to assume solenoidal currents when we count only with electrical measurements.

What really matters is whether or not the a priori information incorporated reflects what is really happening in the brain. The best inverse solution on each particular case is the one associated with the a priori information that resembles most the actual brain functioning. Thus, the ultimate aspect in the evaluation of an inverse solution is the assessment of the validity of the a priori information used. This leads to the conclusion that the development of tools to evaluate the a priori information both theoretically (e.g., using resolution kernels) or experimentally (e.g., intracranial recordings) constitutes an obligatory requisite at the present stage of development.

These facts explain why there is no definite answer to the question of what is better: a dipolar model or a distributed model. Spatio-temporal source models are the result of combining (1) with the constraint that only a few sources with unknown location are active; i.e., the unknown function in (1) is a sum of a few delta functions. As previously described the combination with others constraints will lead to others solutions. Similarly, there is no sense in comparing two different distributed solutions using a fixed set of sources, e.g. dipoles, unless we are sure that these are the sources that actually exist, a topic beyond the scope of this paper.

Even when we should not expect miracles in the solution of this problem, the developments in the last years in the field of distributed solutions are encouraging (see Sects. 2.2, 2.3, 4, and 5 and references therein). Further combination of many of these solutions with physical, anatomical, functional or electrophysiological constraints remains to be explored.

References

Allasia, G. (1995): A class of interpolating positive linear operators: Theoretical and Computational aspects. In: Singh, S.P., Carbone, A. and Watson, B. (eds.): *Approximation theory, Wavelets and Applications.* (Kluwer Academy Publishers, Dordrecht/Boston/London).

Achim, A., Richer, F. and Saint-Hilaire, J. (1991): Methodological considerations for the evaluation of spatio-temporal source models. Electroenceph. clin. Neurophysiol., 1991, 79:227–240.

Backus, G.E. and Gilbert, J.F. (1967): Numerical applications of a formalism for geophysical inverse problems. Geophys. J. Roy. Astron. Soc. 13:247–276.

Backus, G.E. and Gilbert, J.F. (1968): The resolving power of gross earth data. Geophys. J. Roy. Astron. Soc. 16:169–205.

Backus, G.E. and Gilbert, J.F. (1970): Uniqueness in the inversion of gross earth data. Phil. Trans. R. Soc. 266:123–192.

Baillet, S. and Garnero, L. (1997): A bayesian approach to introducing anatomo-functional priors in the EEG/MEG inverse problem. IEEE Trans. Biomed. Eng., 44: 374–385.

Ben-Israel, A., and Greville T.N.E. (1974):*Generalized inverses: Theory and applications.*(John Wiley and Sons, Inc. New York).

Bertero, M., De Mol, C. and Pike, E.R. (1985): Linear inverse problems with discrete data. I: General formulation and singular system analysis. Inverse Problem, 1:301–330.

Bertero, M., De Mol, C., and Pike, E.R. (1988) Linear inverse problems with discrete data. II. Stability and regularization. Inverse Problem 4: 573–594.

Cabrera Fernandez, D., Grave de Peralta, R. and Gonzalez Andino, S. (1995). Some limitations of spatio temporal source models. Brain Topography 7 (3):233–243.

Clarke, C.J.S., Ioannides, A.A. and Bolton, J.P.R. (1989): Localized and distributed source solutions for the biomagnetic inverse problem I. In: Williamson SJ, Hoke M, Stroink G and Kotani M (eds): Advances in Biomagnetism. New York: Plenum Press, pp. 587–590.

Dale, A.M. and Sereno, M.I. (1993): Improved localization of cortical activity by combining EEG and MEG with MRI cortical surface reconstruction: A linear approach. J. Cogn. Neurosci. 5, 162:176.

Fuchs, M., Wischmann, H.A. and Wagner, M. (1994): Generalized minimum norm least squares reconstruction algorithms. In: ISBET Newsletter No. 5, November 1994. Ed: W. Skrandies. 8–11.

Gonzalez Andino, S.L., Grave de Peralta Menenedez, R., Biscay Lirio, R., Jimenez Sobrino, J.C., Pascual Marqui, R.D., Lemagne, J. and Valdes Sosa, P.A.,. (1989): Projective methods for the magnetic direct problem, in: *Advances in Biomagnetism*, edited by S.J. Williamson, M.Hoke, G.Stroink and M. Kotani. Plenum Press, New York, pp. 615–618.

Goronidnitsky, I.F. and Rao, B.D. (1997): Spatial signal reconstruction from limited data using FOCUSS: a re-weighted minimum norm algorithm. IEEE Trans. Sig. Proc., 45 (3): 1–16.

Grave de Peralta Menendez, R. and Gonzalez Andino, S.L. (1994): Single Dipole Localization: Some numerical aspects and a practical rejection criterion for the fitted parameters. Brain Topography 6 (4):277–282.

Grave de Peralta Menendez, R. and Gonzalez Andino S.L. (1998): A critical analysis of linear inverse solutions. IEEE Trans. Biomed. Engn. 45:440–448

Grave de Peralta Menendez, R., Gonzalez Andino, S.L., Cabrera Fernandez, D., Torrez, L. (1993b): A proposal for the inverse solution combining different sources of information. IV International Symposium of the ISBET Society. Havana, Cuba. Abstract published in Brain Topography.

Grave de Peralta Menendez, R., Oropesa Farras, E., Gonzalez Andino, S.L., Cabrera Fernandez, D., Aubert Vazquez, E. (1993a): An approach to phisiollogically meaningful distributed inverse solutions to the neuroelectromagnetic inverse problem. Technical Report 1. July, 1993. Cuban Neuroscience Center, Havana, Cuba.

Grave de Peralta Menendez, R., Gonzalez Andino, S. and Lütkenhöner, B. (1996): Figures of merit to compare linear distributed inverse solutions. Brain Topography. Vol. 9. No. 2:117–124.

Grave de Peralta Menendez, R., Hauk, O., Gonzalez Andino, S., Vogt, H. and Michel, C.M. (1997a) Linear inverse solutions with optimal resolution kernels applied to the electromagnetic tomography. Human Brain Mapping, 5:454–467

Grave de Peralta Menendez, R., Gonzalez Andino, S., Hauk, O., Spinelli, L., Michel, C.M. (1997b): A linear inverse solution with optimal resolution properties: WROP. Biomedizinische Technik, 42:53–56

Grave de Peralta Menendez, R., Gonzalez Andino, S.L., Morand, S., Michel, C.M., (1997c): Imaging the electrical activity of the brain: ELECTRA. submitted.

Greenblatt, R.E. (1993): Probabilistic reconstruction of multiple sources in the neuroelectromagnetic inverse problem. Inverse Problems 9: 271–284.

Hauk, O., Grave de Peralta Menendez, R. and Lütkenhöner, B. (1996): The Backus and Gilbert method and the minimum norm method applied to a simple model of a cortex fold. Proceedings of the Third International Hans Berger Congress, Jena, Germany (to appear).

Hämäläinen, M.S. and Ilmoniemi, R.J. (1984): Interpreting measured magnetic fields of the brain: Estimates of current distributions. Technical Report TKK-F-A559, Helsinski University of Technology.

Hämäläinen, M.S., Hari, R., Ilmoniemi, R.J., Knuutila, J. and Lounasma, O.V. (1993): Magnetoencephalography – theory, instrumentation, and applications to non invasive studies of the working human brain. Rev. Mod. Phys. 65: 413–497. I

Ioannides, A.A., Bolton, J.P.R., Hasson, R. and Clarke C.J.S. (1989): Localised and distributed source solutions for the biomagnetic inverse problem II. In: Williamson SJ, Hoke M, Stroink G and Kotani M (eds): *Advances in Biomagnetism.* New York: Plenum Press, pp. 587–590.

Knösche, T. (1997): Three-dimensional reconstruction from EEG, the performance of linear algoritmhs. Biomedizinische Technik, 42:205–208.

Lanz, G., Michel, C.M., Pascual Marqui, R.D., Spinelli, L., Seeck, M., Seri, S., Landis, T. and Rosem, I. (1997) Extracranial localization of intracranial interictal epileptiform activity using LORETA (low resolution electromagnetic tomography). Electroenc. Clin. Neurophysiol., 102:414–422.

Lawson C.L. and Hanson, R.J. (1974): *Solving least squares problems.* (Prentice Hall, Inc., Englewood Cliffs, New Jersey).

Leahy, R., Mosher, J.C., and Phillips, J.W. (1996): A comparative study of minimum norm inverse methods for MEG imaging. Proceedings of the tenth international conference on Biomagnetism, Biomag '96. Santa Fe, New Mexico

Linz, P. (1979): *Theoretical numerical analysis. An introduction to advanced techniques.* (John Wiley & Sons, Inc., New York).

Matsuura, K. and Okabe, Y. (1995): Selective minimum norm solution of the biomagnetic inverse problem. IEEE Trans. Biomed. Engn., 42: 608–615.

Menke, W. (1989): *Geophysical Data Analysis: Discrete inverse theory.* Academic Press, San Diego, California.

Mosher, J.C., Lewis, P.S. and Leahy, L. (1992): Multiple dipole modeling and localization from spatio temporal MEG data. IEEE Trans. Biomed. Eng., 39: 541–557.

Okada, Y., Huang, J., and Xu, C. (1992): A hierarchical minimum norm estimation method for reconstructing current densities in the brain from remotely measured magnetic fields. in: *Biomagnetism: Clinical Aspects*, edited by Hoke, M., Erne, S., Okada, Y. and Romani, G.L. Elsevier. Amsterdam, pp. 729–734.

Pascual Marqui, R.D. and Michel, C.M. (1994): Rejoinder. In: ISBET Newsletter No. 5, November 1994. Ed: W. Skrandies. 21–25.

Pascual Marqui, R.D. (1995): Reply to comments by Hämäläinen, Ilmoniemi and Nunez. In: ISBET Newsletter No. 6, December 1995. Ed: W. Skrandies. 16–28.

Pascual Marqui, R.D., Michel, C.M. and Lehmann D. (1995): Low resolution electromagnetic tomography: a new method for localizing electrical activity in the brain. Int. J. Psychophysiol. 18: 49–65.

Penrose R. (1955): A generalized inverse for matrices. Proc. Cambridge Philos. Soc. 51: 406–413.

Phillips, J.W., Leahy, R., and Mosher, J.C. (1997): MEG-based imaging of focal neuronal sources, IEEE Trans. on Medical Imaging 16:338–348

Plonsey, R. (1982). The nature of the sources of bioelectric and biomagnetic fields. Biophys. J. 39. pp. 309–312.

Rao, C.R. and Mitra, S.K. (1971): Generalized inverse of matrices and its applications. (John Wiley & Sons, Inc., New York).

Riera, J.J., Valdes, P.A., Fuentes, M.A. and Oharriz, Y. (1997a): Explicit Backus and Gilbert EEG Inverse Solutions for a Spherical Symmetry. Biomedizinische Technik, 42:216–219

Riera, J.J., Fuentes, M.A., Valdes, P.A.and Oharriz, Y. (1997b): Theoretical basis of the EEG Spline Inverse Solutions for a spherical Head Model. Biomedizinische Technik, 42:219–223

Roach, G.F. (1970): *Green functions. Introductory theory with applications.* Van Nostrand Reinhold Co.

Sarvas, J. (1987): Basic mathematical and electromagnetic concepts of the bioelectromagnetic inverse problem. Phys. Med. Biol. 32: 11–22.

Scherg, M., von Cramon, D. (1990): Dipole source potentials of the auditory cortex in normal subjects and patients with temporal lobe lesions. In: Grandori, F., Hoke, M., Romani, G.L., eds. *Auditory evoked magnetic fields and electric potentials.* (Karger, Basel)., 165–193.

Sekihara, K. and Scholz, B. (1995): Average-intensity reconstruction and Wiener reconstruction of bioelectric current distribution based on its estimated covariance matrix. IEEE Trans. Biomed. Eng. 42: 149–157.

Sekihara, K. and Scholz, B. (1996): Generalized wiener estimation of three dimensional current distribution from biomagnetic measurements. IEEE Trans. Biomed. Eng. 43: 281–291.

Sekihara, K., Poeppel, D., Marantz, A., Koizumi, H., Miyashita, Y. (1997a): Noise covariance incorporated MEG-Music algorithms: A method for multiple dipole estimation tolerant of the influence of Background Brain activity. IEEE Trans. Biomed. Eng. Accepted for publication.

Sekihara, K., Poeppel, D., Marantz, A., Philips, C., Koizumi, H., Miyashita, Y. (1997b): MEG Covariance difference analysis: A method to extract target source activities by using task and control measurements. Preprint.

Sezan, M.I., Stark, H. (1982): Image restoration by the method of convex projections: Part II-Application and Numerical results. IEEE Trans. Med. Imaging, vol M1-1: 95–101.

Shimogawara, M.K. and Higuchi, H.M. (1992): Magnetic source imaging by current element distribution, in Biomagnetism: Clinical Aspects, edited by Hoke, M., Erne, S., Okada, Y. and Romani, G.L. Elsevier. Amsterdam, pp. 757–760.

Simard, P.Y. and Mailloux, G.E. (1990): Vector field restoration by the method of convex projection. Computer Vision., Graphics and Image Processing, 52: 360–385.

Spinelli, L., Pascual Marqui, R., Grave de Peralta Menendez, R., and Michel, C.M. (1997): Effect of the number and the configuration of electrodes on distributed source models. VIII World Congress of the ISBET. Zurich, Switzerland.

Srebro, R. (1996): An iterative approach to the solution of the inverse problem. Electroenceph. clin. Neurophysiol., 98: 349–362.

Tarantola, A. (1987): *Inverse Problem Theory.* Elsevier.

Valdes, P.A., Grave de Peralta Menendez, R. and Gonzalez Andino, S.L., (1994): Comments on LORETA. In: ISBET Newsletter No. 5, November 1994. Ed: W. Skrandies. 18–21.

Wahba, G. (1990): *Spline models for observational data.* Society for Industrial and applied mathematics. Philadelphia. Pennsylvania.

Wagner, M. Fuchs, H.-A. Wischmann, R. Drenckhahn, Th. Köhler (1996): Current Density Reconstructions Using the L1 Norm. Proc. of the 10th Int. Conf. of Biomagnetism; BIOMAG 96, Santa Fe, 16.-21.2.1996, to be published (Abstracts, pp. 168).

Wikswo J.P., Jr., Malmivuo, J.A.V., Barry, W.H., Leifer, M.C. and Fairbank W.M. (1979): The theory and application of magnetocardiography. Adv. Cardiovasc. Phys, 2:1–67.

Youla, D.C. and Webb, H. (1982): Image restoration by the method of convex projections: Part I-Theory. IEEE Trans. Med. Imaging, vol M1-1: 81–94.

Ivert B. (1996): Modélisation réaliste de l'activité électrique cérébrale. Ph.D. thesis.

The Spatial Distribution of Spontaneous EEG and MEG

Jan C. De Munck, and Bob W. Van Dijk

MEG Center KNAW, AZVU Rec. C, De Boelelaan 1118, 1081 HV Amsterdam, The Netherlands, email: JC.MUNCK@AZVU.NL

1 Introduction

A subject sitting at rest with electrodes attached to his or her head produces EEG-signals, even when he or she is not performing a specific task. These signals are irregular and it is impossible to predict how the signal will evolve in time with only the knowledge of their behavior in the past. This impossibility does not imply that the signals are completely random, because often characteristic rhythms can be observed when the condition of the subject is changed. A well known example is the appearance of the alpha-rhythm when the subject closes his eyes (e.g. Lopes da Silva, 1987). The precise shape of these EEG signals can not be predicted and the exact knowledge is not very useful. A meaningful description of spontaneous EEG signals can only be given in terms of their statistical properties. The same is true for the behavior of the EEG signals in the space domain. EEG maps can vary from very regular and dipolar patters to very irregular patterns that do not easily reveal where the underlying generators are located. To characterize the maps of spontaneous EEG one has to use statistics.

It is important to study the spatial characteristics of EEG and MEG (E/MEG) data for several reasons. Many researchers are interested in the correlation or coherence of signals from different parts of the brain (e.g. Gevins et al., 1989). When E/MEG is used to record these correlations, it should be taken into account that part of the correlation, measured with different sensors, is caused by the fact that electrical and magnetic fields spread out from a point source. With EEG the spread of activity is worse than with MEG, due to the smearing effect of the low conducting skull. The point spread functions of electric and magnetic fields are responsible for the fact that different E/MEG sensor signals are correlated, even if the underlying sources are completely independent.

Another reason to study the spatial characteristics of the E/MEG is that it may give information on the origin of the underlying generators. The dipole fitting techniques applied on evoked E/MEG cannot generally be used with spontaneous E/MEG. The models used with evoked E/MEG data are based on the assumption that the data is generated by a limited number of deterministic sources (e.g. Scherg and Von Cramon, 1985; De Munck, 1990). Spontaneous E/MEG is generated with the same physical mechanism as evoked E/MEG, but generally there are far more independent sources active. It seems much more realistic to describe the generators of the spontaneous E/MEG using a statistical model, with a statistical distribution of the sources.

A more practical reason to study the spatial characteristics of the E/MEG is to improve source localization techniques for evoked brain activity. In most of the methods presented until now a least squares fit is performed of a model to the data. The rational of using a least squares fit is that it computes the most likely distribution of the noise, assuming it is Gaussian and uncorrelated. Since for both EEG and MEG the most important source of noise is the background activity (it is also the only source of noise that cannot be eliminated experimentally) the use of the least squares fit becomes questionable, when it appears that the background noise is not uncorrelated.

The purpose of this section is to present a statistical description of the spatial distribution of the spontaneous E/MEG. First, some basic understandings will be reviewed which clarify the relation between spatial and temporal characteristics of the E/MEG. This short introduction in statistics is meant to elucidate the derivation and meaning of a mathematical model, that gives an interpretation of spontaneous E/MEG data. Part of this model was published previousley in De Munck (1989) and De Munck et al. (1992).

2 Basic Understandings in Statistics

A stochastic signal is an ordered set of random variables. Therefore, a basic treatment of a stochastic signal should start with a definition of a random variable. One of the key concepts of probability theory is the *probability density function* (pdf) $f_\nu(x)$, corresponding to the random variable $\underline{\nu}$. The pdf is defined in such a way that the probability $P(\underline{\nu} < \nu_0)$ that $\underline{\nu} < \nu_0$ equals

$$P(\underline{\nu} < \nu_0) \equiv \int_{-\infty}^{\nu_0} f_\nu(x)\mathrm{d}x. \tag{1}$$

By adding and subtracting integrals like (1) the probability that $\nu \in V$ can be computed for any (well chosen) subset V of the real numbers $\mathbb{R}$. In this chapter random variables, i.e. variables which have a pdf, are indicated as underlined symbols. The pdf itself will be denoted with the symbol f and it has the random variable as lower index. So, for example, $f_a(x)$ is the pdf of the random variable $\underline{a}$ and x is a dummy variable.

Very often the complete shape of the pdf is not available and one uses the *mean* $E\{\underline{\nu}\}$ and the *variance* $E\{\underline{\nu}^2\}$ to characterize $f_\nu(\nu)$:

$$\begin{aligned} E\{\underline{\nu}\} &\equiv \textstyle\int_{-\infty}^{\infty} \nu f_\nu(\nu)\mathrm{d}\nu \\ E\{\underline{\nu}^2\} &\equiv \textstyle\int_{-\infty}^{\infty} \nu^2 f_\nu(\nu)\mathrm{d}\nu. \end{aligned} \tag{2}$$

The mean of a random variable is the average value that will be assumed when it is drawn infinitely many times from a distribution with pdf $f_\nu(\nu)$. The variance can be used to determine the *standard deviation* σ_ν, which is the average deviation from the mean,

$$\sigma_\nu \equiv \sqrt{(E\{\underline{\nu}\})^2 - E\{\underline{\nu}^2\}} \tag{3}$$

These concepts can easily be extended to a vector $\underline{\boldsymbol{\nu}} = (\underline{\nu}_1, \underline{\nu}_2, \ldots, \underline{\nu}_N)^T$. The pdf is now a multivariate function $f_\nu(\boldsymbol{\nu})$. If one is only interested in the behavior of the first component of $\boldsymbol{\nu}$ one can consider $f_{\nu_1}(\nu_1)$, which is the pdf of the first component:

$$f_{\nu_1}(\nu_1) \equiv \int_{-\infty}^{\infty}\int_{-\infty}^{\infty}\cdots\int_{-\infty}^{\infty} f_{\boldsymbol{\nu}}(\nu_1, \nu_2, \ldots, \nu_N)\mathrm{d}\nu_2\mathrm{d}\nu_3\ldots\mathrm{d}\nu_N. \tag{4}$$

The mean of a multivariate variable is defined per component:

$$\begin{aligned} E\{\underline{\boldsymbol{\nu}}\} &\equiv \int_{\mathrm{IR}^N} \boldsymbol{\nu} f_\nu(\boldsymbol{\nu})\mathrm{d}^N\boldsymbol{\nu} \\ &= \int_{-\infty}^{\infty}\int_{-\infty}^{\infty}\cdots\int_{-\infty}^{\infty} \begin{pmatrix} \nu_1 \\ \nu_2 \\ \vdots \\ \nu_N \end{pmatrix} f_{\boldsymbol{\nu}}(\nu_1, \nu_2, \ldots, \nu_N)\mathrm{d}\nu_1\mathrm{d}\nu_2\ldots\mathrm{d}\nu_N \\ &= \begin{pmatrix} \int_{-\infty}^{\infty} \nu_1 f_{\nu_1}(\nu_1)d\nu_1 \\ \vdots \\ \int_{-\infty}^{\infty} \nu_N f_{\nu_N}(\nu_N)d\nu_N \end{pmatrix}. \end{aligned} \tag{5}$$

The extension of the variance to multiple dimensions is the *variance-covariance* matrix $E\{\underline{\nu}_i\underline{\nu}_j\}$:

$$E\{\underline{\nu}_i\underline{\nu}_j\} \equiv \int_{\mathrm{IR}} \nu_i\nu_j f_\nu(\boldsymbol{\nu})\mathrm{d}^N\boldsymbol{\nu} = \int_{-\infty}^{\infty}\int_{-\infty}^{\infty} \nu_i\nu_j f_{\nu_i\nu_j}(\nu_i,\nu_j)\mathrm{d}\nu_i\mathrm{d}\nu_j, \tag{6}$$

with

$$f_{\nu_i\nu_j}(\nu_i,\nu_j) = \int_{-\infty}^{\infty}\int_{-\infty}^{\infty}\cdots\int_{-\infty}^{\infty} f_{\boldsymbol{\nu}}(\nu_1, \nu_2, \ldots, \nu_N)\mathrm{d}\nu_1\ldots\mathrm{d}\hat{\nu}_i\ldots\mathrm{d}\hat{\nu}_j\ldots\mathrm{d}\nu_N. \tag{7}$$

Here, the symbols $\mathrm{d}\hat{\nu}_i$ and $\mathrm{d}\hat{\nu}_j$ indicate that these variables are omitted. If $i = j$, only one integration variable is omitted. So, the distributions $f_{\nu_i\nu_j}(\nu_i,\nu_j)$ are derived from the general distribution $f_\nu(\boldsymbol{\nu})$ in a similar way as $f_{\nu_i}(\nu_i)$ in (4). If one is only interested in the statistics of combinations of two components of $\underline{\boldsymbol{\nu}}$ the distributions $f_{\nu_i\nu_j}(\nu_i,\nu_j)$ are sufficient. Since $E\{\underline{\nu}_i\underline{\nu}_j\} = E\{\underline{\nu}_j\underline{\nu}_i\}$ the covariance matrix is symmetric. The magnitude of the covariance $E\{\underline{\nu}_i\underline{\nu}_j\}$ indicates the tendency of the components $\underline{\nu}_i$ and $\underline{\nu}_j$ to assume the same value. This tendency can better be expressed in their *correlation coefficient* c_{ij}:

$$c_{ij} = \frac{E\{\underline{\nu}_i \underline{\nu}_j\} - E\{\underline{\nu}_i\}E\{\underline{\nu}_j\}}{\sigma(\underline{\nu}_i)\sigma(\underline{\nu}_j)}, \tag{8}$$

because the effect of the means of the variables is eliminated. Furthermore, the correlation coefficient is a normalized quantity and assumes values between -1 and 1. The definitions (5) and (6) can be generalized to higher order moments, or to more general functions of the random variable. If $\mathbf{g}(\underline{\boldsymbol{\nu}})$ is a function of the random variable $\underline{\boldsymbol{\nu}}$ then the expected value of $\mathbf{g}(\underline{\boldsymbol{\nu}})$ is

$$E\{\mathbf{g}(\underline{\boldsymbol{\nu}})\} = \int_{\mathbb{R}^N} \mathbf{g}(\boldsymbol{\nu}) f_\nu(\boldsymbol{\nu}) \mathrm{d}^N \boldsymbol{\nu} \tag{9}$$

Equation (9) will be used to express the statistics of the sources into the statistics of the E/MEG measurements in the derivation of our model.

A very important concept in statistics is *statistical independence.* Two variables $\underline{\nu}_1$ and $\underline{\nu}_2$ are called statistically independent if $f_{\nu_1\nu_2}(\nu_1, \nu_2) = f_{\nu_1}(\nu_1) f_{\nu_2}(\nu_2)$. The physical interpretation of statistical independence is that the realization of the variable $\underline{\nu}_1$ does not give any information about the realization of variable $\underline{\nu}_2$, and vice versa. If two variables are statistically independent, they are also uncorrelated as follows from (8). The opposite is not necessarily true, however. For instance, if $f_{\nu_1\nu_2}(\nu_1, \nu_2) = 1/2$ for $|\nu_1|+|\nu_2| \leq 1$ and $f_{\nu_1\nu_2}(\nu_1, \nu_2) = 0$ for $|\nu_1| + |\nu_2| > 1$, then $f_{\nu_1\nu_2}(\nu_1, \nu_2) \neq f_{\nu_1}(\nu_1) f_{\nu_2}(\nu 2)$ (ν_1 and ν_2 are dependent), $E\{\nu_1\nu_2\} = E\{\nu_1\} = E\{\nu_2\} = 0$ (but they are uncorrelated).

If the mean and the variance of a random variable are to be estimated from measured data, one needs to make statistical assumptions about its generation. One typical assumption is that the data can be considered as statistical independent realizations of a random variable. Based on these assumptions, one can derive *estimators* (like the arithmetic average of the data), which approximate the mean $E\{\underline{\nu}\}$ as good as possible. The quality of this approximation is expressed in statistical terms, i.e. the probability is given that the statistics deviates more than a certain error threshold from the true value. Therefore, optimal estimators of the mean, the variance and other statistical properties, always depend on the underlying statistical assumptions made. A major unsatisfying aspect of using statistical descriptions of a data set is that the estimators depend on assumptions that cannot be verified experimentally without making new subjective assumptions.

A stochastic signal $\underline{\nu}(t)$ is more involved than a multivariate variable. Physically, it means that on each moment in time t, the value of the signal is the result of the realization of a random variable. Therefore, the pdf $f_\nu(\nu; t)$ of a stochastic signal depends both on ν and on t. The pdf $f_{\nu_1\nu_2}(\nu_1, \nu_2; t_1, t_2)$ of two time samples of $\underline{\nu}(t)$, taken at t_1 and t_2 is a four parameter function. Since for any set of time instances there is a multivariate stochastic variable, a stochastic signal can mathematically be considered as an extension of a multivariate random variable to infinitely many dimensions, Papoulis (1984). If the pdfs f_ν and $f_{\nu_1\nu_2}$ are given, the mean and the covariance can be computed as

$$\begin{aligned} E\{\underline{\nu}(t)\} &\equiv \int_{-\infty}^{\infty} \nu f_\nu(\nu;t)\mathrm{d}\nu \\ E\{\underline{\nu}(t_1)\underline{\nu}(t_2)\} &\equiv \int_{-\infty}^{\infty}\int_{-\infty}^{\infty} \nu_1\nu_2 f_{\nu_1\nu_2}(\nu_1,\nu_2;t_1,t_2)\mathrm{d}\nu_1\mathrm{d}\nu_2. \end{aligned} \tag{10}$$

Higher order moments of a stochastic signal depend on more than two parameters t. Therefore, the complete characterization of a stochastic signal is extremely complex. Even if only the first and second order moments are considered, the situation is involved. If one desires to estimate these moments, one should have a sufficient number of independent realizations of the signal. For that purpose, one has to assume that the signal at one moment is statistically independent from the signal at a much later or earlier moment. Alternatively, one could assume that the signal taken from one site is an independent realization from that taken at another site, so that the average of these signals could be used as an estimate of $E\{\nu(t)\}$. However, this would be a very unrealistic assumption in the case of E/MEG signals because we know that signals at different sites are correlated due to the point spread function of electromagnetic fields.

An important assumption which is often made is that a stochastic signal is *stationary*. This means that the statistical properties of the signal only depend on the time difference between the moments that the samples are taken, and not on the moments itself. In particular, the expectation value $E\{\underline{\nu}(t)\}$ of a stationary signal is independent of time. The covariance of the signal at t_1 and t_2 is only a function one parameter $\tau = t_1 - t_2$:

$$E\{\underline{\nu}(t_1)\underline{\nu}(t_2)\} = \int_{-\infty}^{\infty}\int_{-\infty}^{\infty} \nu_1\nu_2 f_{\nu_1\nu_2}(\nu_1,\nu_2;t_1-t_2)\mathrm{d}\nu_1\mathrm{d}\nu_2 \equiv \mathrm{Cov}_\nu(\tau). \tag{11}$$

The function $\mathrm{Cov}_\nu(\tau)$ indicates the tendency of two samples, separated a distance τ on the time axis, to assume the same value. Therefore, $\mathrm{Cov}_\nu(\tau)$ is typically a decreasing function of τ which tends to zero for $\tau \to \infty$. If the underlying generators of the signal work at specific frequencies, it is useful to express this tendency in the frequency domain. Another reason to represent the covariance function in the frequency domain instead of in the time domain is to study the behavior of the signal, when it is passed through a linear analogous filter. The transformation of $\mathrm{Cov}_\nu(\tau)$ into the frequency domain is called the *power spectrum* $S_\nu(\omega)$:

$$S_\nu(\omega) \equiv \frac{1}{2\pi}\int_{-\infty}^{\infty} \mathrm{Cov}_\nu(\tau)\mathrm{e}^{i\omega\tau}\mathrm{d}\tau \tag{12}$$

Mathematically, the definition and the existence of the power spectrum is much more involved than presented here (e.g. Kedem, 1994), but (12) suits our purposes.

If the signal $\nu(t)$ is passed through a linear time invariant filter with frequency characteristic $H(\omega)$, one can express the mean of the output signal $\underline{\nu}'(t)$ as

$$E\{\underline{\nu}'(t)\} \equiv \hat{H}(\omega)\int_{-\infty}^{\infty} \nu f_\nu(\nu;t)\mathrm{d}\nu = \int_{-\infty}^{\infty} \nu f'_\nu(\nu;t)\mathrm{d}\nu, \tag{13}$$

where the pdf of the output signal $f'_\nu(\nu;t) = \hat{H}(\omega)f_\nu(\nu;t)$. Therefore, if the E/MEG-signals are stationary and if they are passed through the high-pass filter, the mean of the filtered signals will vanish. The effect of the filter on the power spectrum can be expressed as follows:

$$S'_\nu(\omega) = H(\omega)H^*(\omega)S_\nu(\omega), \tag{14}$$

where $H^*(\omega)$ is the complex conjugation of $H(\omega)$. So, the frequency characteristic acts quadratically on the spectrum.

If we want to describe the spatial characteristics of the E/MEG we need to consider the relationship between two stochastic signals $\underline{\nu}_1(t)$ and $\underline{\nu}_2(t)$, registrated with sensors placed at different positions. This relationship can be characterized in the time domain by the *cross covariance*

$$\mathrm{Cov}_{\nu_1\nu_2}(t_1,t_2) \equiv \int_{-\infty}^{\infty}\int_{-\infty}^{\infty} \nu_1\nu_2 f_{\nu_1\nu_2}(\nu_1,\nu_2;t_1,t_2)\mathrm{d}\nu_1\mathrm{d}\nu_2, \tag{15}$$

or, assuming $\nu_1(t)$ and $\nu_2(t)$ are stationary, in the frequency domain by the *cross spectrum*:

$$S_{\nu_1\nu_2}(\omega) \equiv \frac{1}{2\pi}\int_{-\infty}^{\infty} \mathrm{Cov}_{\nu_1\nu_2}(\tau)\mathrm{e}^{i\omega\tau}\mathrm{d}\tau \tag{16}$$

3 Spatial Distribution of Noise

In the previous section, stochastic signals in the time domain were taken to illustrate some definitions, but the same can be done in the space domain. If $\underline{V}(\mathbf{x})$ is a spatial random variable taken at the position $\mathbf{x} \in \mathrm{I\!R}^3$ then we can define the mean and variance of $\underline{V}$ analogously to (10)

$$\begin{aligned} E\{\underline{V}(\mathbf{x})\} &\equiv \textstyle\int_{-\infty}^{\infty} V f_V(V;\mathbf{x})\mathrm{d}V \\ E\{\underline{V}(\mathbf{x}_1)\underline{V}(\mathbf{x}_2)\} &\equiv \textstyle\int_{-\infty}^{\infty}\int_{-\infty}^{\infty} V_1V_2 f_{V_1V_2}(V_1,V_2;\mathbf{x}_1,\mathbf{x}_2)\mathrm{d}V_1\mathrm{d}V_2. \end{aligned} \tag{17}$$

The concept of stationarity in the time domain can be generalized to stationarity in the space domain. Stationarity in the time domain implies that the statistics are invariant for shifts along the time axis. Another way of expressing this is to say that the statistics have *translation symmetry*. In the case of E/MEG, we are dealing with an almost spherical head and a sensor grid which fits on a spherical shell. Therefore, it seems logical to define spatial stationarity using *spherical symmetry*. A spatial random variable is called *spatially stationary*, if the statistics depend only on the angles $\beta_{ij} = \arccos\left(\frac{\mathbf{x}_i \cdot \mathbf{x}_j}{|\mathbf{x}_i||\mathbf{x}_j|}\right)$ between the sample points $\mathbf{x}_i$, viewed from the origin. In particular, the mean of a spatially stationary signal is constant, and the covariance is a function of the angle β_{12} between the two sample points $\mathbf{x}_1$ and $\mathbf{x}_2$, viewed from the origin (which is assumed to be the symmetry point):

$$E\{\underline{V}(\mathbf{x}_1)\underline{V}(\mathbf{x}_2)\} = \int_{-\infty}^{\infty}\int_{-\infty}^{\infty} V_1 V_2 f_{V_1 V_2}(V_1, V_2; \beta) \mathrm{d}V_1 \mathrm{d}V_2 \equiv \mathrm{Cov}_V(\beta_{12}). \tag{18}$$

Similar to the Fourier transform of the time covariance function, the spatial covariance can be expressed as a sum of spatial frequencies. The "natural" functions to express stationary spatial frequencies on a sphere are Legendre polynomials $P_n(\cos\beta)$ (Morse and Feshbach, 1953):

$$\mathrm{Cov}_V(\beta) = \sum_{i=0}^{\infty} b_n P_n(\cos\beta). \tag{19}$$

The coefficients of the Legendre polynomials b_n depend on point spread function of the sources and on the spatial filtering of the volume conductor, as will be derived in the following section. In the time domain, the variable τ extends from $-\infty$ to $+\infty$ and therefore an integral instead of a simple sum appears in the transformation to frequencies. Another consequence is that in the power spectrum $S_V(\omega)$ depends continuously on ω, whereas the b_n are only defined for discrete numbers n.

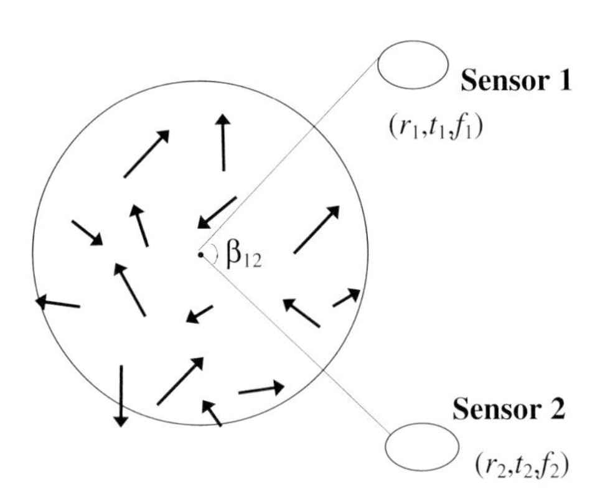

Fig. 1. Graphical presentation of random dipole model. In this model it is assumed that the spontaneous E/MEG is generated by a set of randomly distributed dipole sources, which are embedded in a (spherically symmetric) volume conductor. In this figure the sensors are MEG pick up coils, but they could be just as well EEG electrodes attached to the sphere.

4 A Mathematical Model for the Spatial Distribution of Brain Noise

In this section the definitions presented above will be applied in the derivation of a mathematical model of the spatial distribution of E/MEG signals. The purpose of this model is to pose physiological relevant restrictions on the sources and to predict their implications on the measured E/MEG data. Similar to the stationary dipole model used to describe evoked brain activity (Scherg and Von Cramon, 1985; De Munck, 1990), it will be assumed that spontaneous E/MEG is generated by dipole sources that stay at a fixed position and orientation and of which only the amplitude varies in time. Contrary to evoked brain activity, the sources have a statistical distribution and their amplitudes are considered as stochastic signals.

The derivation will first be presented for the electric potential $\underline{\psi}(\mathbf{x}, t)$. Then the corresponding derivation for the magnetic field will be given, using the same assumptions. For stationary sources the potential can be expressed as:

$$\underline{\psi}(\mathbf{x}, t) = \sum_{l=1}^{L} \underline{s}_l(t)\psi_{\text{dip}}(\mathbf{x}, \underline{\mathbf{p}}_l). \tag{20}$$

Here, the source time functions $\underline{s}_l(t)$ are stochastic signals and $\psi_{\text{dip}}(\mathbf{x}, \underline{\mathbf{p}}_l)$ is the potential at position $\mathbf{x}$, caused by a (random) unit dipole with dipole parameters $\underline{\mathbf{p}}_l$.

Using the same reasoning as in (5) we find for the mean of the potential:

$$E\{\underline{\psi}(\mathbf{x}, t)\} = \sum_{l} \int \mathrm{d}s \int \mathrm{d}\mathbf{p} f_{s_l p_l}(s_l, p_l; t) s_l \psi_{\text{dip}}(\mathbf{x}, \mathbf{p}_l). \tag{21}$$

Here, as well as in the sequel, integration limits which extend from $-\infty$ to ∞ are omitted. If the source time functions are stationary, and if the E/MEG-signals are passed through a high pass filter, the mean of the E/MEG-signal will be zero, as was shown with (13). Since stationarity of the source time functions is a very realistic theoretical assumption and the use of high pass filters is a very common experimental practice we will from now on assume that the source time functions are stationary and that $E\{\underline{\psi}(\mathbf{x}, t)\} = 0$.

The covariance of two potential measurements, one at position $\mathbf{x}_1$ and time t_1 and the other at position $\mathbf{x}_2$ and time t_1 can formally be expressed as

$$\begin{aligned} E\{\underline{\psi}(\mathbf{x}_1, t_1)\underline{\psi}(\mathbf{x}_2, t_2)\} = \; & \textstyle\sum_{l_1 l_2} \int \mathrm{d}s_{l_1} \int \mathrm{d}\mathbf{p}_{l_1} \int \mathrm{d}s_{l_2} \int \mathrm{d}\mathbf{p}_{l_2} \\ & \times f_{s_{l_1} p_{l_1}, s_{l_2} p_{l_2}}(s_{l_1}, p_{l_1}, s_{l_2}, p_{l_2}; t_1 - t_2) \\ & \times s_{l_1} s_{l_2} \psi_{\text{dip}}(\mathbf{x}_1, \mathbf{p}_{l_1}) \psi_{\text{dip}}(\mathbf{x}_2, \mathbf{p}_{l_2}). \end{aligned} \tag{22}$$

In order to compare theory and experiment assumptions need to be made on the distribution of the sources and their amplitudes. Most of the assumptions made imply that a pair of variables is statistically independent. In this way the pdf splits into a product and the remaining integrals are much easier to perform. The assumption of statistical independence might seem as a very strong restriction on the model. However, any multivariate function can be expressed as a sum of products of functions that depend only on one variable. This is a consequence of the fact that the space of multivariate functions is the direct product of spaces of single variate functions. In our application this means that a multivariate distribution can be expressed as the superposition of statistically independent states. By simply adding the expressions for the mean and covariances of different independent states, the model can be generalized afterwards so that statistical dependence of variables are properly described.

First we will assume that *the source time functions are independent of the dipole positions and orientations.* In a formula, this is expressed as

$$f_{s_{l_1} p_{l_1}, s_{l_2} p_{l_2}}(s_{l_1}, p_{l_1}, s_{l_2}, p_{l_2}; t_1 - t_2) = f_{s_{l_1} s_{l_2}}(s_{l_1}, s_{l_2}; t_1 - t_2) f_{p_{l_1} p_{l_2}}(p_{l_1}, p_{l_2}) \tag{23}$$

This assumption will separate the integrals over the source time function from the integrals over the dipole parameters:

$$E\{\underline{\psi}(\mathbf{x}_1, t_1)\underline{\psi}(\mathbf{x}_2, t_2)\} = \sum_{l_1 l_2} \mathrm{Cov}_{s_{l_1} s_{l_2}}(t_1 - t_2) \tag{24}$$

$$\times \int \mathrm{d}\mathbf{p}_{l_1} \int \mathrm{d}\mathbf{p}_{l_2} f_{p_{l_1} p_{l_2}}(p_{l_1}, p_{l_2}) \psi_{\mathrm{dip}}(\mathbf{x}_1, \mathbf{p}_{l_1}) \psi_{\mathrm{dip}}(\mathbf{x}_2, \mathbf{p}_{l_2}),$$

where $\mathrm{Cov}_{s_{l_1} s_{l_2}}(t_1 - t_2)$ is the cross covariance of the amplitude time functions of dipoles l_1 and l_2. The next assumption is that *each pair of dipoles is statistically independent*, i.e.

$$f_{p_{l_1} p_{l_2}}(p_{l_1}, p_{l_2}) = \begin{cases} f_{p_{l_1}}(p_{l_1}) f_{p_{l_2}}(p_{l_2}) & \text{if } l_1 \neq l_2 \\ f_{p_{l_1}}(p_{l_1}) & \text{if } l_1 = l_2 \end{cases} \tag{25}$$

If (25) is substituted into (24), the double sum over l_1 and l_2 is splits into a double sum with $l_1 \neq l_2$ and a single sum with $l_1 = l_2$:

$$\begin{aligned} E\{&\underline{\psi}(\mathbf{x}_1, t_1)\, \underline{\psi}(\mathbf{x}_2, t_2)\} \\ &= \textstyle\sum_{l_1 \neq l_2} \mathrm{Cov}_{s_{l_1} s_{l_2}}(t_1 - t_2) E\{\underline{\psi}_{\mathrm{dip}}(\mathbf{x}_1, \mathbf{p}_l)\} E\{\underline{\psi}_{\mathrm{dip}}(\mathbf{x}_2, \mathbf{p}_l)\} \\ &+ \textstyle\sum_l \mathrm{Cov}_{s_l}(t_1 - t_2) \mathrm{Cov}_{p_l}\{\underline{\psi}_{\mathrm{dip}}(\mathbf{x}_1, \mathbf{p}_l), \underline{\psi}_{\mathrm{dip}}(\mathbf{x}_2, \mathbf{p}_l)\}, \end{aligned} \tag{26}$$

where $\mathrm{Cov}_{p_l}(\,)$ is the covariance of two potential measurements at $\mathbf{x}_1$ and $\mathbf{x}_2$. Ignoring the first order moments, we obtain

$$E\{\underline{\psi}(\mathbf{x}_1, t_1)\underline{\psi}(\mathbf{x}_2, t_2)\} = \sum_l \mathrm{Cov}_{s_l}(t_1 - t_2) \mathrm{Cov}_{p_l}\{\underline{\psi}_{\mathrm{dip}}(\mathbf{x}_1, \mathbf{p}_l), \underline{\psi}_{\mathrm{dip}}(\mathbf{x}_2, \mathbf{p}_l)\}, \tag{27}$$

with

$$\mathrm{Cov}_{p_l}\{\underline{\psi}_{\mathrm{dip}}(\mathbf{x}_1,\mathbf{p}_l),\underline{\psi}_{\mathrm{dip}}(\mathbf{x}_2,\mathbf{p}_l)\} = \int \mathrm{d}\mathbf{p}_l f_{p_l}(\mathbf{p}_l)\psi_{\mathrm{dip}}(\mathbf{x}_1,\mathbf{p}_l)\psi_{\mathrm{dip}}(\mathbf{x}_2,\mathbf{p}_l). \tag{28}$$

In the following this expression will be worked out into detail. It will be assumed that *the dipoles are located in a spherically symmetric volume conductor*, consisting of a set of concentric spherical shells with different conductivities. These shells represent for instance the brain, the cerebrospinal fluid layer, the skull and the skin. The index l will be dropped for convenience. The dipole parameter vector $\mathbf{p}_l$ consists of the following components $\mathbf{p}_l = (\rho_0, \theta_0, \phi_0, \theta, \phi)^T$, where $(\rho_0, \theta_0, \phi_0)$ denotes the dipole position in spherical coordinates and (θ, ϕ) is the dipole orientation, also in spherical coordinates, see Fig. 2.

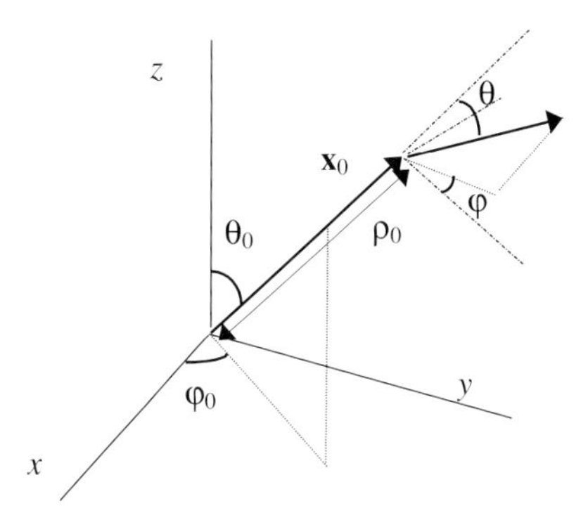

Fig. 2. Definition of the mathematical symbols used to describe a dipole in spherical coordinates. The triple $(\rho_0, \theta_0, \phi_0)$ denotes the dipole position in spherical coordinates, θ is the angle between the dipole vector and its position vector and ϕ is the dipole orientation in the tangential plane.

The angle θ is the angle of the dipole orientation with the dipole position vector and ϕ is the orientation in the tangential plane. A spherical symmetric model is obtained by assuming that the *pdf of the dipole source only depends on ρ_0 and ϕ*. When furthermore *the dipole position is independent of its orientation*, we have for the pdf:

$$f_p(\mathbf{p}) = f_{\rho_0}(\rho_0) f_\theta(\theta). \tag{29}$$

With these five assumptions it is possible to express $\mathrm{Cov}_{pl}()$ in a simple series of Legendre polynomials. For that purpose, the dipole potential ψ_{dip} is expressed as the gradient of the monopole potential ψ_{mon} (De Munck, 1988; De Munck and Peters, 1993)

$$\psi_{\mathrm{dip}}(\mathbf{x}_1, \mathbf{p}) = \left(\cos\theta \frac{\partial}{\partial \rho_0} + \frac{\sin\theta\cos\phi}{\rho_0}\frac{\partial}{\partial\theta_0} + \frac{\sin\theta\sin\phi}{\rho_0\sin\theta_0}\frac{\partial}{\partial\phi_0}\right)\psi_{\mathrm{mon}}(\mathbf{x}_1, \mathbf{x}_0). \tag{30}$$

When (29) and (30) are substituted into (28), the following expression is obtained:

$$\begin{aligned} &\mathrm{Cov}_p\{\underline{\psi}_{\mathrm{dip}}(\mathbf{x}_1,\mathbf{p}),\underline{\psi}_{\mathrm{dip}}(\mathbf{x}_2,\mathbf{p})\} \\ &= \int \mathrm{d}\mathbf{x}_0 f_{\rho_0}(\rho_0)\left(M_r^2 \frac{\partial\psi_1}{\partial\rho_0}\frac{\partial\psi_2}{\partial\rho_0} + \frac{M_\theta^2}{\rho_0^2}\left(\frac{\partial\psi_1}{\partial\theta_0}\frac{\partial\psi_2}{\partial\theta_0} + \frac{1}{\sin^2\theta_0}\frac{\partial\psi_1}{\partial\phi_0}\frac{\partial\psi_2}{\partial\phi_0}\right)\right) \end{aligned} \tag{31}$$

where ψ_i is an abbreviation for $\psi_{\mathrm{mon}}(\mathbf{x}_i, \mathbf{x}_0)$ and

$$\begin{aligned} M_r^2 &= 2\pi \int_0^\pi f_\theta(\theta)\sin\theta\cos^2\theta \mathrm{d}\theta \\ M_\theta^2 &= 2\pi \int_0^\pi f_\theta(\theta)\sin^3\theta \mathrm{d}\theta. \end{aligned} \tag{32}$$

Here M_r^2 denotes the variance of the radial component of the dipole, and M_θ^2 denotes the tangential variance. The factor 2π in (31) and (32) results from the integration over ϕ.

Equation (31) can be evaluated further, by expanding ψ_{mon} in a series of spherical harmonics, and by expressing the electrode positions $\mathbf{x}_1$ and $\mathbf{x}_2$ in spherical coordinates, $\mathbf{x}_1 \simeq (r_1, t_1, f_1)^T$ and $\mathbf{x}_2 \simeq (r_2, t_2, f_2)^T$:

$$\begin{aligned} \psi_{\mathrm{mon}}(\mathbf{x}_1, \mathbf{x}_0) &= \frac{1}{4\pi}\sum_{n=0}^{\infty} a_n \left(\frac{\rho_0}{r_1}\right)^n P_n(\cos\beta_{10}) \\ &= \sum_{n=0}^{\infty} a_n \left(\frac{\rho_0}{r_1}\right)^n \frac{1}{2n+1}\sum_{m=-n}^{n} Y_{nm}(t_1 f_1) Y_{nm}(\theta_0\phi_0) \end{aligned} \tag{33}$$

and similarly for $\psi_{\mathrm{mon}}(\mathbf{x}_2, \mathbf{x}_0)$. In (33) β_{10} is the angle between the electrode and the source point, $Y_{nm}()$ are normalized spherical harmonics, which have the property:

$$\int_0^{2\pi} \mathrm{d}\phi \int_0^\pi \cos\theta \mathrm{d}\theta Y_{nm}(\theta,\phi) Y_{n'm'}(\theta,\phi) = \delta_{nn'}\delta_{mm'}, \tag{34}$$

and $P_n(\cos\beta_{01})$ are Legendre polynomials and $\delta_{nn'}$ is the Kronecker delta. Finally, the coefficients a_n are dependent on the radii and conductivities of the spherical volume conductor model (De Munck and Peters, 1993). When it is assumed that the dipole is in the innermost shell and all shells are isotropic, the coefficients a_n are independent of ρ_0. It can be observed that for constant a_n, e.g. $a_n = 1/r_1$, (33) represents the expansion of the monopole potential in Legendre polynomials. Therefore, one may consider a_n as the spatial frequency characteristic of a spherically symmetric filter that smoothes out the infinite medium potential, which has $(\rho_0/r_1)^n$ as spatial frequencies.

After substitution of (33) into (31) one finds for the radial part of the covariance:

$$\mathrm{Cov}_{P,\mathrm{RAD}}\{\underline{\psi}_{\mathrm{dip}}(\mathbf{x}_1), \underline{\psi}_{\mathrm{dip}}(\mathbf{x}_2)\} = M_r^2 \int_0^{\infty} \mathrm{d}\rho_0 \int_0^{2\pi} \mathrm{d}\phi_0 \int_0^{\pi} \mathrm{d}\theta_0 \rho_0^2 f_{\rho_0}(\rho_0)$$
$$\times \sum_{n=0}^{\infty} \frac{n}{2n+1} \frac{a_n}{\rho_0} \left(\frac{\rho_0}{r_1}\right)^n \sum_{m=-n}^{n} Y_{nm}(t_1, f_1) Y_{nm}(\theta_0, \phi_0)$$
$$\times \sum_{n'=0}^{\infty} \frac{n'}{2n'+1} \frac{a_{n'}}{\rho_0} \left(\frac{\rho_0}{r_2}\right)^{n'} \sum_{m'=-n'}^{n'} Y_{n'm'}(t_2, f_2) Y_{n'm'}(\theta_0, \phi_0)$$
$$= \frac{M_r^2}{4\pi} \sum_{n=0}^{\infty} \frac{a_n^2 n^2}{2n+1} \int_0^{\infty} f_{\rho_0} \left(\frac{\rho_0^2}{r_1 r_2}\right)^n \mathrm{d}\rho_0 P_n(\cos\beta_{12}). \tag{35}$$

Here we have used the orthonormality of the spherical harmonics, equation (34). With the tangential part of the covariance the spherical harmonics are differentiated with respect to θ_0 and ϕ_0. Therefore, instead of the orthonormality of the spherical harmonics, we have to use the identity

$$\int_0^{\pi} d\cos\theta \left(\frac{\mathrm{d}P_{nm}(\cos\theta)}{\mathrm{d}\cos\theta} \frac{\mathrm{d}P_{n'm}(\cos\theta)}{\mathrm{d}\cos\theta} \sin^2\theta + \frac{m^2}{\sin^2\theta} P_{nm}(\cos\theta) P_{n'm}(\cos\theta)\right)$$
$$= \delta_{nn'} \frac{2n(n+1)}{2n+1} \frac{(n+m)!}{(n-m)!}, \tag{36}$$

which can be found in Smythe (1950). Here the factor m^2 originates from taking the derivative of $\cos m\phi$ and $\sin m\phi$ with respect to ϕ. After some manipulation one finds, similar to (35) a series expansion of the covariance in Legendre polynomials of β_{12}, the angle between both electrodes. The expansion has the form of (19), with

$$b_n = \frac{a_n^2}{4\pi(2n+1)} \int_0^{\infty} f_{\rho_0}(\rho_0) \left(\frac{\rho_0^2}{r_1 r_2}\right)^n \mathrm{d}\rho_0 (M_r^2 n^2 + M_\theta^2 n(n+1)). \tag{37}$$

Finally, if the dipoles are concentrated on a spherical shell wit radius r_0, we find:

$$\mathrm{Cov}(\underline{\psi}_{\mathrm{dip}}(\mathbf{x}_1), \underline{\psi}_{\mathrm{dip}}(\mathbf{x}_2)) \tag{38}$$
$$= \frac{1}{4\pi} \sum_{n=0}^{\infty} \frac{a_n^2}{2n+1} \left(\frac{r_0^2}{r_1 r_2}\right)^n (M_r^2 n^2 + M_\theta^2 n(n+1)) P_n(\cos\beta_{12}).$$

Despite its complicated derivation, (38) has a rather simple interpretation. First we note that, due to the spherically symmetric volume conductor and the spherically symmetric distribution of the dipoles, the covariance is spatially stationary because it only depends on the "distance" β_{12} between the sensors. The factor n^2 with the radial component of the covariance is due to the fact that in (31) the derivative of $(\rho_0/r_1)^n$ with respect to ρ_0 was taken. The factor $n(n+1)$ accompanying the tangential part of the covariance similarly originates from taking the derivative with respect to θ and ϕ. As noted before, we may consider the smearing effect of the skull, the cerebrospinal fluid layer and the skin as a linear spherically symmetric spatial filter, with frequency characteristic a_n. Then we observe from (38) that this spatial filter acts quadratically on the input frequencies, similar to the effect of a linear time invariant filter on the spectrum of the input signal (see equation 14). Finally, in (38) there appears a factor $1/(2n+1)$ which has no counterpart in the time domain. The origin of this factor is related to the fact that there are $2n+1$ linearly independent spherical harmonics of degree n. In the time domain, we have for all harmonics e^{iwt} the same number of linearly independent functions, i.e. $\cos\omega t$ and $\sin\omega t$. The quadratic effect of the spatial filter ensures that the series expansion in (38) converges rapidly, in particular for deep sources, if $(\rho_0/\rho_1) \ll 1$. In the limiting case of $\rho_0 \to 0$ the first term dominates and the covariance function tends to $\cos(\beta_{12})$.

For the covariance of two MEG measurements on the same dipole we can derive a similar formula. It is very difficult to start from the usual expression for the magnetic field outside a sphere. Instead, we use the fact that outside the head the magnetic field is the gradient of a magnetic potential (Sarvas, 1987):

$$\mathbf{B}(\mathbf{x}) = \nabla U(\mathbf{x}), \tag{39}$$

and we express the magnetic potential as a series of spherical harmonics,

$$U(\mathbf{x}_1, \mathbf{x}_0) = \frac{-\mu_0}{4\pi}\left(\frac{\sin\theta\cos\phi}{\rho_0\sin\theta_0}\frac{\partial}{\partial\phi_0} - \frac{\sin\theta\sin\phi}{\rho_0}\frac{\partial}{\partial\theta_0}\right) \times \sum_{m=0}^{\infty}\frac{1}{n+1}\left(\frac{\rho_0}{r_1}\right)^n P_n(\cos\beta_{10}). \tag{40}$$

Note that compared to the tangential part of (30) the place of $\sin\phi$ and $\cos\phi$ are swapped and that there is a minus sign in front of $\sin\phi$. This difference reflects the well known fact that the maps of EEG and MEG are rotated 90° with respect to one another. Nonetheless, the computation of the covariance between two magnetic potentials is very similar to the computations carried out before for tangential dipoles. We find

$$\mathrm{Cov}\{\underline{U}_{\mathrm{dip}}(\mathbf{x}_1), \underline{U}_{\mathrm{dip}}(\mathbf{x}_2)\} = \frac{M_\theta^2}{4\pi}\sum_{n=0}^{\infty}\frac{n}{n+1}\frac{1}{2n+1}\left(\frac{r_0^2}{r_1 r_2}\right)^{n+1} P_n(\cos\beta_{12}), \tag{41}$$

where again it has been assumed that the dipoles are confined to a spherical shell of radius r_0.

In order to compute the covariance between two magnetic field measurements, we have to differentiate (41) with respect to $\mathbf{x}_1$ and $\mathbf{x}_2$. This differentiation is complicated by the fact that in (41) $\mathbf{x}_1$ and $\mathbf{x}_2$ appear in spherical coordinates. If both sensor coils have a radial orientation, we only have to differentiate (41) with respect to r_1 and r_2. However, if the center of the spherical conductor does not exactly coincide with the center of the grid the sensors pick up tangential parts of the field also, and we have to take the directional derivatives of (41). We may use,

$$\nabla_1 r_1 = \frac{\mathbf{x}_1}{r_1} \equiv \hat{\mathbf{x}}_1 \quad \text{and } \nabla_1 \cos(\beta_{12}) = \nabla_1 \frac{\mathbf{x}_1 \mathbf{x}_2}{r_1 r_2} = \frac{\hat{\mathbf{x}}_2 - \cos(\beta_{12})\hat{\mathbf{x}}_1}{r_1} \tag{42}$$

and we find for the magnetic covariance, measured in the directions $\mathbf{n}_1$ and $\mathbf{n}_2$,

$$\begin{aligned}
&\mathrm{Cov}\ \{\underline{B}^n_{\mathrm{dip}}(\mathbf{x}_1), \underline{B}^n_{\mathrm{dip}}(\mathbf{x}_2)\} \qquad (43)\\
&= (\mathbf{n}_1\nabla_1)(\mathbf{n}_2\nabla_2)\mathrm{Cov}\{\underline{U}_{\mathrm{dip}}(\mathbf{x}_1), \underline{U}_{\mathrm{dip}}(\mathbf{x}_2)\}\\
&= (\mathbf{n}_1\hat{\mathbf{x}}_1)(\mathbf{n}_2\hat{\mathbf{x}}_2)S_1\\
&\quad - (\mathbf{n}_1\hat{\mathbf{x}}_1(\mathbf{n}_2\hat{\mathbf{x}}_1 - (\hat{\mathbf{x}}_1\hat{\mathbf{x}}_2)(\mathbf{n}_1\hat{\mathbf{x}}_2)) + \mathbf{n}_2\hat{\mathbf{x}}_2(\mathbf{n}_1\hat{\mathbf{x}}_2 - (\hat{\mathbf{x}}_2\hat{\mathbf{x}}_1)(\mathbf{n}_2\hat{\mathbf{x}}_1)))\ S_2\\
&\quad + (\mathbf{n}_1\mathbf{n}_2 - (\mathbf{n}_1\hat{\mathbf{x}}_2)(\mathbf{n}_2\hat{\mathbf{x}}_2) - (\mathbf{n}_2\hat{\mathbf{x}}_1)(\mathbf{n}_1\hat{\mathbf{x}}_1) + (\hat{\mathbf{x}}_1\hat{\mathbf{x}}_2)(\mathbf{n}_1\hat{\mathbf{x}}_1)(\mathbf{n}_2\hat{\mathbf{x}}_2))\ S_3,\\
&\quad + (\mathbf{n}_1\hat{\mathbf{x}}_2 - (\hat{\mathbf{x}}_1\hat{\mathbf{x}}_2)(\mathbf{n}_1\hat{\mathbf{x}}_1))(\mathbf{n}_2\hat{\mathbf{x}}_1 - (\hat{\mathbf{x}}_2\hat{\mathbf{x}}_1)(\mathbf{n}_2\hat{\mathbf{x}}_2))S_4
\end{aligned}$$

with

$$\begin{aligned}
S_1 &= \frac{\mu_0^2 M_\theta^2}{4\pi}\sum_{n=0}^{\infty}\frac{n(n+1)}{2n+1}\left(\frac{r_0^2}{r_1 r_2}\right)^{n+1} P_n(\cos\beta_{12})\\
S_2 &= \frac{\mu_0^2 M_\theta^2}{4\pi}\sum_{n=0}^{\infty}\frac{n}{2n+1}\left(\frac{r_0^2}{r_1 r_2}\right)^{n+1} P'_n(\cos\beta_{12})\\
S_3 &= \frac{\mu_0^2 M_\theta^2}{4\pi}\sum_{n=0}^{\infty}\frac{n}{n+1}\frac{1}{2n+1}\left(\frac{r_0^2}{r_1 r_2}\right)^{n+1} P'_n(\cos\beta_{12})\\
S_4 &= \frac{\mu_0^2 M_\theta^2}{4\pi}\sum_{n=0}^{\infty}\frac{n}{n+1}\frac{1}{2n+1}\left(\frac{r_0^2}{r_1 r_2}\right)^{n+1} P''_n(\cos\beta_{12}).
\end{aligned} \tag{44}$$

Here $P'_n(\cos\beta_{12})$ and $P''_n(\cos\beta_{12})$ are the first and the second derivative of the Legendre Polynomial with respect to $\cos\beta_{12}$. Equation (43) looks complicated, but it is very general and it can be implemented easily in a computer program. If the sensors have a radial orientation, which means that they are all directed towards the center of the volume conductor, only the first sum remains, the others vanish:

$$\mathrm{Cov}\{\underline{B}^{\mathrm{rad}}_{\mathrm{dip}}(\mathbf{x}_1), \underline{B}^{\mathrm{rad}}_{\mathrm{dip}}(\mathbf{x}_2)\} = \frac{M_\theta^2}{4\pi}\sum_{n=0}^{\infty}\frac{n(n+1)}{2n+1}\left(\frac{r_0^2}{r_1 r_2}\right)^{n+1} P_n(\cos\beta_{12}). \tag{45}$$

One other interesting case occurs if EEG and MEG are measured simultaneously. It follows from (30) and (40) that the covariance between electric and magnetic fields vanishes, so that the EEG and MEG are uncorrelated:

$$\mathrm{Cov}\{\underline{\psi}_{\mathrm{dip}}(\mathbf{x}_1), \underline{U}_{\mathrm{dip}}(\mathbf{x}_2)\} = \int \mathrm{d}\mathbf{x}_0 \int_0^{\pi} \mathrm{d}\theta \sin\theta \int_0^{2\pi} \mathrm{d}\phi f_{\rho_0}(\rho_0) f_\theta(\theta)$$
$$\times \left(\cos\theta \frac{\partial}{\partial \rho_0} + \frac{\sin\theta\cos\phi}{\rho_0}\frac{\partial}{\partial\theta_0} + \frac{\sin\theta\sin\phi}{\rho_0\sin\theta_0}\frac{\partial}{\partial\phi_0} \right) \psi_{\mathrm{mon}}(\mathbf{x}_1, \mathbf{x}_0)$$
$$\times \frac{\mu_0}{4\pi} \left(\frac{\sin\theta\sin\phi}{\rho_0}\frac{\partial}{\partial\theta_0} - \frac{\sin\theta\cos\phi}{\rho_0\sin\theta_0}\frac{\partial}{\partial\phi_0} \right) \sum_{n=0}^{\infty} \frac{1}{n+1} \left(\frac{\rho_0}{r_2}\right)^n P_n(\cos\beta_{20})$$
$$= \int \mathrm{d}\mathbf{x}_0 \int_0^{\pi} \mathrm{d}\theta \sin\theta \int_0^{2\pi} \mathrm{d}\phi f_{\rho_0}(\rho_0) f_\theta(\theta) \frac{\mu_0 \sin^2\theta}{4\pi\sin\theta_0} \left(\frac{\sin^2\phi}{\rho_0^2} - \frac{\cos^2\phi}{\rho_0^2} \right)$$
$$\times \frac{\partial^2}{\partial\theta_0\partial\phi_0} \psi_{\mathrm{dip}}(\mathbf{x}_1, \mathbf{x}_0) \sum_{n=0}^{\infty} \frac{1}{n+1} \left(\frac{\rho_0}{r_2}\right)^n P_n(\cos\beta_{20}) = 0. \tag{46}$$

Here the radial part of the electric potential disappears because in combination with the tangential part of the magnetic potential one only obtains terms with $\cos\phi$ and $\sin\phi$, of which the ϕ-integral vanishes. Similarly, the cross-terms vanish because the ϕ-integral over $\cos\phi\,\sin\phi$ vanishes. The remaining two terms do not vanish, but they cancel. Note that this result is true, independent of the assumption made for the distribution of the dipoles position parameters, $f_{\mathbf{x}_0}(\mathbf{x}_0)$.

5 Model Predictions

In the previous section it is shown that a spherical symmetric distribution of dipoles in a spherical volume conductor results in a spatially stationary covariance function. Furthermore, the shapes of the electric and magnetic covariance functions are predicted. Here, these model predictions are examined, in order to make it possible to verify them experimentally. In practice, the magnetic field is not measured directly, but *gradiometers* are used. A gradiometer consists of two coils with opposite windings, a pick up coil close to the head and a compensation coil, some distance away. In this way the background noise, which has a long wave length compared to spatial patterns of the brain signals, will be suppressed. The covariance of two gradiometer measurements can be expressed in the covariance of magnetic field measurements:

$$\begin{aligned} \mathrm{Cov}_{\mathrm{Grad}}(\mathbf{x}_1, \mathbf{x}_2) = {} & \mathrm{Cov}_B(\mathbf{x}_1^P, \mathbf{x}_2^P) - \mathrm{Cov}_B(\mathbf{x}_1^P, \mathbf{x}_2^C) \\ & - \mathrm{Cov}_B(\mathbf{x}_1^C, \mathbf{x}_2^P) + \mathrm{Cov}_B(\mathbf{x}_1^C, \mathbf{x}_2^C). \end{aligned} \tag{47}$$

Here the upper indices P and C refer to the pick up and the compensation coils, respectively. Instead of the covariance, we here consider the correlation squared, which is a normalized quantity between 0 and 1:

$$Cor^2(\mathbf{x}_1, \mathbf{x}_2) = \frac{\mathrm{Cov}^2(\mathbf{x}_1, \mathbf{x}_2)}{\mathrm{Var}(\mathbf{x}_1)\mathrm{Var}(\mathbf{x}_2)}. \tag{48}$$

In Fig. 3 the correlation squared is plotted as a function of gradiometer distance, for different depths of the dipoles. It was assumed that the pick up coils and the compensation coils have a radial coordinate of 11.5 cm and 16.5 cm respectively. These figures roughly coincide with the dimensions of the CTF-whole helmet system.

We observe that for superficial dipoles, with a radial coordinate r_0 of 0.8 times the head radius, the gradiometers are almost uncorrelated when the gradiometers are separated more than 30°. For deeper dipoles, this correlation distance increases and becomes less well defined. For dipoles with a radial coordinate of 0.5 times the head radius the correlation increases for large gradiometer separation. This behavior can be understood because in the limit of $r_0 \to 0$ only the first term of the Legendre series survives, which is $\cos \beta$, so that the correlation squared is $\cos^2 \beta$.

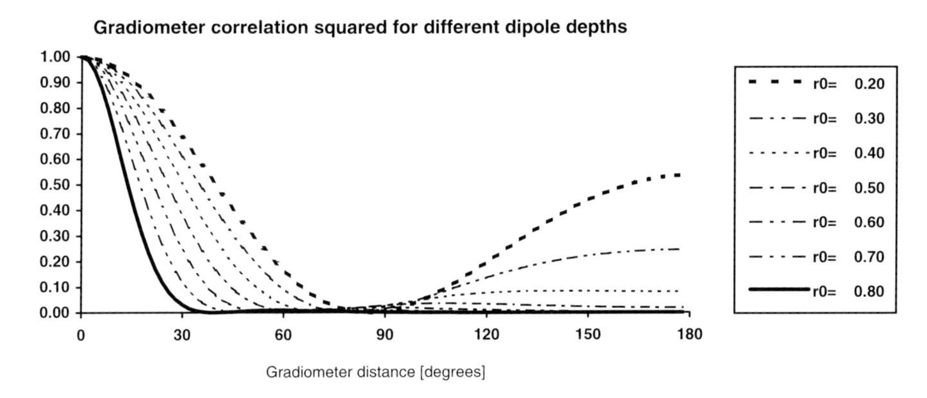

Fig. 3. The correlation squared as a function of gradiometer distance, for dipoles on spherical shells of different radii. These radii are expressed as a fraction of the head radius. A radius of 0.8 roughly corresponds to the location of the cortex. The pick up coils have a radial coordinate of 1.1 times the head radius.

The correlation length depends, at least in principle, on the baseline. Therefore, the random dipole model could be used to optimize the design of a gradiometer system. In order to quantify the dependence of the correlation length on the baseline, we plotted in Fig. 4 the correlation function for different baselengths. We observe that the shorter the baseline, the shorter is the correlation length. The effect is however quite small. Therefore, the proper choice of the baseline should be led by other considerations (Vrba, 1997), such as the spatial characteristics of the environmental noise inside a magnetically shielded room. The thick line in the middle represents the baselength of the CTF system that we have in Amsterdam.

The covariance of two electric measurements can be plotted similarly, using equation (38). This is done in Fig. 5. The sensor distance for which the

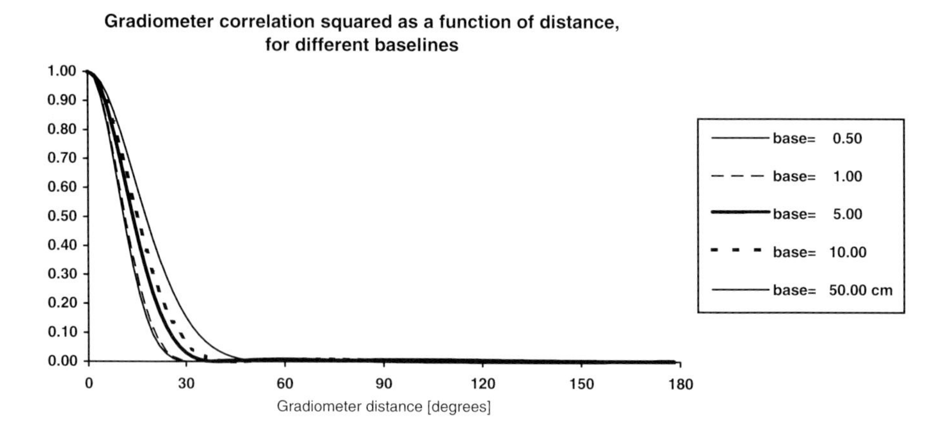

Fig. 4. The correlation as a function of gradiometer distance for different baselines. The radial coordinate of the pick up coils was set at 11 cm and the dipoles are distributed over of spherical shell with a radius of 8 cm.

correlation vanishes is about 70°, which is more that twice the MEG correlation distance. This behavior of the correlation function is due to the smearing effect of the skull. For deep dipoles the correlation increases with distance, for distances larger than 90°. The large correlation over large distances is caused by the fact that deep dipoles have their minimum and maximum potentials at opposite sites on the sphere.

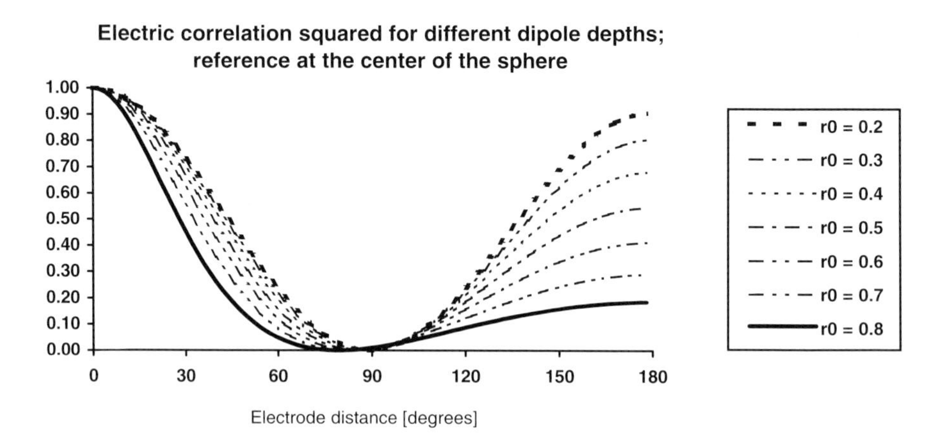

Fig. 5. The electric correlation squared as a function of gradiometer distance, for dipoles on spherical shells of different radii. These are theoretical curves, because the potentials are defined with respect to a reference in the center of the sphere.

The problem with Fig. 5 is however, that it can never be tested experimentally because the reference electrode is not adequately taken into account. This is a consequence of the fact that in Fig. 5, we have ignored the arbitrary constant in the right hand side of (30), which was silently set at zero. A zero constant in (30) means that the potential ψ_{dip} should be considered as the potential difference between a surface electrode and an electrode at the center of the sphere. However, in practice the EEG is always measured with respect to a reference electrode attached to the skin and the EEG reflects the potential difference between two points on the skin. Therefore, in practice, the EEG-correlation function is dependent on three electrode positions, the two "active" electrodes and one reference:

$$Cor(\underline{\psi}(\mathbf{x}_1), \underline{\psi}(\mathbf{x}_2)) = \frac{\text{Cov}(\underline{\psi}(\mathbf{x}_1) - \underline{\psi}(\mathbf{x}_{\text{ref}}), \underline{\psi}(\mathbf{x}_2) - \underline{\psi}(\mathbf{x}_{\text{ref}}))}{\sqrt{\text{Var}(\underline{\psi}(\mathbf{x}_1) - \underline{\psi}(\mathbf{x}_{\text{ref}}))\text{Var}(\underline{\psi}(\mathbf{x}_2) - \underline{\psi}(\mathbf{x}_{\text{ref}}))}}. \quad (49)$$

It appears that this dependence of the EEG-correlation function is quite strong. To demonstrate this, we plotted the EEG-correlation function squared for different references in Fig. 7. In these curves, the reference was placed on the z-axis (Fig. 6) and both active electrodes were placed on a great circle making an angle χ with the z-axis.

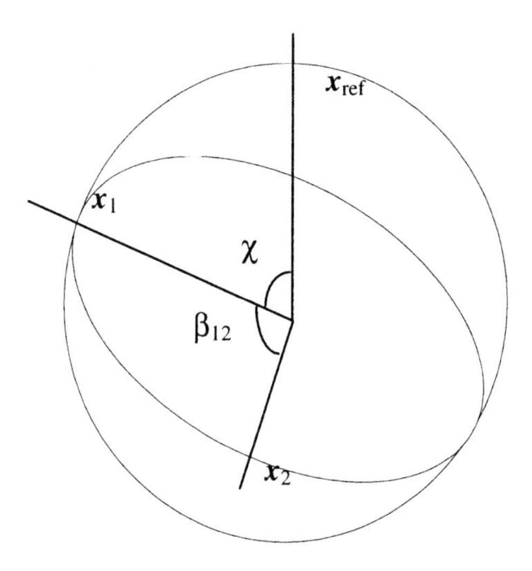

Fig. 6. The relative positions of the electrodes used demonstrate the dependence of the EEG correlation on the reference electrode.

It appears that in all cases, the correlation function decreases monotonically with electrode distance, but the rate of decrease is strongly dependent on the position of the reference electrode with respect to the active electrodes. Therefore, if the spatial stationarity of experimentally determined

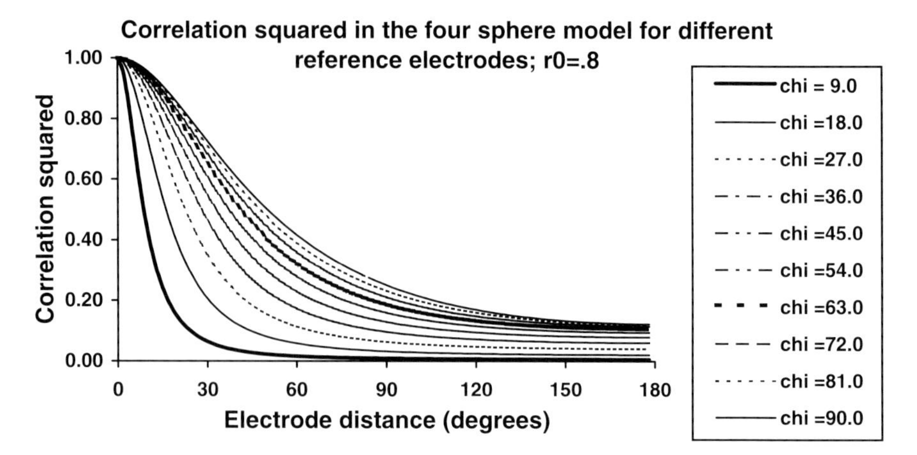

Fig. 7. The EEG correlation function squared is shown for different relative positions of the reference and both active electrodes, see Fig. 6.

correlations were tested by simply plotting these correlations against electrode distance, a big cloud of points would be obtained, irrespective of the true stationary nature of the data.

It could be attempted to reduce the effect of the reference electrode by computing the EEG data with respect to some artificial reference, like the average or the weighted average reference (Nunez et al., 1997). However, one could argue that in this way the problem would only increase in complexity because the new plot will depend on even more parameters than only the position of the reference. Instead, we propose to eliminate the effect of the reference by simply considering the variance of the EEG signal as a function of electrode distance. This variance is a second order statistic, which only depends on the positions of the two active electrodes and the effect of the reference is eliminated:

$$\mathrm{Var}(\underline{\psi}(\mathbf{x}_1)-\underline{\psi}(\mathbf{x}_2)) = \mathrm{Var}(\underline{\psi}(\mathbf{x}_1)) + \mathrm{Var}(\underline{\psi}(\mathbf{x}_2)) - 2\mathrm{Cov}(\underline{\psi}(\mathbf{x}_1), \underline{\psi}(\mathbf{x}_2)). \quad (50)$$

The first two terms are constant (because of stationarity) and the second term is the covariance put upside down. Therefore, the variance function is an increasing function of electrode distance, which starts in the origin because for nearby electrodes the variance of the difference signal vanishes. If the variance is plotted as a function of electrode distance, one obtains, using the assumptions of the random dipole model, the curves presented in Fig. 8. Each curve is normalized with the largest value, i.e. the variance measured with two diametrical electrodes. Note that, contrary to the MEG case, there is hardly a difference in the variance function of deep and superficial curves. The cause of the resemblance is the smearing effect of the low conducting skull. To demonstrate this effect, we plotted in Fig. 9 the variance function for the homogeneous sphere.

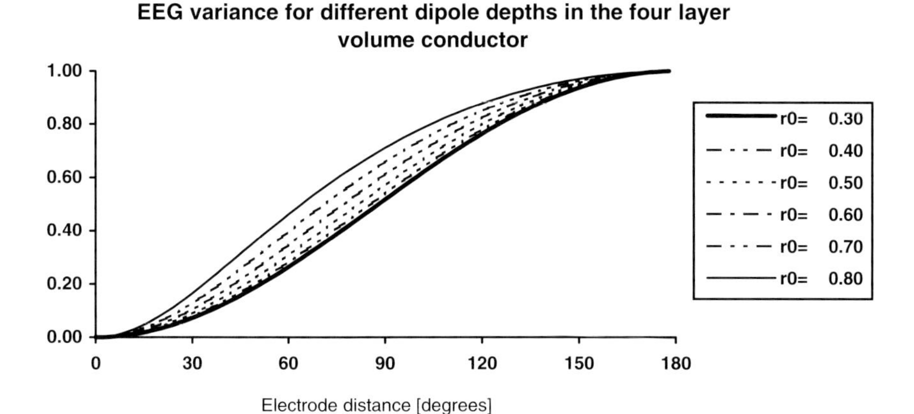

Fig. 8. The EEG variance as a function of electrode distance, for a set of dipoles randomly distributed over a spherical shell with radius r_0, embedded in a layered spherical volume conductor. Each curve is normalized with the largest value present on the curve.

It can be observed that for superficial dipoles the variance obtains its maximum value already for electrodes separated 30° or more. The reason why the skull effect is so strong is that it acts quadratically in (38). Mathematically this means that the series converges fast and that the convergence is determined by the spatial filtering factors a_n (instead of the ratio $r_0^2/(r_1 r_2)$, because all curves in Fig. 8 are very similar to the first term, $1 - \cos\beta$.

One might think that a characterization of the EEG correlations does not give a complete description because cross-terms are ignored. However, it can be shown that if the variance of each pair of electrodes is given, one can easily compute the covariance of differentially measured signals:

$$\mathrm{Cov}(\psi_1 - \psi_{\mathrm{ref}}, \psi_2 - \psi_{\mathrm{ref}}) = \frac{1}{2}(\mathrm{Var}(\psi_1 - \psi_{\mathrm{ref}}) + (\mathrm{Var}(\psi_2 - \psi_{\mathrm{ref}}) - (\mathrm{Var}(\psi_1 - \psi_2)). \tag{51}$$

Once these covariances are determined, one can also compute the correlation using (49).

To illustrate the physiological relevance of the random dipole model and to test the model predictions we show some data in Figs. 10 and 11. Figure 10 presents the correlation of the MEG data of a sleeping subject, as a function of gradiometer distance. This correlation is plotted in different frequency bands, 2–6 Hz, 6–10 Hz, 10–14 Hz and 14–18 Hz. There appears to be a remarkable resemblance between the data in Fig. 10 and the theoretical curve in Fig. 3, with $r_0 = 0.8$. Note that there are 61 channels, so that there are 1830 channel pairs for which the correlation has been determined. Of all these pairs there are only a very few which do not agree with the theoretical predictions. These few exceptions are potentially interesting points for further inspection.

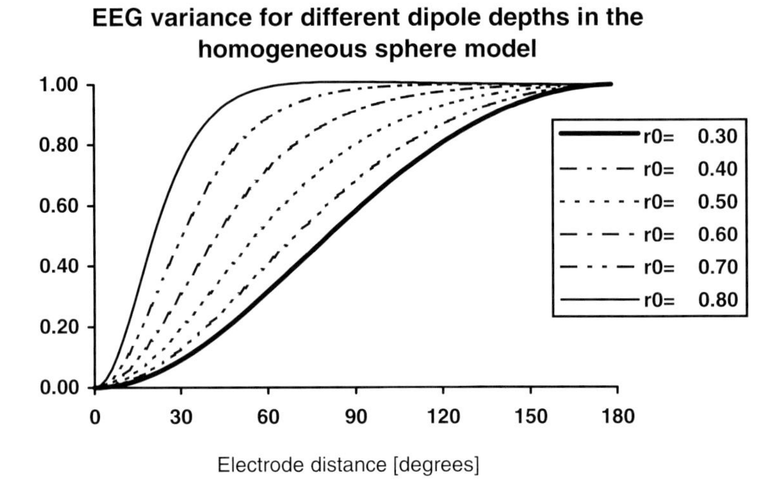

Fig. 9. The EEG variance as a function of electrode distance, for a set of dipoles randomly distributed over a spherical shell with radius r_0, embedded in a homogeneously conducting sphere.

The underlying correlation cannot simply be explained by the point spread function of the magnetic field. Therefore, provided that artifacts can be ruled out, these points represent connected brain structures.

Note furthermore that a complete agreement between theory and experiment cannot be expected because the MEG helmet is not perfectly spherically symmetric and the center of the subject's head does not necessarily coincide with the symmetry center of the helmet. These sub-optimal experimental conditions do not explain the exceptional correlations, but they do explain why there is a spread in the correlations at a gradiometer distance of 30°.

Figure 11 shows the EEG variance as a function of electrode distance, of a subject sitting at rest. These EEG data, recorded from 32 electrodes, was band pass filtered from 5 to 8 Hz, from 9 to 12 Hz and from 13 to 40 Hz. The variance in each of these bands was determined for all pairs of electrodes. The electrode positions were determined using the method presented in (De Munck et al., 1991) and for all pairs of electrodes the distance, expressed as the angle viewed from the center of the best fitting sphere, was determined. Figure 10 shows the variance as a function of electrode distance, for each of the three frequency bands. Similar to the theoretical curves presented in Figs. 8 and 9, the variance is normalized with the largest variance present in the data set.

In the lower frequency band we observe that for each electrode distance there is an almost unique variance, despite the variance estimation errors and the finite precision of the electrode position determination. This observation implies that the variance is spatially stationary, and indicates that

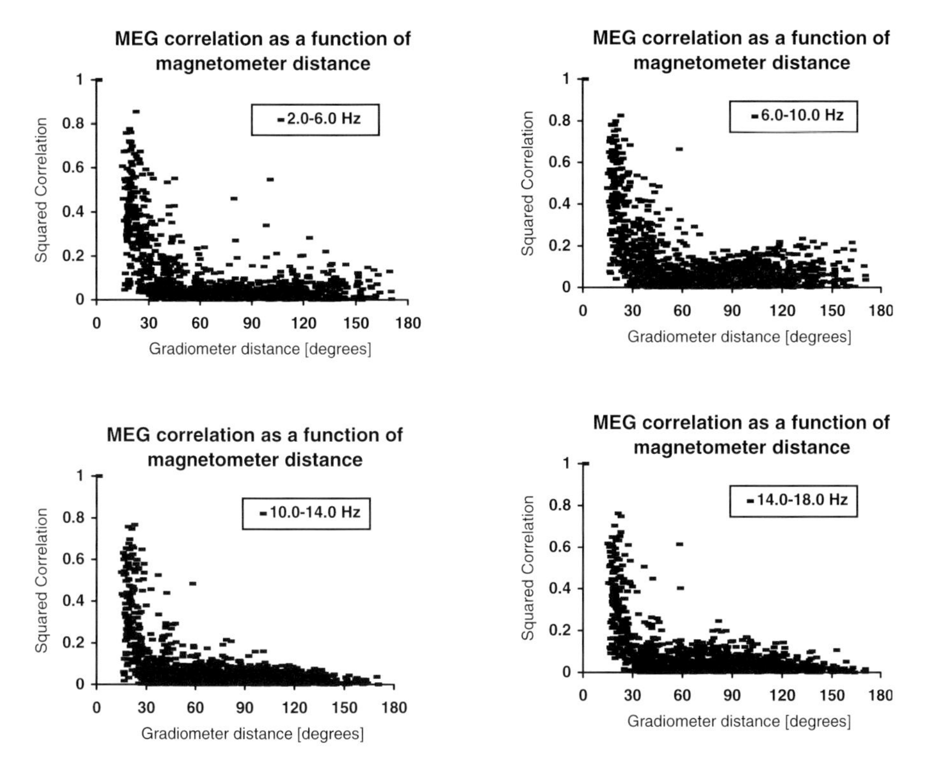

Fig. 10. The squared correlation of MEG data as a function of gradiometer distance for a sleeping subject.

the underlying generators have a spherically symmetric distribution. For the alpha-band the situation is more complicated. There we have for each electrode distance a large spread in variance. Apparently, the alpha band is not spatially stationary and the generators of the alpha band do not have a spherically symmetric distribution. This finding is in agreement with e.g. Grummich et al. (1992). For the higher frequency band the situation is less clear. If one considers the variance as a function of electrode position and if one keeps one of the electrodes fixed then one obtains nice increasing curves as predicted by the model. However, if all these curves are plotted in one graph, as was done in Fig. 11, a cloud of points is obtained, which is not so nicely stationary as in the lower frequency band.

6 Discussion

In the previous sections a theoretical model was presented which explains and quantifies the correlations of spontaneous brain activity. There are several potential applications of the model. One could use it as a description of the

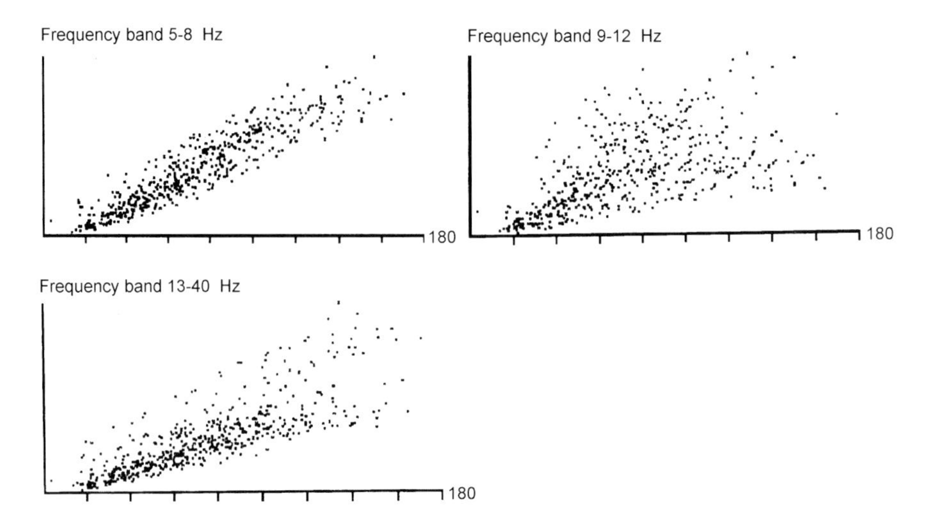

Fig. 11. Experimental data showing the variance as a function of electrode distance for three frequency bands, from 5 to 8 Hz, from 9 to 12 Hz and from 13 to 40 Hz.

spatial distribution of brain noise, or as a description of the generators of certain rhythms present E/MEG recordings in terms of distributed sources. Furthermore, the formulation of the model in terms of spatial frequencies has implications for mapping techniques which are based on spatial filtering. In the following subsections these topics are addressed in more detail.

6.1 Statistically Realistic Models

Source localization techniques, applied on evoked potential (EP) or evoked magnetic field (EF) data, are based on the minimization of a cost function. This cost function is a measure to compare the EP/EF data with the model predictions. Many choices of the cost function are possible, but in practice the sum of squared differences is chosen almost always. The reason why this choice is made, is that the sum of squared differences is easy to handle mathematically, and in particular it is easy to find the minimum when (part of) the model parameters are linear. From a theoretical point of view, it can be argued that the minimization of the sum of squared differences maximizes the likelihood of the noise (i.e. the difference between data and model), assuming it is Gaussian and uncorrelated (e.g. Bard, 1974). Although this theoretical argument may be based on unrealistic assumptions, it provides a framework to make an objective choice for a proper cost function.

If the data is represented by the matrix R, where each row represents the time trace of one channel, and if the corresponding model is represented by the matrix $\tilde{R}$, the ordinary sum of squared differences cost can be expressed as:

$$H = \sum_{i,j} (\tilde{r}_{ij} - r_{ij})^2 = Tr\left\{(\tilde{R} - R)(\tilde{R} - R)^T\right\}. \tag{52}$$

When it is assumed that background noise is correlated Gaussian noise, which is a very reasonable assumption (Knuutila and Hämäläinen, 1987), the maximum likelihood estimator is obtained by minimizing (53) instead of (52):

$$H = \sum_{ij,i'j'} C^{\text{inv}}_{ij,i'j'} \left(\tilde{r}_{ij} - r_{ij}\right)\left(\tilde{r}_{i'j'} - r_{i'j'}\right), \tag{53}$$

where $C_{ij,i'j'}$ is the covariance between the noise on channel i, time sample j and channel i', time sample j'. In (53) these covariances are put in a matrix with ij as one index and $i'j'$ as the other index. Of that matrix the inverse has to be determined. Although theoretically straight forward, a practical application of (53) is hampered by the fact that the dimensions of the covariance matrix may become huge. For instance, if there are 100 channels and 50 time samples, the dimensions of C are 5000×5000 and it becomes computationally expensive to invert C.

If the assumption is made that the spatial distribution of the background E/MEG is the same for all frequency bands, then we can express the spatio-temporal covariance as the Kronecker product of a temporal part T and a spatial part X. The Kronecker product of two matrices is obtained by multiplying the matrix elements:

$$C_{ij,i'j'} = T_{j,j'} X_{i,i'}, \tag{54}$$

Using some matrix manipulations (Magnus and Neudecker, 1988), it can be found that the cost function simplifies to

$$H = Tr\left\{X^{\text{inv}}(\tilde{R} - R)T^{\text{inv}}(\tilde{R} - R)^T\right\}. \tag{55}$$

We observe that instead of inverting one big matrix C, we have to invert two matrices X and T separately, which is computationally much less demanding. The validity of underlying assumption still remains to be investigated. Our preliminary MEG sleep data (Fig. 10) show that, at least in some conditions, a separation of spatial and temporal correlation is present.

Recently several studies have been published in which the effect of including the background noise into the cost function has been examined (e.g. Sekihara et al., 1994; Huizenga and Molenaar, 1994 and Yamazaki et al., 1994). These studies show that ignoring the covariance decreases the stability and the reliability of the dipole inverse solution. However, these studies are restricted to the space domain and a logical next step would be to generalize these results to spatio temporal models, with a complete spatio-temporal description of the background noise.

6.2 Distributed Sources

The random dipole model might be considered as a distributed source model because the E/MEG sources are not confined to a point, but instead these sources spread out over a larger region. However, contrary to the conventional distributed source models presented in this chapter, the sources do not describe the EEG on a single time point, but they give a (statistical) description of a large period of time (typically 1 to 10 min). Nonetheless, there is an interesting mathematical relationship between the random dipole model and the distributed source model, as it was proposed by Hämäläinen and Ilmoniemi (1994).

In the formulation of these authors, the inverse problem of E/MEG is solved by finding that distribution of current dipoles $\mathbf{J}(\mathbf{x})$, which minimizes

$$\iiint_V f(\mathbf{x})\mathbf{J}(\mathbf{x}) \cdot \mathbf{J}(\mathbf{x}) d\mathbf{x}, \tag{56}$$

in such a way that the potential (or magnetic field) corresponding to $\mathbf{J}(\mathbf{x})$ passes through the measurements ψ_i (or $\mathbf{B}_i$). An expression for the potential corresponding to $\mathbf{J}(\mathbf{x})$ can simply be obtained by multiplying $\mathbf{J}(\mathbf{x})$ with the point dipole potential, which equals $\nabla\psi_{\text{mon}}(\mathbf{x}, \mathbf{x}_i)$, and integrating over the whole volume conductor:

$$\psi_i \equiv \psi(\mathbf{x}_i) = \iiint_V \mathbf{J}(\mathbf{x}) \cdot \nabla\psi_{\text{mon}}(\mathbf{x}_i, \mathbf{x}) \mathrm{d}\mathbf{x}. \tag{57}$$

We can find the find the minimum of (56) under the constraints posed by the data, using Lagrange multipliers μ_j

$$H(\mathbf{J}) = \min_{\mathbf{J}(\mathbf{x})} \iiint_V f(\mathbf{x})\mathbf{J}(\mathbf{x}) \cdot \mathbf{J}(\mathbf{x}) \mathrm{d}\mathbf{x} + 2\sum_j \mu_j (\psi_j - \iiint_V \mathbf{J}(\mathbf{x}) \cdot \nabla\psi_{\text{mon}}(\mathbf{x}, \mathbf{x}_j) \mathrm{d}\mathbf{x}). \tag{58}$$

It follows that the solution of (58) is

$$\mathbf{J}(\mathbf{x}) = \sum_j \mu_j \nabla\psi_{\text{mon}}(\mathbf{x}, \mathbf{x}_j) \tag{59}$$

where the coefficients μ_j can be found by solving the system:

$$\psi(\mathbf{x}_i) = \sum_j \mu_j \iiint_V f(\mathbf{x}) \nabla\psi_{\text{mon}}(\mathbf{x}, \mathbf{x}_j) \cdot \nabla\psi_{\text{mon}}(\mathbf{x}, \mathbf{x}_i) \mathrm{d}\mathbf{x} = \sum_j \mu_j \operatorname{Cov}_P(\mathbf{x}_j, \mathbf{x}_i). \tag{60}$$

Here the last equality follows from (31). Therefore, we can use (38) to compute the matrix elements required in the minimum norm estimation method according to Hämäläinen and Ilmoniemi. For the magnetic field, a similar result can be derived. Instead of time consuming 3D integrals, one can use (43) and (44) to compute the matrix elements. Moreover, if one has a combination of EEG and MEG measurements, it follows from (46) the E/MEG cross-terms in the matrix vanish, so that EEG and MEG measurements independently contribute to the solution (60).

In the short derivation presented here, use was made of a mathematical resemblance between the random dipole model and the minimum norm estimation problem posed by Hämäläinen and Ilmoniemi. Both models are based on distributed sources, but whether there also is a precise physical or physiological analogy, e.g. in terms of Bayesian estimators, is not easy to say.

7 Conclusion

We have presented a mathematical model to describe spontaneous EEG and MEG activity. The model predictions are in good agreement with some experimentally determined data sets. The model itself has applications in the improvement of dipole localization techniques and in the spatial analysis of spontaneous brain activity.

References

Y. Bard (1974): *Nonlinear parameter estimation*, Academic Press, New York, San Francisco, London, pp 83–140.

J.C. de Munck, P.C.M Vijn and F. Lopes da Silva (1992): A random dipole model for spontaneous brain activity, IEEE Trans. Biomed. Eng. BME 39(8): 791–804.

J.C. de Munck (1988): The potential distribution in a layered anisotropic spheroidal volume conductor, J. Appl. Phys. 64(2):464–470.

J.C. de Munck (1989): A mathematical and physical interpretation of the electromagnetic field of the brain, PhD-thesis, University of Amsterdam.

J.C. de Munck (1990): The estimation of time varying dipoles on the basis of evoked potentials, Electroenceph. Clin. Neurophysiol. 77:156–160.

J.C. de Munck, P.C.M. Vijn and H. Spekreijse (1991): A practical method for determining electrode positions on the head, Electroenceph. Clin. Neurophys. 78:85–87.

J.C. de Munck and M.J. Peters (1993): A fast method to compute the potential in the multisphere model, IEEE Trans. Biomed. Eng. BME 40(11): 1166–1174.

M.S. Hämäläinen and R.J. Ilmoniemi (1994): Interpreting magnetic fields of the brain: minimum norm estimates, Med.&Biol. Eng. & Comp. 32, 35–42.

H.M. Huizenga and P.C.M. Molenaar (1994): Equivalent source estimation of scalp potential fields contaminated by heteroscedastic and correlated noise, Brain Topography 8(1):13–33.

A.S. Gevins, S.L. Bressler, N.H. Morgan, B.A. Cutillo, R.M. White, D.S. Greer and J. Illes (1989): Event-related covariances during a bimanual visuomotor task. I. Methods and analysis of stimulus- and response-locked data, Electroenceph. Clin. Neurophys. 74:58–75.

P. Grummich, J. Vieth, H. Kober and T. Scholtz (1992): Separation of sources of alpha activity in multichannel MEG. in: Biomagnetism: Clinical Aspects. M. Hoke et al. (eds.), Elsevier, Amsterdam.

B. Kedem, (1994): Time series analysis by higher order crossings, IEEE Press, New York.

J. Knuutila and M.S. Hämäläinen, (1987): Characterization of brain noise using a high sensitivity 7-channel magnetometer, proc. 6-th Int. Conf. Biomag. (Tokyo, 1987), ed. K.Atsumi et al. (Tokyo: Tokyo Denki University) pp 186–189.

F. Lopes da Silva (1987): EEG analysis: theory and practice In: *Electroencephalography: Basic principles, clinical applications and related fields*, E. Niedermeyer and F. Lopes da Silva (eds.), 2nd edition, Urban & Schwarzenberg, Baltimore-Munich, pp 871–897.

J.R. Magnus and H. Neudecker (1988): *Matrix differential calculus with applications in statistics and econometrics*, Wiley & Sons, Chichester, New York.

P.M. Morse and H. Feshbach (1953): *Methods of theoretical phyiscis, I and II*, McGraw-Hill, New York.

A. Papoulis (1984): *Probability, random variables and stochastic processes*, McGraw-Hill, Aukland-Tokyo.

P.L. Nunez, R. Shrinivasan, A.F. Westdorp, R.S. Wijsinghe, D.M. Tucker, R.B. Silberstein and P.J. Cadusch (1997): EEG Coherency I: statistics, reference electrode, volume conduction, Laplacians, cortical imaging and interpretation of multiscales, Electroenceph. Clin. Neurophysiol. 103:499–515.

J. Sarvas (1987): Basic mathematical and electromagnetic concepts of the biomagnetic inverse problem, Phys. Med. Biol. 32(1):11–22.

M. Scherg and D. Von Cramon (1985): A new interpretation of the generators of BAEP waves in I-V: Results of a spatio-temporal dipole model, Electroenceph. Clin. Neurophysiol. 62:290–299.

K. Sekihara, F. Takeuchi, S. Kuriki and H. Koizumi(1994): Reduction of brain noise influence in evoked neuromagnetic source localization using noise spatial correlation, Phys. Med. Biol. 39: 937–946.

W.P. Smythe (1950): *Static and dynamic electricity*, McGraw-Hill, New York, pp. 129–155.

J. Vrba (1997): Baseline optimization for noise cancellation systems, Proceedings - 19th International Conference - IEEE/EMBS Oct.30 - Nov. 2, 1997, Chicago, IL. USA.

T. Yamazaki, B.W. Van Dijk and H. Spekreijse (1994): Confidence limits for the parameter estimation in the dipole localisation method on the basis of spatial correlation of background EEG, Brain Topography 5(2):195–198.

Neurophysiological Brain Function and Synchronization Processes

Jürgen Kurths[1], Peter Tass[2], and Bärbel Schack[3]

[1] Department of Theoretical Physics, University of Potsdam, Am Neuen Palais 19, PF 601553, D-14415 Potsdam, Germany
[2] Department of Neurology, Heinrich-Heine-University, Moorenstr. 5, D-40225 Düsseldorf, Germany
[3] Institute of Medical Statistics, Computer Science and Documentation, University of Jena, Jahnstr. 3, D-07740 Jena, Germany

Synchronization of neuronal activity plays an important role in neurophysiology. Data from biological systems, for instance, obtained from magnetoencephalography (MEG) and electroencephalography (EEG) recordings are nonstationary. For this reason it is imperative to design methods aiming at investigating the time evolution of synchronization processes. In this chapter two methods are presented which are based on different concepts and, thus, address different aspects of synchronization:

Schack's approach is founded on the notion of instantaneous *coherence*. The latter is a correlation coefficient per frequency describing the momentary level of coupling between two signals for a certain frequency or frequency band. Schack proves the functional relationship between instantaneous EEG and MEG coherence and elementary cognitive processes and applies her method here in order to investigate these coherences during word processing and the Stroop task.

Tass et al. understand synchronization in terms of $n : m$ *phase locking* of noisy oscillators. By means of the Hilbert transform instantaneous phases of oscillatory signals are determined. The $n : m$ phase differences are analysed with methods from statistical physics. This approach is applied here to investigate synchronization processes generating parkinsonian tremor.

Dynamic Topographic Spectral Analysis of Cognitive Processes

Bärbel Schack

Institute of Medical Statistics, Computer Science and Documentation, University of Jena, Jahnstraße 3, D-07740 Jena, Germany

1 Method

One approach for considering information processing is the use of EEG spectral analysis methods. Because of the fast changes in the EEG signal during cognitive processes, methods are necessary that allow a high time resolution. Principally, there are two approaches to time frequency analysis. The first is based on the convolution of the Fourier transform with special windows. Examples of this method are short-time Fourier transformation and Wavelet transformation. The second approach uses adaptive filtering methods for a parametric estimation of the instantaneous spectral matrix. In the one-dimensional case the theory of adaptive filters is well developed (see e.g. Haykin (1986)); examples are the Kalman algorithm, the RLS and the LMS algorithms. The extension of these methods to the multi-dimensional case is not trivial. In the following Sect. 1.1 a modified and generalized LMS algorithm for the estimation of time-dependent parameters of an ARMA model will be presented. The signal-dependent adaptive determination of the stepwidth of the algorithm leads to a stable estimation procedure. The time-dependent model parameters are then used for the parametric estimation of a time-dependent spectral density matrix as will be shown in Sect. 1.2.

1.1 Adaptive Fit of Linear Models with Time-Dependent Parameters

Let $\boldsymbol{X}= \{ \boldsymbol{X}_n = \left[X_n^1, ..., X_n^d\right]^T \}_{n=0,1,...}$ be a d-dimensional stochastic process with the expectation vector zero. Further, let $\boldsymbol{x}= \{ \boldsymbol{x}_n = \left[x_n^1, ..., x_n^d\right]^T \}_{n=0,1,...}$ be the observed d-dimensional signal values. This d-dimensional signal will be fitted by a d-dimensional ARMA(p,q) model which is able to react to structure changes of the signal. This condition is fulfilled by an autoregressive moving average model $\boldsymbol{Y}= \{ \boldsymbol{Y}_n = \left[Y_n^1, ..., Y_n^d\right]^T \}_{n=0,1,...}$ with time-varying parameters of the form

$$\boldsymbol{Y}_n + \sum_{k=1}^{p} A^k(n)\boldsymbol{Y}_{n-k} = \boldsymbol{Z}_n - \sum_{j=1}^{q} B^j(n)\boldsymbol{Z}_{n-j}. \qquad (1)$$

Thereby, p and q are the orders of the model, $\boldsymbol{Z}_n$ is a d-dimensional noise process, $\boldsymbol{Y}_n$ is the d-dimensional model process, and $A^k(n)$ and $B^j(n)$ are $d * d$ matrices of the autoregressive and moving average parameters

$$A^k(n) = \begin{pmatrix} a^k_{11}(n) \; a^k_{12}(n) \cdots a^k_{1d}(n) \\ a^k_{21}(n) \; a^k_{22}(n) \cdots a^k_{2d}(n) \\ \cdots\cdots\cdots\cdots\cdots\cdots\cdots \\ a^k_{d1}(n) \; a^k_{d2}(n) \cdots a^k_{dd}(n) \end{pmatrix} \tag{2}$$

and

$$B^j(n) = \begin{pmatrix} b^j_{11}(n) \; b^j_{12}(n) \cdots b^j_{1d}(n) \\ b^j_{21}(n) \; b^j_{22}(n) \cdots b^j_{2d}(n) \\ \cdots\cdots\cdots\cdots\cdots\cdots\cdots \\ b^j_{d1}(n) \; b^j_{d2}(n) \cdots b^j_{dd}(n) \end{pmatrix} . \tag{3}$$

In the case of stationary processes several estimation procedures of the parameter matrices are known (Brockwell and Davis (1991)). For one-dimensional instationary signals different techniques of adaptive filtering were used for fitting a time-changing autoregressive model to the signal (Haykin (1986)). The LMS algorithm may be generalized and modified for adaptive fitting ARMA models in the multi-dimensional case, as follows: the parameter matrices are changed at every sample point in a manner that minimizes the squared prediction error of the model using the stochastic gradient method. This approach leads to the following **Adaptive Estimation Procedure:**

$$\boldsymbol{e}_0 = 0 \tag{4}$$

$$\boldsymbol{e}_n = \boldsymbol{x}_n + \sum_{k=1}^{p} \hat{A}^k(n-1) \cdot \boldsymbol{x}^T_{n-k} + \sum_{j=1}^{q} \hat{B}^j(n-1) \cdot \boldsymbol{e}^T_{n-j} \tag{5}$$

$$\hat{A}^k(n) = 0 \text{ (null matrix), for } n \le k, \;\; k = 1, ..., p \tag{6}$$

$$\hat{B}^j(n) = 0 \text{ (null matrix), for } n \le j, \;\; j = 1, ..., q \tag{7}$$

$$\hat{A}^k(n) = \hat{A}^k(n-1) - c_n \cdot \boldsymbol{e}_n \cdot \boldsymbol{x}^T_{n-k} \;\; ; k = 1, ..., p \tag{8}$$

$$\hat{B}^j(n) = \hat{B}^j(n-1) - c_n \cdot \boldsymbol{e}_n \cdot \boldsymbol{e}^T_{n-j} \;\; ; j = 1, ..., q. \tag{9}$$

The stepwidth c_n is decisive for the stability and the adaptation speed of the model fitting procedure. In the case of the LMS algorithm ($c_n = c =$ const) for a stationary autoregressive process ($q = 0$) it can be shown that the condition

$$0 < c < \frac{2}{p \cdot d \cdot \sum_{i=1}^{d} \sigma_i^2}, \tag{10}$$

where σ_i^2 is the variance of the ith component of the signal, is sufficient for the mean convergence of the parameters of the model (Schack (1997)). In the case of instationarity the condition (10) cannot be guaranteed because of possible changes in the variance vector. But condition (10) may be displaced by the momentary analogon

$$0 < c(n) < \frac{2}{p \cdot d \cdot \sum_{i=1}^{d} \hat{\sigma}_i^2(n)}, \tag{11}$$

where $\hat{\sigma}_i^2(n)$ denotes the adaptive variance estimation of the ith component

$$\hat{\sigma}_i^2(0) = 0 \tag{12}$$
$$\hat{\sigma}_i^2(n) = \hat{\sigma}_i^2(n-1) - c_s \cdot \left(\hat{\sigma}_i^2(n-1) - (x_n)^2\right) \ , i = 1, ..., d, \ \ n = 1, 2, ... \tag{13}$$

with $0 < c_s < 1$. Because of (11) the time-varying stepwidth $c(n)$ for the adaptive estimation algorithm will be chosen in the following form:

$$c(n) = \frac{f}{1 + \sum_{i=1}^{d} \hat{\sigma}_i^2(n)}, \tag{14}$$

with the adaptation factor f fulfilling the condition

$$f < \frac{2}{(p+q) \cdot d}. \tag{15}$$

For the adaptive filtering of instationary signals besides the choice of the momentary stepwidth the problem of parameter drift arises (Sethares et al. (1986)). The model parameters can increase infinitely if no boundaries are defined for them. In the stationary case a natural limitation of the model parameters is given by the assumption of causality and invertibility of the ARMA(p,q) model. If the ARMA(p, q) model is a causal process then the following unequations hold for the model parameters (Schack (1997)):

$$\left| \sum_{j=1}^{d} a_{ij}^k \right| < \binom{p}{k}, \quad k = 1, ..., p \tag{16}$$
$$\left| \sum_{i=1}^{d} a_{ij}^k \right| < \binom{p}{k}, \quad k = 1, ..., p. \tag{17}$$

Further, the condition

$$\sum_{j=1}^{q} \sum_{i=1}^{d} \sum_{k=1}^{d} \left| b_{ik}^j \right| < 1 \tag{18}$$

guarantees the invertibility of the model process (Schack (1997)). Thus, a natural way to avoid parameter drift of the momentary model parameters in the instationary case is the assumption of the conditions (16) or (17) and (18) for every time point. Choosing the adaptive step width (14) with the adaptation factor f corresponding to (15) and taking into consideration the conditions (16), (17) and (18) we get a stable algorithm for the adaptive fitting of a multi-dimensional ARMA model with time-dependent parameters to instationary signals. A suitable choice of model orders p and q is important for the quality of the model fit. In the stationary case, there are different criteria for consistent estimations of the model orders (Hannan and Kavalieris (1984)). Because of the changes of the signal structure in the instationary case a consistent estimation of the model orders is not possible. However, it is important to choose the orders high enough for a sufficiently accurate model fit. A practicable way is to estimate the model orders for different intervals of the measured signal and to fix the maximal values for further analysis.

1.2 Multivariate Dynamical Spectral Analysis

The adaptive fitting procedure of linear models with time-dependent parameters enables the dynamic estimation of the spectral density matrix and in this way the adaptation of all spectral parameters to structural changes of the signal. For each sample point, the momentary transfer matrix $\boldsymbol{H}_n(\lambda)$ of the fitted ARMA model may be calculated by the formula

$$\boldsymbol{H}_n(\lambda) = \boldsymbol{A}_n^{-1}(\lambda) * \boldsymbol{B}_n(\lambda) \tag{19}$$

where

$$\boldsymbol{A}_n(\lambda) = \boldsymbol{I} + \sum_{k=1}^{p} \hat{A}^k(n) \cdot \exp(-ik\lambda) \tag{20}$$

and

$$\boldsymbol{B}_n(\lambda) = \boldsymbol{I} - \sum_{j=1}^{q} \hat{B}^j(n) \cdot \exp(-ik\lambda) \ . \tag{21}$$

At the same time, an adaptive estimation of the momentary covariance matrix

$$\boldsymbol{S}(n) = \begin{pmatrix} \hat{s}_{11}(n) & \hat{s}_{12}(n) & \cdots & \hat{s}_{1d}(n) \\ \hat{s}_{21}(n) & \hat{s}_{22}(n) & \cdots & \hat{s}_{2d}(n) \\ \cdots & \cdots & \cdots & \cdots \\ \hat{s}_{d1}(n) & \hat{s}_{d2}(n) & \cdots & \hat{s}_{dd}(n) \end{pmatrix} \tag{22}$$

of the prediction error vector $\boldsymbol{e}= \{\boldsymbol{e}_n = \left[e_n^1, ..., e_n^d\right]^T\}_{n=0,1,...}$ is possible in the following way:

$$\hat{s}_{ij}(0) = 0, i, j = 1, ..., d \tag{23}$$

$$\hat{s}_{ij}(n) = \hat{s}_{ij}(n-1) - c_s \cdot (\hat{s}_{ij}(n-1) - e_n^i \cdot e_n^j), \quad i, j = 1, ..., d \tag{24}$$

where c_s is a constant with $0 < c_s < 1$. Taking into consideration the adaptive model estimation and (19)–(24), it is possible to calculate the spectral density matrix at every sample point n by

$$\boldsymbol{f}_n(\lambda) = \boldsymbol{H}_n(\lambda) * \boldsymbol{S}_n(\lambda) * \boldsymbol{H}_n^{*T}(\lambda), \tag{25}$$

where $\boldsymbol{H}_n^{*T}(\lambda)$ denotes the complex conjugate and transpose of $\boldsymbol{H}_n(\lambda)$. Thus

$$\boldsymbol{f}_n(\lambda) = \begin{pmatrix} f_{11,n}(\lambda) & f_{12,n}(\lambda) & \cdots & f_{1d,n}(\lambda) \\ f_{21,n}(\lambda) & f_{22,n}(\lambda) & \cdots & f_{2d,n}(\lambda) \\ \cdots & \cdots & \cdots & \cdots \\ f_{d1,n}(\lambda) & f_{d2,n}(\lambda) & \cdots & f_{dd,n}(\lambda) \end{pmatrix} \tag{26}$$

denotes the momentary spectral density matrix. In such a way the parametric calculation (25) of the spectral density matrix of the adaptively fitted ARMA model enables the dynamic estimation of the spectra of a non-stationary signal. For $d = 1$ in particular we get the instantaneous auto power spectra for each signal component,

$$\hat{f}_{ii,n}(\lambda) \tag{27}$$

and for $d = 2$ the instantaneous coherence spectra and the instantaneous cross phase spectra for each pair of signal components:

$$\hat{\rho}^2_{ij,n}(\lambda) = \frac{\hat{f}_{ij,n}(\lambda)}{\hat{f}_{ii,n}(\lambda) \cdot \hat{f}_{jj,n}(\lambda)} \tag{28}$$

$$\hat{\phi}_{ij,n}(\lambda) = \arg \mathrm{Re}[\hat{f}_{ij,n}(\lambda)] + \mathrm{i} \cdot \mathrm{Im}[\hat{f}_{ij,n}(\lambda)], \tag{29}$$

where $\mathrm{Re}[\cdot]$ denotes the real part and $\mathrm{Im}[\cdot]$ denotes the imaginary part of a complex value. The detailed time-frequency analysis for signal power (27), coherence (29) and cross phase (29) may be used for finding frequency bands with characteristic changes in these spectral parameters. The limitation to sensitive frequency bands enables a necessary data reduction. As a result we get time curves of mean frequency band parameters for power, coherence and cross phase. For a chosen frequency band $[\lambda_{\mathrm{low}}, \lambda_{\mathrm{upper}}]$ the mean frequency parameters are computed by

$$\hat{f}_{ii}(n) = \frac{1}{\mathrm{card}} \cdot \sum_{\lambda_{\mathrm{low}} \leq \lambda_k \leq \lambda_{\mathrm{upper}}} \hat{f}_{ii,n}(\lambda_k) \tag{30}$$

$$\hat{\rho}^2_{ij}(n) = \frac{1}{\mathrm{card}} \cdot \sum_{\lambda_{\mathrm{low}} \leq \lambda_k \leq \lambda_{\mathrm{upper}}} \hat{\rho}^2_{ij,n}(\lambda_k) \tag{31}$$

$$\hat{\phi}_{ij}(n) = \frac{1}{\mathrm{card}} \cdot \sum_{\lambda_{\mathrm{low}} \leq \lambda_k \leq \lambda_{\mathrm{upper}}} \hat{\phi}_{ij,n}(\lambda_k) \tag{32}$$

for each sample point n and arbitrary signal component i or signal component pair i, j. With card the number of discrete frequency points with $\lambda_{\mathrm{low}} \leq \lambda_k \leq \lambda_{\mathrm{upper}}$ is denoted. The calculation of the instantaneous spectral densiy matrix (25) on the basis of the adaptive estimation procedure also allows the estimation of spectral parameters with dimensions higher than 2. If the components of a higher-dimensional signal have a determined spatial order, mapping procedures were usually used. Well-known examples are power and coherence maps. A dynamic topographic analysis is made possible by combining these mapping procedures and the estimation of frequency band parameters with high time resolution.

2 Possibilities for Investigating Cognitive Processes by Instantaneous Frequency Parameters

The EEG or MEG signals of cognitive processes are characterized by fast changes of the signal structure, especially in the frequency domain. The dynamic parametric approach for estimating the spectral parameters of a multi-dimensional signal continuously in time opens up new possibilities for considering such information processing. The principial methods of analysis are demonstrated in the following example: healthy volunteers had to memorize auditorily presented concrete nouns and to recall them. The EEG was recorded from 19 scalp electrodes (international 10/20 system with mean ear lobe reference) with a sampling frequency of 256 Hz. The duration of a single word representation was shorter than 1 sec. The end of the word presentation was marked and the pre- and postintervals both of 1 sec length were analysed.

The two curves at the bottom of Fig. 1 show the EEG signal on the electrode positions P4 and T6 during the presentation of the word "knife". The beginning of the presentation was 220 ms and the end was 1000 ms. In the upper images the corresponding two time-power analyses are illustrated. Fast changes of the lower frequency components with high amplitudes may be observed. Besides this, frequency components with small amplitudes within

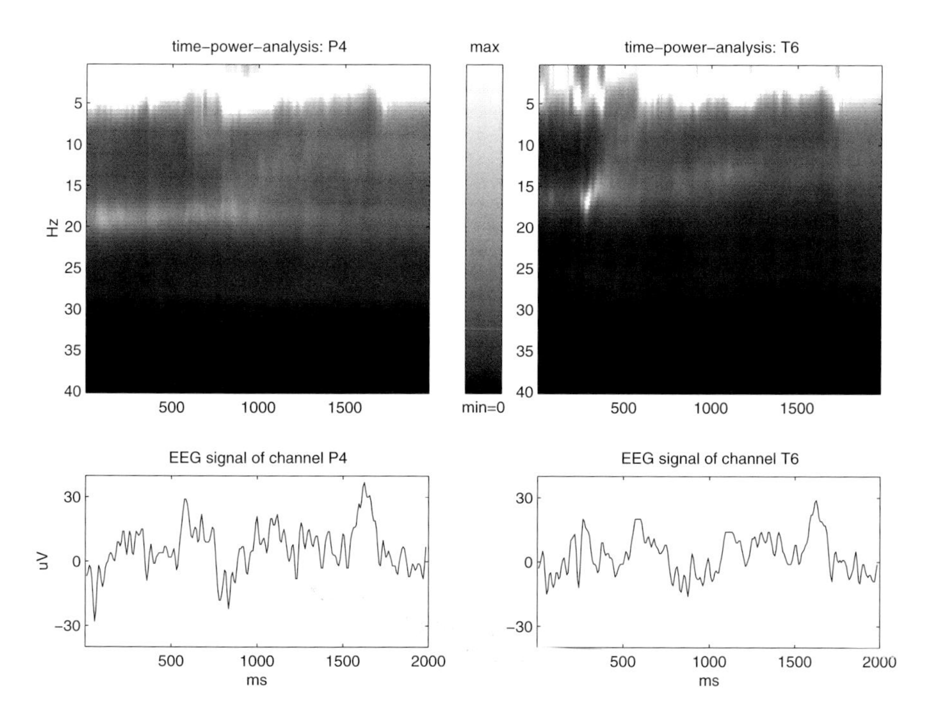

Fig. 1. Dynamic power analysis during word processing

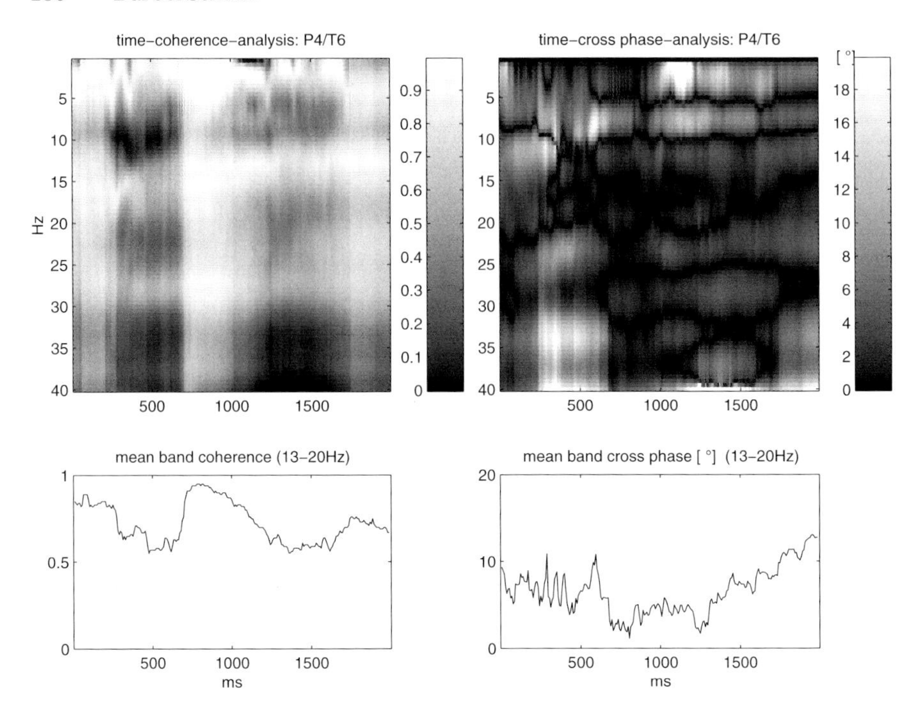

Fig. 2. Dynamic coherence and phase analysis during word processing

the beta1-frequency band (13–18 Hz) may be pointed out in the second half of the time interval. The question arises of possible synchronization of these two signals within different frequency bands. This problem can be solved by providing a time-coherence-analysis of the EEG signals at electrode positions P4 and T6, shown in the upper left image of Fig. 2.

The coherence function increases within the frequency bands 1–5 Hz, 13–18 Hz and 25–30 Hz in the time interval from 750 ms up to the end of the task. The curve at the bottom left shows the time evolution of the mean band coherence of the beta1-band (13–18 Hz). In the end of the word presentation there is a strong increase in band coherence. This phenomenon speaks for a high level of synchronization in the right parietotemporal cortex area. This supposition may be supported by considering the cross phase of the two EEG signals. The right upper image of Fig. 2 presents the time-cross phase-analysis. Within the dark areas there is a very low phase shift which can be seen especially in the beta1-frequency band in the second half of the time interval analysed. This suppression will be supported by the time curve of the mean frequency band cross phase (13-18 Hz) illustrated at the bottom right of the image. Because of the proximity of the electrodes O4 and T6 the low cross phase during the time interval with high coherences (nearly 750-1250 ms) points to a high level of phase synchronization. Using mapping procedures based on band powers and local band coherences the examinations may be extended to the topography of the whole cortex. For the local coherence maps the coherences between adjacent electrodes along the transverse and longitudinal electrode rows were calculated and fictive coherence electrode positions were placed in the middle of the electrode pairs. The linear interpolation results in the functional local coherence map (Rappelsberger and Petsche (1988)).

Because of the continuous estimation of the spectral parameters in time we get mapping sequences for band power and band coherence with a time resolution which depends on the sampling frequency. Figure 3 shows the sequences of power (A) and of mean local coherences (B) for the beta1-frequency band for the whole single task. At the end of the presentation of the word "knife" an increase of both the power and the coherence in the right parietotemporal cortex may be seen. Moreover, in the whole interval of word presentation there are high coherences in the frontal area, which vanish afterwards. These coherences in the frontocentral area are not accompanied by high band powers. This example demonstrates the fast changes in the EEG during information processing.

3 The Sensitivity of Instantaneous Coherence for Considering Thinking Processes

3.1 The Relationship Between the Time Evolution of Information Processing and Coherent Events

In EEG analysis of cognitive processes, the coherence function is used for describing the functional relationship between different areas of the cortex by several authors (e.g., Rappelsberger and Petsche (1988), Petsche et al. (1992), Weiss and Rappelsberger (1996),Thatcher et al. (1986)). Most of the authors show that EEG coherence more or less adequately reflects mental activity. One reason for this phenomenon may be the close connection between the coherence function and the negative of entropy. Gelfand and Yaglom (1959) developed the quantitative measure of the amount of information between two vectors of continuously distributed random variables X and Y. This measure is defined by

$$I_{x,y} = \int_{-\infty}^{+\infty}\int_{-\infty}^{+\infty} f_{X,Y}(x,y)\ln[\frac{f_{X,Y}(x,y)}{f_X(x)\cdot f_Y(y)}]\mathrm{d}x\cdot\mathrm{d}y, \tag{33}$$

where $f_X(x), f_Y(y)$ are the probability density functions of the individual random variables and $f_{X,Y}(x,y)$ is the joint probability density function. The form (33) of this measure is well known in statistics as the negative of the entropy of the true distribution $f_{X,Y}(x,y)$ with respect to an assumed distribution of independent variables, $f_X(x)\cdot f_Y(y)$. The measure $I_{x,y}$ between the two vectors of stationary, jointly normally distributed time series may be expressed by terms of the coherence function

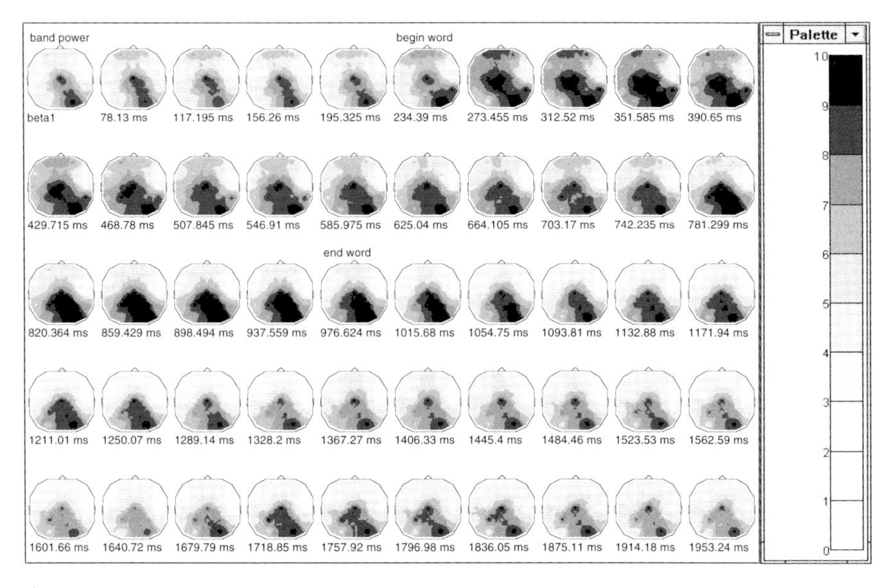

A

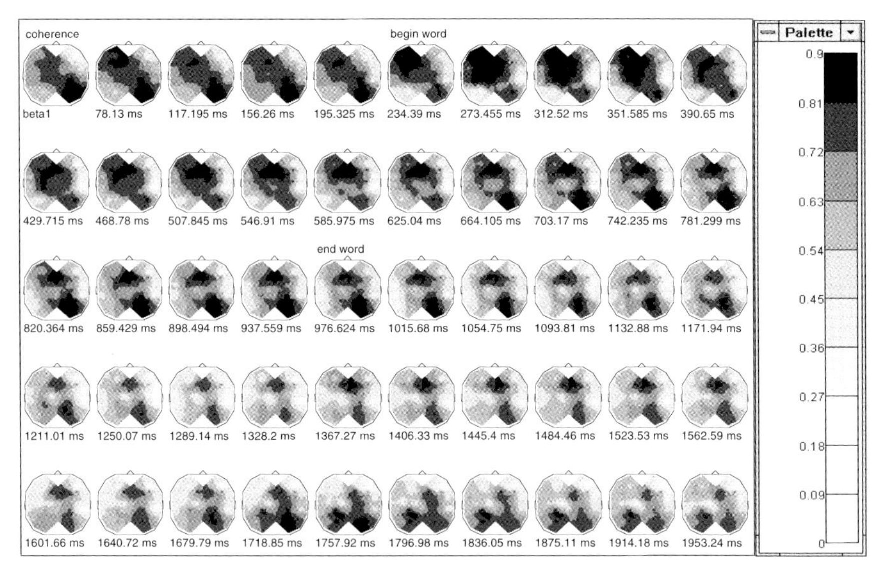

B

Fig. 3. Dynamic mapping of power (A) and coherence (B) within the frequency band (13–20 Hz)

$$I_{x,y} = - \int_{0}^{f_{Nyq}} \ln[1 - \rho^2_{x,y}(\lambda)] \mathrm{d}\lambda, \tag{34}$$

whereby f_{Nyq} denotes the Nyquist frequency and $\rho^2_{x,y}(\lambda)$ the quadratic coherence function between two signals x and y (Gelfand and Yaglom (1959)). The form (34) of the negative of entropy includes the possibility of restriction to sensitive frequency bands. Because of the monotony of the logarithm function the negative of entropy and the coherence function describe an increase in synchronization between two signals in a very similar manner. Furthermore, the coherence function is a normalized measure. This property is advantageous for comparing different tasks or results for different subjects. In this section the connection between cognitive processes and the coherence function will be shown. Because of the processual character of mental activity the dynamic approach explained in the two Sects. 1.1 and 1.2 will be applied.

Elementary thinking processes are used for demonstrating the sensitivity of the instantaneous coherence. These processes represent different modalities of human information processing and furthermore, the time evolution of these cognitive tasks may be influenced by an external parameter.

For elementary cognitive processes of concept activation and pattern comparison the following experiment was conducted (Krause et al. (1997)). The volunteer presses a button, and after 600 ms a question "Identical name?" or "Identical pattern?" and a first item -a letter or a point pattern (Garner figure)- are to be seen on a screen. The second item is given after a variable interstimulus interval (ISI). The volunteer has to answer "yes" or "no" as quickly as possible by pressing the corresponding button. Four different situations are considered: In the first case, comparing two letters written in different fonts (a small letter and a large one) and answering the question "Identical name?", a category concept must be activated. We call it concept activation with forced choice (c.a. forced choice). In the second case, comparing two letters written in the same font and answering the question "Identical name?", it is not necessary to activate a category concept. Because this problem may be solved by pattern comparison this situation is called concept activation with free choice (c.a. free choice). In the third case the volunteer has to answer the question "Identical pattern?", comparing two letters written in the same font. He/she can solve the problem by comparing patterns or activating a name. This case is called pattern comparison with free choice (p.c. free choice). The tasks of these three conditions were randomly mixed. In the fourth situation the volunteer has to compare two simple point patterns. This task was used as a reference situation and is called pattern comparison with forced choice (p.c. forced choice).

The ISI was changed in steps of 100 ms from 100 ms up to 600 ms. The aim of the following study was to consider a possible dependence of the time evolution of the coherence function on the length of the interstimulus interval and the discrimination of the different situations.

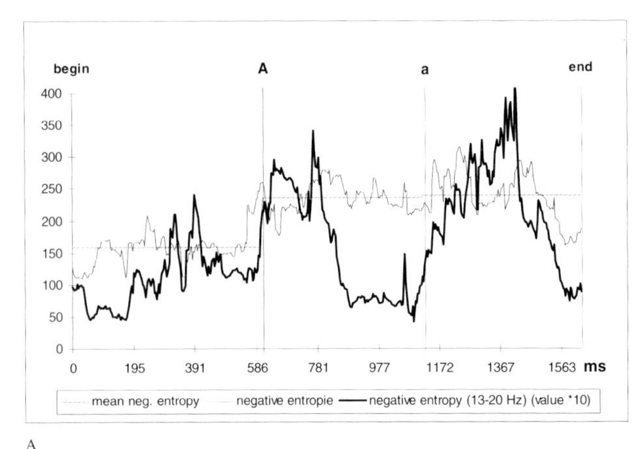

A

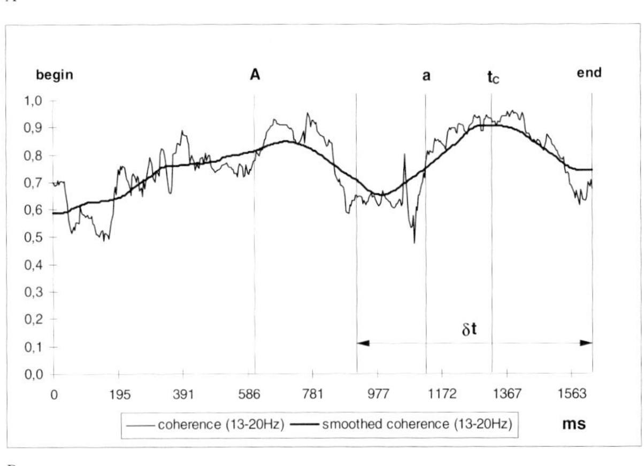

B

Fig. 4. Description of synchronization phenomena by instantaneous negative entropy (A) and adaptive coherence (B) during a single task of concept activation with forced choice

Detailed time-coherence analysis has shown characteristic changes within the frequency band of 13–20 Hz (Schack and Krause (1995)) during cognitive tasks. The importance of the high time resolution and of the restriction to sensitive frequency bands is demonstrated in the upper image (A) of Fig. 4. Different calculations of the negative entropy for the EEG of the electrode pair T5/P3 are shown during a whole single task of concept activation with forced choice. The gray step function consists of the mean values of the negative entropy within the interval from the start up to the first item, within the time interval between showing the first letter and showing the second letter and within the time interval between showing the second letter and the end of the task. An increase in the negative entropy may be observed clearly after the appearance of the first letter. The negative entropy function with high time evolution (thin curve) shows dynamic behaviour. This dynamic behaviour of the negative entropy increases rapidly after restriction to the frequency band 13–20 Hz (thick curve: time evolution of the negative entropy restricted to 13–20 Hz and multiplied by 10). The instantaneous coherence function of the frequency band 13–20 Hz for the electrode pair T5/P3 shows the same dynamic behaviour as the corresponding entropy (thin curve in the lower image (B) of Fig. 4). If a functional relationsship exists between the external parameter ISI of the task and the evolution of the coherence, the time point of the increase of the mean band coherence $\rho^2_{[13,20]}(t)$ after the presentation of the second item must change with the change of the ISI. In order to exclude a random increase in the coherence after showing the second item, the following calculations for 30 local coherences, for all four situations, and for all ISI were made.

Firstly, the time curve of the coherence function was smoothed by a rectangular window with a width of 300 ms (thick curve in image (B) of Fig. 4). After that the time point of maximal coherence t_C

$$\rho^2_{[13,20]}(t_C) \geq \rho^2_{[13,20]}(t) \ , \ \forall t \in \delta t \ , \delta t = [900\,\text{ms, end of the task}], \qquad (35)$$

was calculated from 300 ms after the appearance of the first item on the screen up to the end of the task. This procedure was done for all single trials, after which the time points of maximal coherence were averaged. An increase of the time point of maximal band coherence (13–20 Hz) which depends on increasing the ISI could be observed for all 30 neighboured electrode pairs and all four comparison procedure situations. Furthermore, in the case of concept activation with forced choice the time points of maximal coherence are later than in all other cases, especially in the left pariototemporal area. Image (A) in Fig. 5 shows this functional relationship between the ISI and the averaged results (from four volunteers) of the time points of maximal coherence in the left parietotemporal area (mean of the electrode pairs T5/P3, P3/Pz, T5/O1 and P3/O1) for all four comparison tasks.

If the increase in coherence reflects synchronization processes for activation phases during the comparison procedures it is meaningful to compare the maximum values of the smoothed coherence corresponding to the time points of maximum coherence for the different situations. Image (B) in Fig. 5 shows the maps of mean maximal local coherences for the four comparison procedures situations. Obviously, the highest synchronization processes may be observed in the frontocentral and the left and right parietotemporal areas of the cortex for all cases. Further, in the situation of concept activation with forced choice there are higher coherences in the left parietotemporal area in comparison with all of the other three situations (see image (C) in Fig. 5). Discrimination of the different situations by means of maximal coherences for the frequency band 13–20 Hz could be ensured statistically. Table 1 shows the result of comparing maximal local coherences of the situation concept activation with forced choice with the other three situations on the basis of the Wilcoxon test with paired samples for 7 subjects and the ISI of 300 ms.

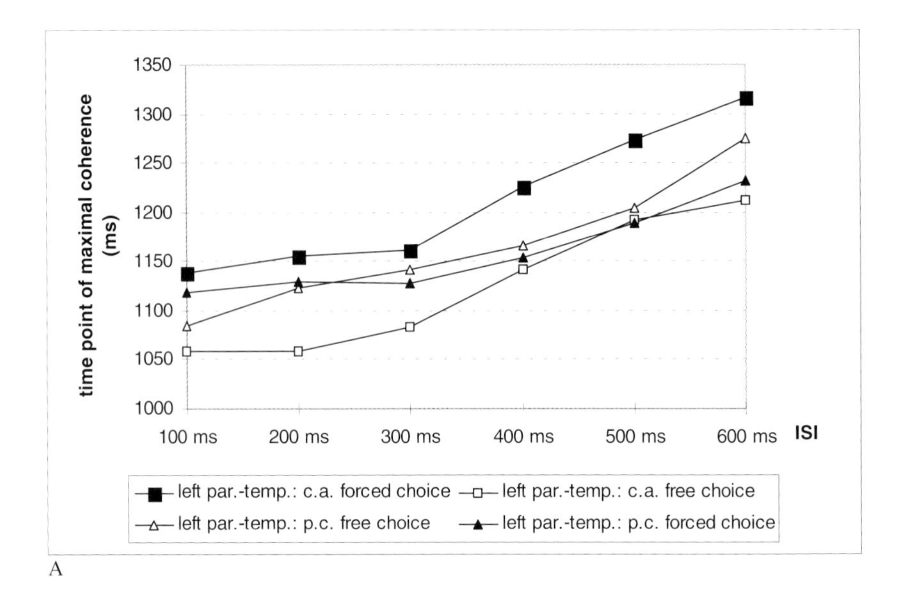

A

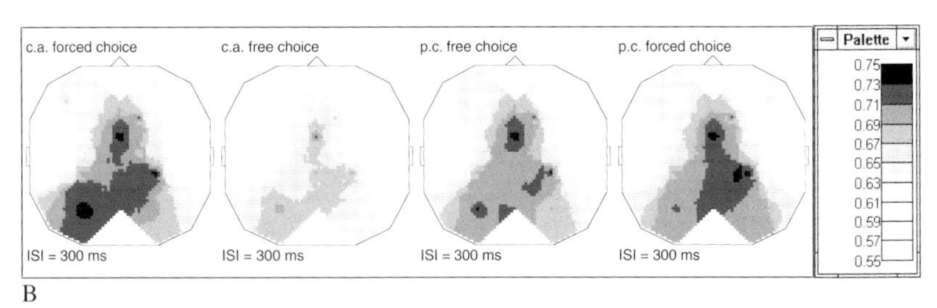

B

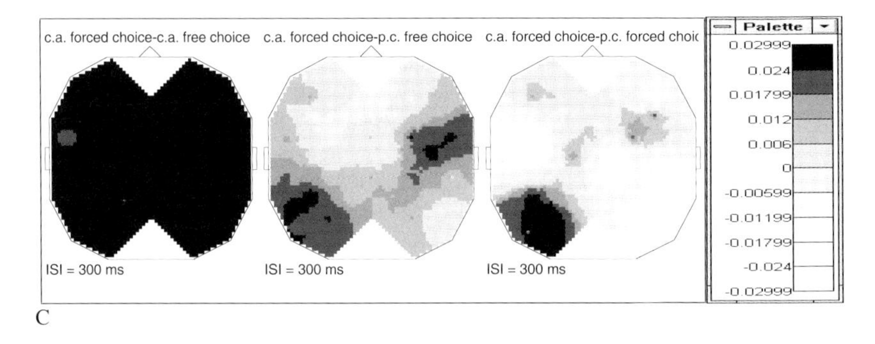

C

Fig. 5. Dependence of time points of maximal coherence in the left parietotemporal area on the ISI (A). Local coherence maps with areas of high synchronization (B) and difference maps between c.a. with forced choice and the other three situations (C) for the ISI = 300 ms

Table 1. Electrode pairs with significant differences between maximal coherences of concept activation with forced choice and the other situations

ISI = 300 ms	significance level: 0,05	significance level: 0,1
concept activation with free choice	T5/P3,T5/O1,P3/O1,T5/T3, P3/C3,P3/Pz,C3/Cz,Cz/C4 Pz/P4,P4/C4,C4/F4,Fp1/F3 F3/Fz,F4/F8,F4/Fp2,F8/Fp2	Fz/Cz,Cz/Pz,P4/O2
pattern comparison with free choice	T5/P3,T5/T3,P3/O1,C4/T4	T5/O1,F8/T4
pattern comparison with forced choice	T5/P3	T5/O1,F4/C4

In the case of concept activation with forced choice there are higher coherences in the left parietotemporal area in comparison with all of the other three situations where the comparison procedure may be solved by pattern comparison. Discrimination between such elementary cognitive processes is only possible because a spatiotemporal approach is used.

3.2 Similarities and Differences Between EEG and MEG Coherences

The results mentioned in Sects. 2 and 3.1 were obtained on the basis of ear lobe reference. Several authors (Fein et al. (1988), Nunez et al. (1997), Essl and Rappelsberger (1998)) have investigated the influence of the reference position on the value of the coherence. In Sect. 3.1 a strong relationship between the time evolution of information processing and the dynamical changes in instantaneous EEG coherence with ear lobe reference was shown. In order to have a reference-independent parameter for studying information processing we calculated the coherence for pairs of MEG channels in the left parietotemporal area. The MEG with a dewar with 31 channels located over the left parietotemporal region of the cortex were registered simultaneously with 19 channels (international 10/20 system with linked ears reference) of the EEG during the elementary comparison procedures for the interstimulus intervals 300 ms and 500 ms as explained in Sect. 3.1. The similar dynamic behaviour of the EEG coherences (ear lobe reference) and the MEG coherence within the frequency band (13–20 Hz) is demonstrated by their time evolution during a single task (image (A) of Fig. 6).

The increase of the local EEG and MEG coherences after the presentation of the second letter in the case of concept activation can be seen very clearly and the time evolution between the presentation of the first letter and the end of the task is similar. For a statistical evaluation, the time points of maximal coherence for all four tasks and the ISIs of 300 ms and 500 ms (3 subjects) were calculated for both the EEG and the MEG. Image (B) in Fig. 6 shows the result. The time points of maximal local synchronization in the

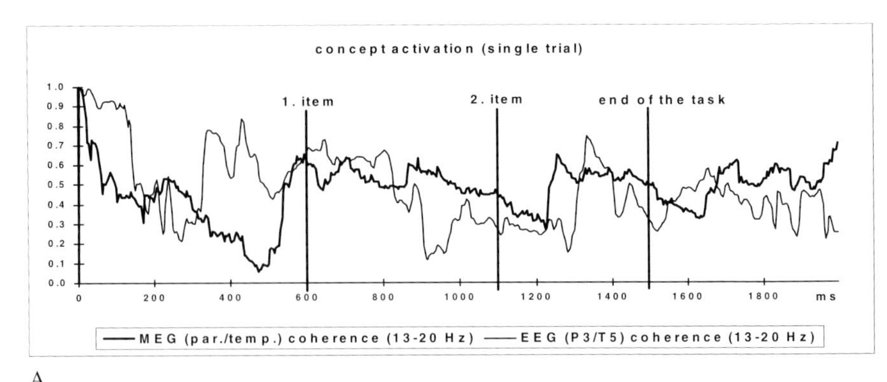

A

B

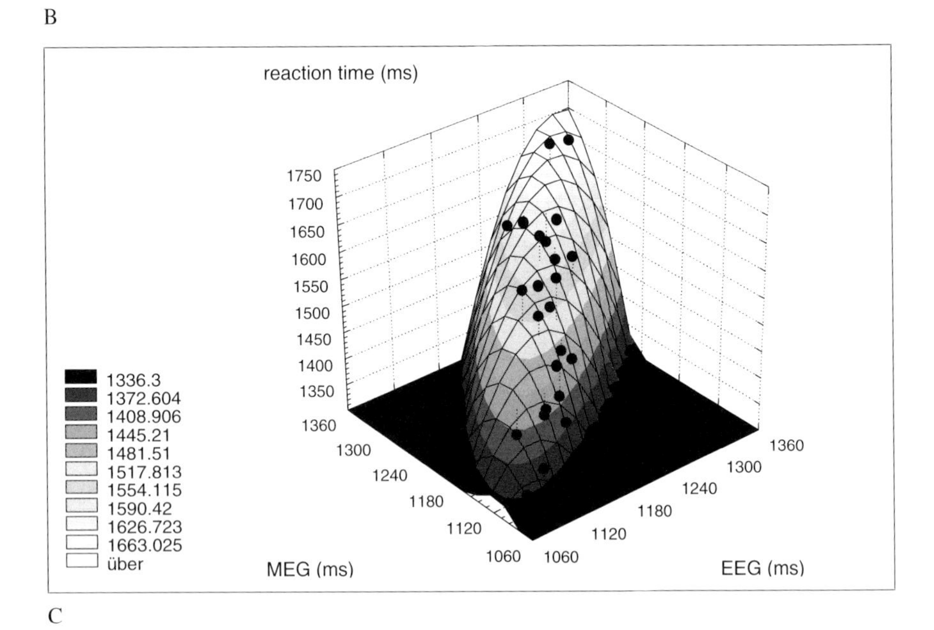

C

Fig. 6. Time evolution of adaptive EEG and MEG coherences (13–20 Hz) for a single task (A) and their mean time points of maximal coherence (B). Correlations between reaction time and time points of maximal EEG and MEG coherences (C).

left parietotemporal area are equal in a statistical sense. Further, there seems to be a strong relationship between the time points of maximal coherence and the reaction times. Indeed the three-dimensional scatterplot in image (C) of Fig. 6 illustrates the significant correlation between the reaction time and the time points of maximal EEG and MEG coherence within the frequency band 13–20 Hz.

Further, the simultaneous examination of EEG and MEG coherence have shown that there are many more disturbances in the EEG coherence which may be caused by volume conduction.

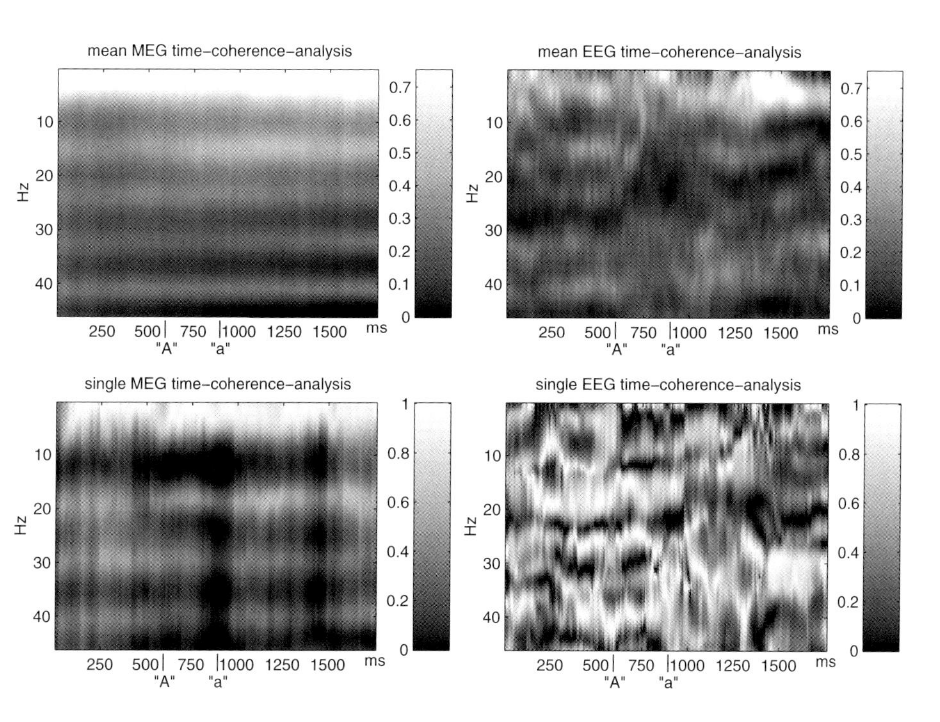

Fig. 7. Comparison of time-coherence-analyses of EEG and MEG

These properties of EEG and MEG coherence can be clearly observed by comparing images in Fig. 7. The two upper images are averaged time coherence analyses (3 subjects, 180 single trials) for the MEG and EEG in the case of concept activation with forced choice (ISI = 300 ms). Within the MEG plot the sensitive frequency bands may be seen very clearly. Besides an increase in coherence within the band 13–20 Hz after the second item similar dynamics may be seen in the lower frequencies. For the EEG there are increasd coherences within the same frequency bands. The blurring effect for the EEG can be explained by considering the time-coherence-analysis of a single trial (two lower images in Fig. 7). Both the blurring effect within the

EEG and the appearance of higher values of local EEG coherences may be caused by the volume conduction of the EEG. However, both EEG and MEG instantaneous coherence reflect the time evolution of a cognitive process in a similar manner.

4 Dynamic Topographical Coherence Analysis

The dynamic approach for considering cognitive processes by instantaneous coherence will be applied now to the Stroop test for colors in a global topographical manner.

The experimental situation was the following: the word of a color appears on a screen where the color in which the word is written either agrees with the word or not. The volunteer has to say the color of the word. Two situations differ: the congruent situation where the color and the word agree and the incongruent situation where the color and the word do not agree. 10 male volunteers participated in the study. During the experiment the EEG was registered according to the international 10/20 system with linked ears reference. Further, the speech was recorded to prove the truth of the answers and to registrate the reaction time.

The aim of the study was to detect areas of the cortex where synchronization effects are generated during the information processing and to find synchronization parameters which allow differences between the congruent and the incongruent situations. From many psychologists' examinations (e.g., Glaser (1992)) it is well-known that the reaction time is longer in the incongruent case. Different psychophysiological investigations have shown that the prefrontal area is involved during the Stroop task (e.g., Wendrell et al. (1995)). David (1992) has studied the influence of hemisphere specialization on the analysis of Stroop stimuli. For this reason a global analysis strategy concerning the whole cortex and the relationships between different areas of the cortex is necessary. Firstly, the reaction times were proved. We got a mean reaction time of 793 ms in the incongruent case and 708 ms in the congruent case, which agree with the results known from the literature. Using the experiences of the Stroop task mentioned above, time-coherence-analysis was firstly made for different neighboured electrode pairs in the prefrontal and left and right parietotemporal areas. An increase in coherence function during the task could be observed again in the frequency band 13–20 Hz. For this reason further considerations were restricted to this frequency domain. For a topographic survey the time functions of mean band coherences (13–20 Hz) for all neighboured electrode pairs along horizontal and longitudinal directions (so-called local coherences) were calculated. Using the special mapping procedure for local coherences we get mapping sequences of instantaneous band coherences. The upper image (A) in Fig. 8 shows the sequence of local band coherences (13–20 Hz) in the incongruent case (averaged over all persons). Obviously, there are centres of local synchronization within the

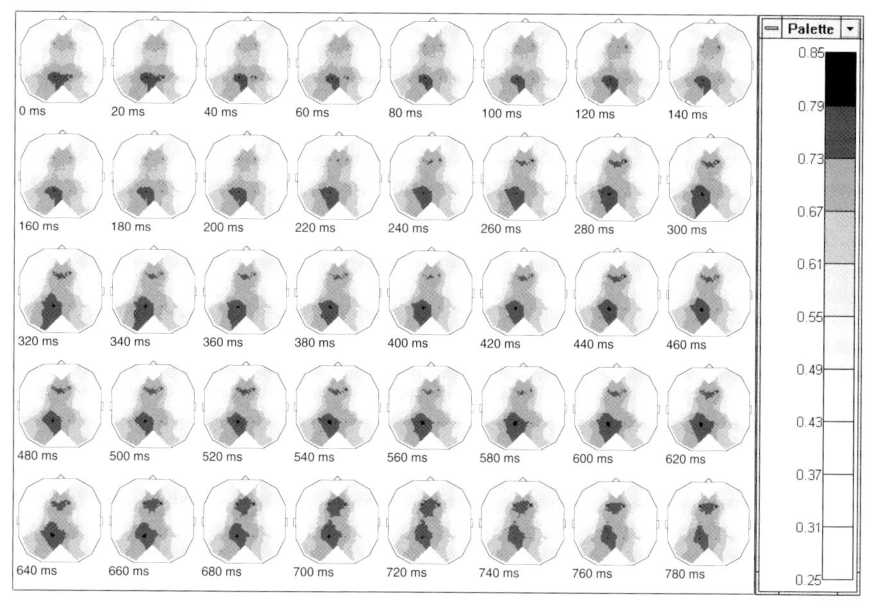

A

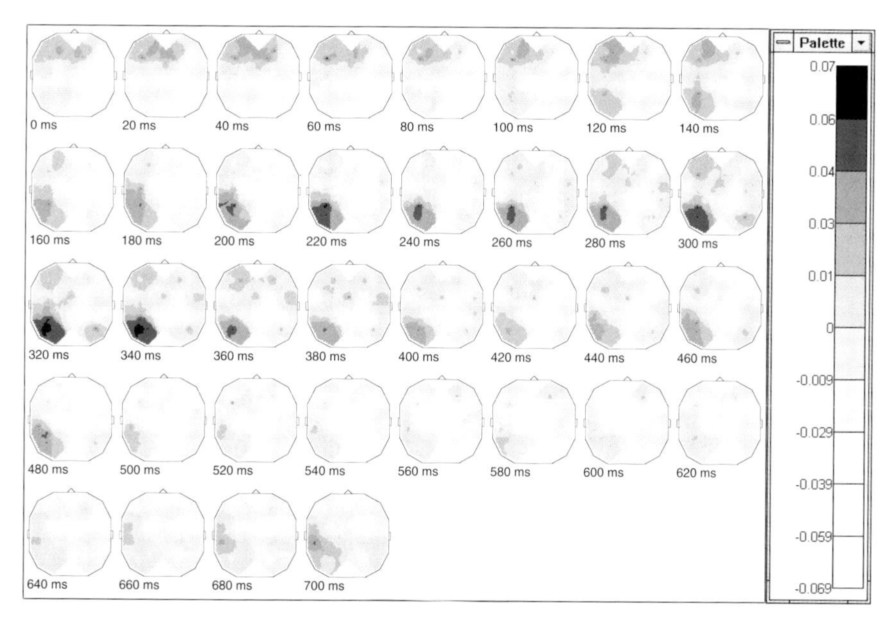

B

Fig. 8. Map sequence of local coherences (13–20 Hz) during the incongruent situation (A) and the corresponding difference sequence to the congruent case (B)

prefrontal and the parietal areas. Furthermore, in a defined time interval from approximately 200 ms up to 380 ms the band coherence in the left parietotemporal cortex is higher in the incongruent case than in the congruent case, as can be seen by the difference sequence (incongruent case - congruent case) of local band coherences in the image (B) of Fig. 8. From these results it is interesting to ask about synchronization between the frontal and parietal areas. For this reason the frontoparietal coherences between the electrode pairs F3/P3, F3/Pz, F3/P4, Fz/P3, Fz/Pz, Fz/P4, F4/P3, F4/Pz and F4/P4 were calculated.

In order to finde the time interval of maximal synchronization phenomena the time points of maximal coherence within the whole time interval of the task (from presentation of the word up to the beginning of speech) were calculated in the manner explained in Sect. 3.1. The mean time point of maximal frontoparietal coherences was later in the incongruent case, as is shown in Table 2.

Table 2. Time points of maximal band coherence (13–20 Hz) during the Stroop task

el.pairs	F3/P3	F3/Pz	F3/P4	Fz/P3	Fz/Pz	Fz/P4	F4/P3	F4/Pz	F4/P4
incongr.(ms)	394	389	412	367	381	412	351	383	416
congr.(ms)	317	308	336	326	321	329	316	327	307

Obviously, this result conforms with the corresponding relationship of the reaction times. A first differentiation between the congruent and incongruent situations is possible by means of the values of maximal coherence. There are higher maximal coherences in the incongruent case for the electrode pairs F3/Pz, F3/P4 and Fz/P4 and a lower maximal coherence for the electrode pair F4/Pz (significance level: 0,05; Wilcoxon test with paired samples). This result points to the influence of hemisphere specialization on the analysis of Stroop stimuli.

Furthermore, it is interesting to look for the correlation between the reaction times and the time points of maximal coherences. The result is illustrated in Fig. 9. There are significantly high correlations for all frontoparietal electrode pairs besides F4/P3 in the incongruent case, whereas in the congruent case there exist high correlations only for the electrode pairs Fz/Pz, Fz/P4, F4/Pz and F4/P4.

The scatterplots in Fig. 9 show the high correlation between the reaction time and synchronization times in the right hemisphere for both cases whereas in the left hemisphere such a high correlation appears only in the incongruent case.

In summary, spatiotemporal coherence analysis enables discrimination between the congruent and incongruent situations of the Stroop task.

significant correlations

between reaction time and time points of maximal frontoparietal coherences

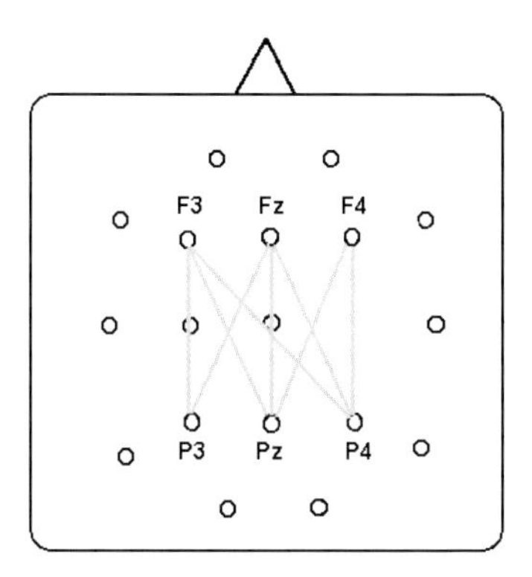

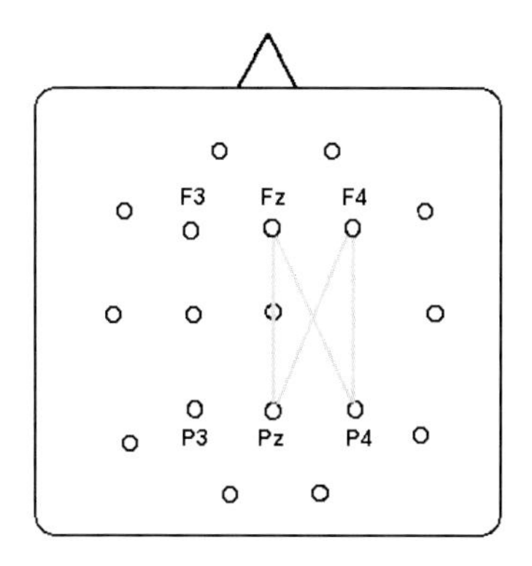

incongruent case
max. correlation coefficient: F3/Pz
time point of max. coherence: 399 ms

congruent case
max. correlation coefficient: Fz/P4
time point of max. coherence: 321 ms

scatterplots

between reaction time and time points of maximal frontoparietal coherences

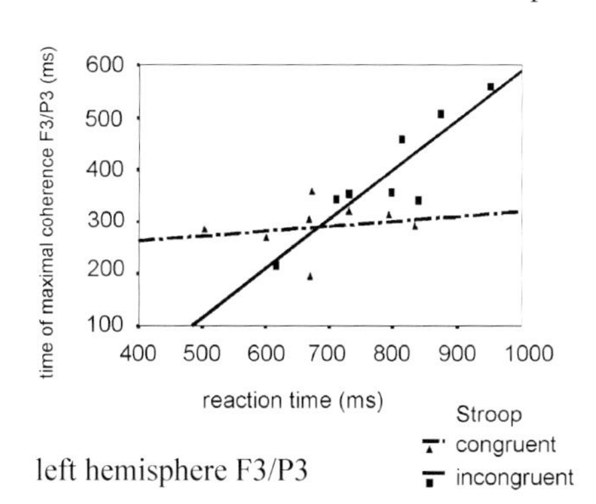

left hemisphere F3/P3

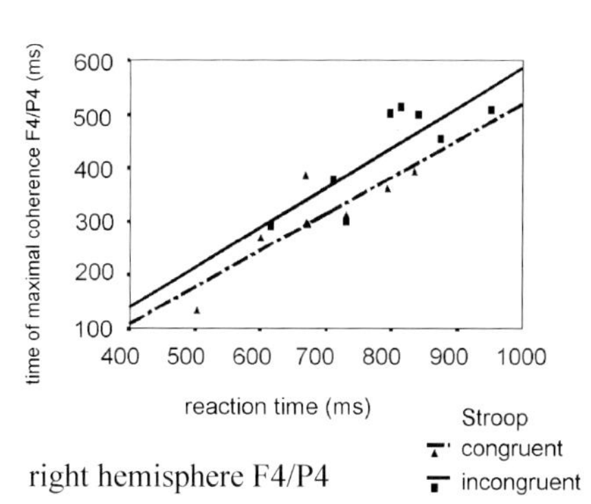

right hemisphere F4/P4

Fig. 9. Frontoparietal electrode pairs with significant high correlations between reaction time and synchronization time. The correlations in the incongruent (thick lines) and the congruent case (dotted lines)

5 Discussion

In Sect. 1 a general approach for dynamic multivariate spectral analysis is presented. This method allows calculation of the time evolution of arbitrary spectral parameters of a multi-dimensional signal. The temporal description of a cognitive process by dynamic power, coherence and cross phase analysis is demonstrated (Sect. 2). Instantaneous coherence in particular seems to be a sensitive parameter for reflecting synchronization phenomena during cognitive processes (Sect. 3). Because of the dependence of EEG coherence on the reference a simultaneous analysis of EEG coherence with ear lobe reference and MEG coherence was executed. The time behaviour of both instantaneous coherences was similar, whereas there are differences between them concerning the height of the coherence values and the clarity of sensitive frequency bands. The results concerning the Stroop task in Sect. 4 demonstrate that the investigation of local adaptive coherences and of adaptive coherences between electrode positions in different regions of interest and the inclusion of mapping procedures allow both a spatial and a temporal analysis of brain functioning. The results presented were obtained by the application of one- and two-dimensional spectral parameters. The method introduced in Sect. 1 also allows the calculation of higher order spectral parameters with high time resolution. Because synchronization effects may contain whole cortex areas the multiple EEG/MEG coherence seems to be such an interesting parameter. Because of the restriction of the EEG and the MEG to measurements on the surface of the head the spatial resolution is low. A combination of imaging methods like fMRI and PET with high spatial resolution and dynamic analyis methods of the EEG/MEG with high time resolution would rapidly increase knowledge about brain functioning.

Acknowledgements: I would like to thank G. Grießbach of the Department of Biomedical Engineering of the university of Ilmenau and J. Bolten of the Institute of Medical Statistics, Computer Sciences and Documentation of the university of Jena for their extensive help in preparing the necessary software. Further, I would like to thank W. Krause and B. Kriese, Institute of Psychology of the university of Jena, P. Rappelsberger and S. Weiss , Institute of Neurophysiology of the university of Vienna, A. Chen, Human Psychology and Pain Research Laboratory of the university of Manchester and H. Nowak, Biomagnetic centre of the university of Jena for recording and preparing electroencephalogram and magnetoencephalogram data and for helpful discussions.

The study was supported by the Deutsche Forschungsgesellschaft (project Scha 741/1-1).

References

Brockwell, P.J., Davis, R.A. (1991): *Time Series: Theory and Methods* (Springer, New York)

David, A.S. (1997): Stroop effects within and between the cerebral hemispheres: studies in normals and acallosals. Neuropsychlogia **30**, 161–175

Essl, M., Rappelsberger, P. (1998): EEG Coherence and reference signals: experimental results and mathematical explanations Med. Biol. Eng. Comput. **36**, 1–8

Fein, G., Raz, J., Brown, F.F., Merrin, E.L.(1988): Common reference coherence data are confounded by power and phase effects. Electroencephal. clin. Neurophysiology **69**, 581–584

Gelfand, I.M., Yaglom, A.M. (1959): Calculation of the amount of information about a random function contained in another such function. Amer. math. Soc. Transl. **12**, 199–24

Glaser, W.R. (1992): Picture naming. Cognition **42**, 61–105

Hannan, E.J., Kavalieris, L. (1984): Multivariate Linear Time Series Models. Biometrika **69**, 81–94

Haykin, S. (1986): *Adaptive Filter Theory* (Prentice Hall, Englewood Cliffs, New Jersey)

Krause, W., Schack, B., Gibbons, H., Kriese, B.,(1997): Über die Unterscheidbarkeit begrifflicher und bildhaft-anschaulicher Repräsentationen bei elementaren Denkanforderungen. Z Psychol **205**, 169–203

Nunez, P.L., Srinivasan, R., Wijesinghe, R.S., Westdorp, A.F., Tucker, D.M., Silberstein, R.B., Cadusch, P.J. (1997): EEG CoherencyI: Statistics, Reference Electrode, Volume Conduction, Laplacians, Cortical Imaging, and Interpretation at Multiple Scales. Electroencephal. clin, Neurophysiology **103**, 499–515

Rappelsberger, P., Petsche, H. (1988): Probability Mapping: Power and Coherence Analyses of Cognitive Processes. Brain Topography **1**, 46–54

Petsche, H., Lacriox, D., Lindner, K., Rappelsberger, P., Schmidt-Heinrich, E. (1992): Thinking with images or thinking with language: a pilot EEG probability mapping study. International Journal of Psychphysiology **12**, 31–39

Schack, B., Bareshova, E., Grießbach, G., Witte, H. (1994): Methods of dynamic spectral analysis by self–exciting ARMA models and their application to analysing biosignals. Med. Biol. Eng. Comput. **33**, 492–498

Schack, B., Grießbach, G., Arnold, M., Bolten, J. (1995): Dynamic cross-spectral analysis of biological signals by means of bivariate ARMA processes with time-dependent coefficients. Med. Biol. Eng. Comput. **33**, 605–610

Schack, B., Krause, W. (1995): Dynamic Power and Coherence Analysis of Ultra Short-Term Cognitive Processes - a Methodical Study. Brain Topography **8**, 127–136

Schack, B. (1997): *Adaptive Verfahren zur Spektralanalyse instationärer mehrdimensionaler biologischer Signale* (Thesis, Technical University Ilmenau)

Sethares, W.A., Lawrence, D.A., Johnson, C.R., Bitmead, R.R. (1986): Parameter Drift in LMS Adaptive Filters. IEEE Trans. ASSP **4**, 868–879

Thatcher, R.W., Krause, P.J., Hrybyk, M. (1986): Cortico-cortical associations and EEG coherence: a two-compartmental model. Electroencephal. clin. Neurophysiology **64**, 123–143

Weiss, S., Rappelsberger, P. (1996): EEG coherence within the 13–18 Hz band as a correlate of a distinct lexical organisation of concrete and abstract nouns in humans. Neuroscience Letters **209**, 17–20

Wendrell, P., Junque, C., Pujol, J., Jurado, M.A., Molet, J., Grafman, J. (1995): The role of prefrontal regions in the Stroop task. Neuropsychologia **33**, 341–352

Complex Phase Synchronization in Neurophysiological Data

Peter Tass[1], Jürgen Kurths[2], Michael Rosenblum[2], Jörg Weule[1], Arkady Pikovsky[2], Jens Volkmann[1], Alfons Schnitzler[1], and Hans-Joachim Freund[1]

[1] Department of Neurology, Heinrich-Heine-University, Moorenstr. 5, D-40225 Düsseldorf, Germany
[2] Department of Theoretical Physics, University of Potsdam, Am Neuen Palais 19, PF 601553, D-14415 Potsdam, Germany

1 Introduction

Synchronization is a general phenomenon that appears in various fields of science (Andronov et al. 1966, Haken 1983, Blekhman 1988, Hayashi 1964, Landa 1996). As a consequence of its generality it lacks a unique definition (Blekhman 1988, Blekhman et al. 1995). In this chapter we understand synchronization in terms of phase synchronization (Rosenblum et al. 1996, Rosenblum et al. 1997a, Rosenblum et al. 1997b, Pikovsky et al. 1997a, Pikovsky et al. 1997b), i.e. adjustment of rhythms of two interacting self-sustained oscillators. This type of synchronization will be explained below. The concept of phase synchronization was already used for the analysis of bivariate data and revealed different synchronization patterns occuring in visually guided tracking movements (Tass et al. 1996) and in the cardiorespiratory interaction (Schäfer et al. 1998).

Several central issues in neuroscience also refer to phase synchronization processes (see, e.g., Singer and Gray 1995). For this reason we have developed a method for the detection of $n{:}m$ phase synchronization in magnetoencephalography (MEG) data and related signals (Tass et al. 1998). We apply our method here in order to analyse MEG data and electromyogram (EMG) data from patients suffering from Parkinson's disease. The spatio-temporal synchronization patterns revealed by our approach will be compared with results of previous studies. This will point out that our single run analysis tool provides us with additional information which is important from the physiological point of view.

2 Synchronization in the Nervous System

This section is divided into two parts. The first part is devoted to the Parkinsonian resting tremor, especially, we study the synchronization between tremor and cerebral rhythmic activity. We have chosen this topic because a variety of findings revealed by different techniques (intracortical recordings, current dipole analysis of evoked MEG data, magnetic field tomography, positron emission tomography (PET)) serve as a sound basis for introducing our data analysis tool. Comparing our results with the below mentioned previous studies we can judge whether our results agree with the known facts and whether our approach reveals additional information about the synchronization processes generating the parkinsonian tremor.

In the second part we will sketch the increasing experimental evidence showing that synchronization is a basic mechanism in the nervous system. This will point out the physiological relevance of detecting phase synchronization in MEG data from a general point of view.

2.1 Parkinsonian Resting Tremor

As yet, the genesis of physiological and pathological tremors is a matter of debate (Elble and Koller 1990). Resting tremor in Parkinson's disease (PD) is an involuntary shaking with a frequency around 3 Hz to 6 Hz which predominantly affects the distal portion of the upper limb. Apart from advanced cases where the tremor persists during movement and steady posture, classically the parkinsonian tremor decreases or vanishes during voluntary action.

Several studies clearly indicate that parkinsonian tremor is caused by a central oscillator. For a detailed discussion we refer to Volkmann et al. (1996). Let us just mention that in the anterior nucleus of the ventrolateral thalamus (VLa) there are the so-called no-response cells which are neither modulated by somatosensory stimuli nor by active or passive movements (Lenz et al. 1994). The synchronized output of these no-response cells acts on the periphery via the motor cortex. This follows from recordings of tremor locked activity from sensorimotor cortex during neurosurgery in PD patients (Alberts et al. 1969) and in monkeys with experimentally induced Parkinson-like tremor (Lamarre and Joffroy 1979).

Volkmann et al. (1996) suggested that the synchronized oscillatory activity of VLa feeds into two loops: The thalamus (VLa) drives the supplementary motor area (SMA) and the premotor cortex (PMC). Both SMA and PMC drive the primary motor cortex (M1). The intrinsic loop is closed as SMA, PMC and M1 feedback into VLa via the basal ganglia. On the other hand peripheral feedback from muscle spindles and joint receptors reaches the motor cortex via the thalamus (extrinsic loop) and may serve to stabilize the intrinsic oscillation.

Parker et al. (1992) studied changes of regional cerebral blood flow (rCBF) associated with resting tremor of PD patients. The regions which showed significant rCBF increases during tremor periods were hand and lower limb sensory, motor and premotor cortex, supplementray motor cortex (SMA) and cerebellum. PET is integrating the cerebral activity over a time frame of several seconds and is therefore obviously not suited. Nor is it capable of resolving the sequence of neuronal activation during a tremor cycle.

MEG and electroencephalography (EEG) are currently the only available techniques which non-invasively allow to monitor brain activity with a time resolution in the ms range that is mandatory to study synchronization. Except for some slowing of background activity, surface EEG is classically normal in PD patients (Soikkeli et al. 1991). Using MEG, Volkmann et al. (1996) observed tremor-related neuromagnetic activity over wide areas of the frontal and parietal cortex. Based on averaged MEG data, the spatial and temporal organization of this activity was investigated by means of single equivalent current dipole (ECD) analysis and fully three-dimensional distributed source solutions (magnetic field tomography, MFT). ECD and MFT solutions were superimposed on high-resolution MRI to localize the tremor related current sources. The findings of Volkmann et al. (1996) indicate that parkinsonian tremor is associated with rhythmic subsequent electrical activation at the diencephalic level and in lateral premotor, somatomotor, and somatosensory cortex.

As the signal-to-noise ratio of MEG data is rather low, the study of Volkmann et al. (1996) was based on averages triggered to the EMG burst onset. However, one has to be aware of two maior drawbacks of averaging techniques: (a) Averaging cancels out dynamical features which are not closely connected to the EMG burst onset. From animal experiments it is well known that, e.g., synchronized oscillatory activity in the visual cortex is typically averaged out as it is not tightly stimulus locked (Eckhorn et al. 1990, Singer and Gray 1995). (b) Averaging is not appropriate for revealing the time course of instationary processes.

Here we apply our method to study the limitations of the previous studies and to answer the following questions: (a) In which MEG channels can we detect activity which is $n{:}m$ synchronized to the tremor? (b) Are these channels synchronized among each other, and if so which type of $n{:}m$ locking can be observed? (c) Because our method is a single run analysis, we moreover are able to study the temporal evolution of synchronization.

2.2 Synchronization: a Basic Physiological Mechanism

A vast number of animal studies revealed that synchronized activity abounds in the nervous system (see, e.g., Freeman 1975, Steriade, Jones, Llinás 1988). Indeed, central issues in neuroscience refer to synchronization of neuronal activity within a certain brain area and between distinct areas. For instance, results of animal experiments indicate that synchronization in the visual cortex is a central physiological mechanism underlying visual pattern recognition (Gray and Singer 1987, Eckhorn et al. 1988, Singer and Gray 1995). Additionally, several studies indicate that synchronized oscillatory activity in the sensorimotor cortex may serve for the integration and coordination of information underlying control and execution of movements (MacKay 1997).

Moreover, synchronization seems to play an important role concerning the interaction of different brain areas. For example, a recent study revealed that during a visuomotor integration task in the awake cat synchronization is observed between areas of the visual cortex and parietal cortex, and between areas of the parietal and motor cortex (Roelfsema et al. 1997). However, as yet, little is known about synchronization processes among brain areas and their functional and behavioural correlates in humans.

On the other hand pathological synchronization processes such as previously discussed in Parkinson's disease or most extremely during epileptic seizures, may interfere with normal brain function. Hence, to understand how the human brain works under normal and pathological conditions, synchronization of neuronal oscillatory activity within a single area and between different areas of the brain has to be analysed in humans, too. Additionally, the synchronized cerebral activity has to be related to peripheral signals, such as the EMG which reflects the bursting activty of peripheral nerves.

Neuronal activity of the human brain can noninvasively be assessed by means of whole-head MEG (Ahonen et al. 1991, Hämäläinen et al. 1993). To this end, the magnetic field generated by active neurons is recorded outside the brain (Cohen 1972) with superconducting detectors, the so-called SQUIDs (superconducting quantum interference device) (Zimmermann 1977). In this study, cortical signals were recorded with Neuromag-122 which consists of 122 SQUIDs arranged in a helmet-shaped array (Ahonen et al. 1991). This type of neuromagnetometer records the two orthogonal tangential derivatives of the magnetic field component perpendicular to the helmet surface at each measurement location. Therefore, the largest magnetic gradient is recorded directly above a cortical current source. For this reason one can distinguish between the cortical activity of distinct areas provided the latter are sufficiently remote (cf. Hämäläinen et al. 1993).

Before we present our analysis of the MEG data let us first explain phase synchronization in the presence of noise and our approach for detecting it.

3 Phase Synchronization: Basic Notions

Classically synchronization of two periodic oscillators is understood as *phase locking*, i.e.

$$|\psi_{n,m}| = |n\phi_1 - m\phi_2| < \text{const} \; , \tag{1}$$

where n and m are integers, ϕ_1 and ϕ_2 are phases of two oscillators and $\psi_{n,m}$ is the generalized phase difference, or relative phase. $\psi_{n,m}$, as well as $\phi_{1,2}$ are defined not on the circle $[0, 2\pi]$ but on the whole real line. In this simplest case condition (1) is equivalent to the notion of *frequency locking* $n\Omega_1 = m\Omega_2$, where $\Omega_j = \langle\dot{\phi}_j\rangle$ and brackets mean time averaging. We note that sometimes definition (1) is understood in a very narrow sense $|n\phi_1 - m\phi_2| =$ const that is only valid for quasi-harmonic oscillations and excludes, e.g., the obvious case of synchronization of two relaxational oscillators and other non-trivial phenomena. Thus, synchronization of periodic oscillators means the appearance of locking, or adjustment of frequencies, due to interaction of

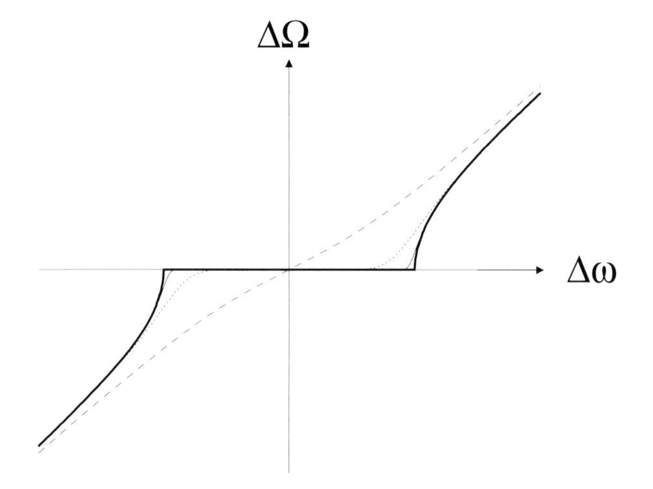

Fig. 1. Qualitative dependence of the frequency difference $\Delta\Omega$ of two coupled periodic oscillators on the parameter mismatch $\Delta\omega$ for a constant coupling strength. The noise free case is depicted by the bold line. Solid, dotted and dashed lines correspond to noisy oscillators with an increasing noise level.

non-identical oscillators for a certain range of the parameters that govern the mismatch $\Delta\omega$ in frequencies of uncoupled systems (Fig. 1). In other words, if the frequency of one oscillator varies, the second one follows this variation.

The definition of synchronization in *noisy systems* is not so trivial. If noise is small, then the condition of frequency locking is fulfilled only approximately and the relative phase fluctuates in a random way (Stratonovich 1963, Rosenblum et al. 1997a). Large noise amplitudes may cause *phase slips*, i.e. rapid 2π jumps of the relative phase. The dependence of the frequency difference $\Delta\Omega$ on the mismatch $\Delta\omega$ is now smeared (Fig. 1), and synchronization appears only as a tendency. In this case the question "synchronous or not synchronous" cannot be answered in a unique way, but only treated in a statistical sense. Following the basic work of Stratonovich (1963) we understand synchronization of noisy systems as appearance of one peak or more peaks in the *distribution of the relative phase* $\psi_{n,m}$ mod 2π. Before proceeding with further illustrations of this case, we introduce the notion of *phase synchronization of chaotic systems* (Rosenblum et al. 1996, Rosenblum et al. 1997a, Rosenblum et al. 1997b, Pikovsky et al. 1997a, Pikovsky et al. 1997b), and explain the common features of these two phenomena. The main idea is that the notion of phase can generally be introduced for chaotic systems as well, and phase locking can be observed, while the amplitudes remain chaotic. It was shown that the phase dynamics is qualitatively similar to that of noisy periodic oscillators, where the chaotic (although purely deterministic) nature of amplitudes plays the role of a noisy perturbation to phases. Thus, the random-walk-like motion of the relative phase and phase slips can be observed for this class of systems as well. As a consequence of this similarity, we can consider synchronization of noisy periodic and chaotic systems from a common viewpoint.

4 Synchronization in the Presence of Noise: a Model Example

In order to illustrate the impact of noise on phase and frequency locking, let us consider a chaotic system with random perturbations. As a model example we choose two coupled non-identical Rössler systems subjected to noisy perturbations:

$$\begin{aligned} \dot{x}_{1,2} &= -\omega_{1,2} y_{1,2} - z_{1,2} + \xi_{1,2} + \varepsilon(x_{2,1} - x_{1,2}), \\ \dot{y}_{1,2} &= \omega_{1,2} x_{1,2} + 0.15 y_{1,2}, \\ \dot{z}_{1,2} &= 0.2 + z_{1,2}(x_{1,2} - 10). \end{aligned} \tag{2}$$

Here we introduce the parameters $\omega_{1,2} = 1 \pm \Delta\omega$ and ε which govern the frequency mismatch and the strength of coupling, respectively; $\xi_{1,2}$ are two Gaussian delta-correlated noise terms

$$\langle \xi_{1,2}(t)\xi_{1,2}(t')\rangle = 2D\delta(t - t')\ , \quad \langle \xi_1(t)\xi_2(t)\rangle = 0\ . \tag{3}$$

The system is simulated by Euler's technique with the time step $\Delta t = 2\pi/1000$. In the following we fix $D = 1$.

First we consider the case of two identical oscillators, i.e. $\Delta\omega = 0$. Here we cannot use the criterion of frequency locking for the description of synchronization, as the averaged frequencies are equal even if the oscillators are uncoupled. Nevertheless, we can distinguish the uncoupled ($\varepsilon = 0$) and coupled cases ($\varepsilon = 0.04$) if we look at the distribution of the relative phase (Fig. 2). In both cases the relative phase performs a random-walk-like motion, but its distribution is uniform in the absence of coupling and has a well-expressed peak if two systems weakly interact. Thus, in the latter case we have both frequency and phase locking.

Now we consider detuned oscillators, $\Delta\omega = 0.015$. In the absence of noisy perturbations, the phase difference oscillates around some constant level, and its distribution obviously has a sharp peak. Therefore we can speak of frequency and phase locking here (Fig. 3). In the presence of noise the relative phase performs a biased random walk, so there is obviously no frequency locking. Nevertheless, the distribution of the phase definitely shows the presence of phase locking (Fig. 3).

To summarize, the notions of phase and frequency locking are no longer equivalent in the case of noisy systems. To reveal the kind of synchronization, one has to analyze the distribution of the relative phase.

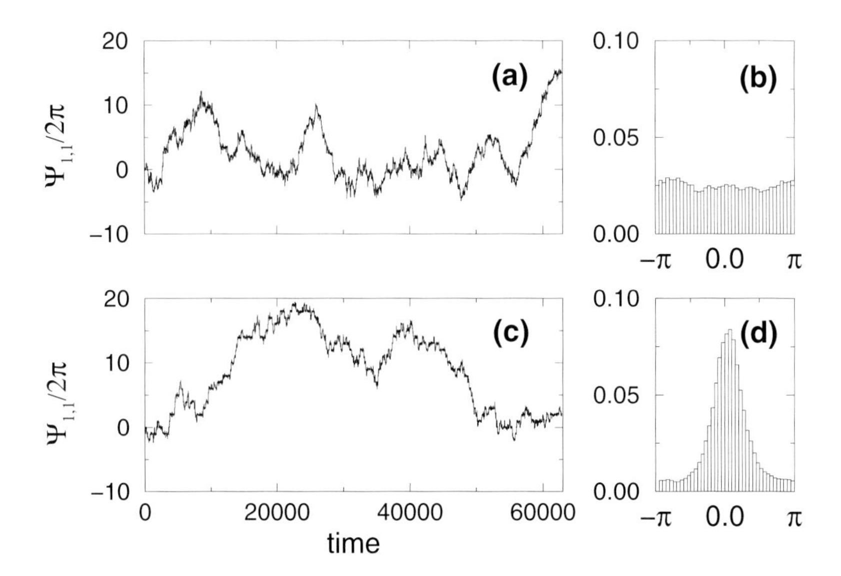

Fig. 2. Relative phase $\psi_{1,1} = \phi_1 - \phi_2$ and distribution of $\psi_{1,1} \bmod 2\pi$ for the case of uncoupled (a, b) and coupled (c, d) identical chaotic Rössler systems (2) perturbed by noise. Although the fluctuations of $\psi_{1,1}$ in both cases seem to be quite alike, the distributions (b) and (d) clearly identify the difference between coupled and uncoupled regimes.

5 Detection of Phase Synchronization

The methods developed in nonlinear dynamics within the last two decades essentially enriched the arsenal of time series analysis. These methods usually allow us to reveal some information about the system, considered as a black box, from a scalar signal measured at its output. *Bivariate data* are still mainly analyzed by means of statistical techniques, such as cross-spectrum analysis, computation of mutual information or maximal correlation (Pompe 1993, Renyi 1970, Voss, Kurths 1997). The essential limitation of these techniques is the requirement of stationarity. Here we show that the ideas of nonlinear dynamics, namely synchronization, can be used to retrieve additional information on the underlying dynamics, especially for non-stationary systems (Rosenblum et al. 1997b).

Suppose we have two simultaneously measured signals. Quite often we have some additional knowledge about the system under study and can therefore suppose that these signals are the outputs of two possibly weakly interacting subsystems. The presence of this interaction can be found by means of the analysis of *instantaneous* phases of these signals. These phases can be obtained by means of the analytic signal concept (Gabor 1946) based on the Hilbert transform. An introduction to this technique was presented elsewhere (Rosenblum et al. 1996, Pikovsky et al. 1997a, Rosenblum, Kurths

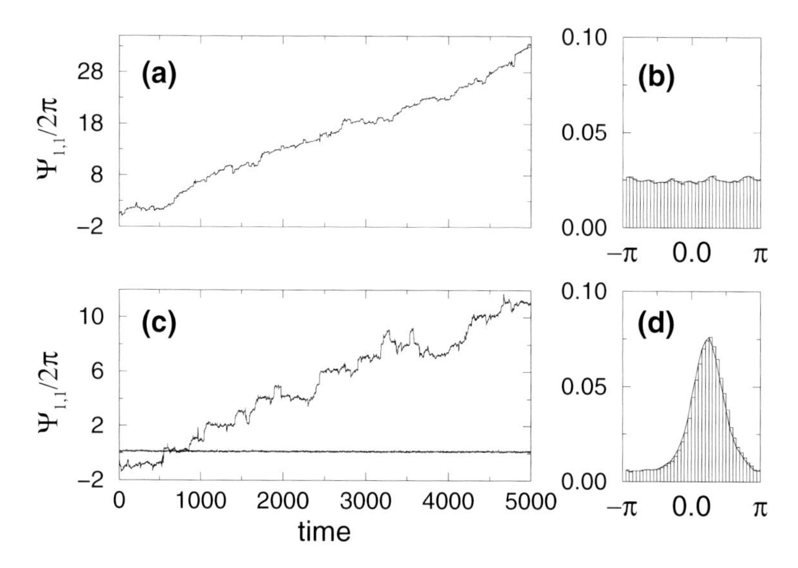

Fig. 3. Relative phase $\psi_{1,1} = \phi_1 - \phi_2$ and distribution of $\psi_{1,1}$ mod 2π for the case of uncoupled (a, b) and coupled (c, d) non-identical chaotic Rössler systems (2) perturbed by noise. The almost horizontal curve in (c) corresponds to the absence of noise: in this case the phase difference fluctuates around some constant value due to influence of chaotic amplitudes. These fluctuations are rather small, and no phase slips are observed (this fact is explained by the high phase coherence properties of the Rössler attractor, see the discussion in Pikovsky et al. 1997a). In contrast to the noisy case, here we observe both frequency and phase locking.

1998). This general approach unambiguously gives the instantaneous phase and amplitude for an arbitrary scalar signal $s(t)$. The analytic signal $\zeta(t)$ is a complex function of time defined as

$$\zeta(t) = s(t) + \mathrm{i}\tilde{s}(t) = A(t)\mathrm{e}^{i\phi(t)} \tag{4}$$

where the function $\tilde{s}(t)$ is the Hilbert transform of $s(t)$

$$\tilde{s}(t) = \pi^{-1}\mathrm{P.V.}\int_{-\infty}^{\infty} \frac{s(\tau)}{t-\tau}\mathrm{d}\tau \tag{5}$$

and P.V. means that the integral is taken in the sense of the Cauchy principal value. The instantaneous amplitude $A(t)$ and the instantaneous phase $\phi(t)$ of the signal $s(t)$ are thus uniquely defined by (4).

Taking into account that data obtained from natural systems, here MEG data, are nonstationary we have to detect quasistationary epochs of phase locking. As already mentioned in the former section synchronization in terms of statistical physics is defined by the existence of one or more preferred values of the relative phase $\psi_{n,m}$ mod 2π. For this reason we perform a sliding window analysis, and within each window the extent of synchronization is estimated statistically.

With this aim in view for every timing point t we determine the distribution of $\psi_{n,m}$ mod 2π within the window $[t - \Delta t, t + \Delta t]$. To characterize the distribution we use an *estimation based on normalized entropy:* In each window $[t - \Delta t, t + \Delta t]$ the entropy

$$e_{nm}(t) = -\sum_{j=1}^{N} p_j \ln p_j \tag{6}$$

of the the distribution is determined, where p_j denotes the relative frequency of finding ψ_{nm} mod 2π within the jth bin. The optimal number of bins is given by $N = \exp[0.626 + 0.4 \ln(M - 1)]$, where M is the number of samples (Otnes and Enochson 1972). The value of e_{nm} lies between 0 and $\hat{e}_{nm} = \ln N$. Normalization of (6) yields

$$\varrho_{n,m}(t) = \frac{\hat{e}_{nm} - e_{nm}(t)}{\hat{e}_{nm}} , \tag{7}$$

which will be called the *n:m locking index*. Hence, $\varrho_{n,m}(t)$ is equal to 0 or 1 provided the distribution is uniform or has a single Dirac peak. In this way $\varrho_{n,m}(t)$ enables us to estimate the difference between the actual distribution of $\psi_{n,m}$ mod 2π and a uniform distribution.

No straightforward method is known for the determination of the integers n and m. Hence, we calculate $\varrho_{n,m}$ for several pairs of n and m, where the range of the ratio n/m is restricted due to the problem under consideration. We will come back to this point below.

To get rid of influences and artifacts which are due to noise and the bandpass filtering, a "synchronization threshold" was determined by using bandpass filtered white noise as surrogate. The bandpass was the same as for the MEG and EMG signals, respectively. For a sufficiently long data set of surrogates the $n{:}m$ locking index was determined, and the 95.4th percentile was chosen as synchronization threshold. Subtracting this (time independent) threshold value $c_{n,m}$ from the $n{:}m$ locking index $\varrho_{n,m}(t)$ finally yields the $n{:}m$ *synchronization index*

$$\rho_{n,m}(t) = \varrho_{n,m}(t) - c_{n,m} \, . \tag{8}$$

In addition to this procedure we determined the "synchronization threshold" in two different ways: (a) Channels which did not show tremor locked activity served as surrogates. (b) Data from an empty room measurement (performed directly before the investigation of the PD patient) served as surrogates. In both cases we obtained practically the same synchronization threshold. For a detailed analysis and discussion we refer to Tass et al. (1998). The difficulties in using surrogates for testing phase synchronization are explained in Rosenblum et al. (1998).

Let us recall the above mentioned similarity of phase dynamics in noisy and chaotic oscillators. A very important consequence of this fact is that, using the synchronization approach to data analysis, we can avoid the hardly solvable dilemma “noise vs chaos”: irrespective of the origin of the observed signals, the approach and techniques of the analysis are unique. For this reason we are allowed to completely focus on phase relations, in this way addressing the dynamical phenomenon which is relevant from the physiological point of view.

6 Experimental Results

We present data from a 36-year-old patient suffering from an early-stage, tremor-dominant Parkinson's disease which was almost exclusively restricted to the right side. MEG recording was performed using a 122-channel MEG system (Ahonen et al. 1991). During the MEG recording the patient was comfortably sitting with eyes closed. Surface EMG recordings were done using pairs of silver cup electrodes which were attached over the flexor digitorum superficialis and the extensor indicis muscles.

6.1 Signal Pre-processing

The pre-processing of EMG and MEG signals is different.

1. *Pre-processing of the EMG:* We want to analyse the phase relationship between cerebral activity and compound motor unit activity, that is why we perform a standard EMG pre-processing procedure as illustrated in Fig. 5. First a high-pass filtering (> 60 Hz) is carried out to extract the EMG's burst activity (Fig. 4). In this way we get rid of low frequency artifacts such as movement artifacts. Next, in order to determine the envelope of the bursts we rectify the signal, where rectifying x means calculating $\mid x \mid$ (Fig. 5). Finally, by means of a bandpass filter the signal's principle frequency component is

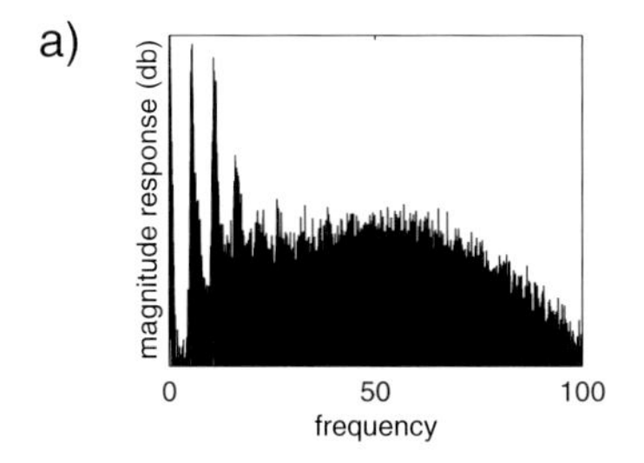

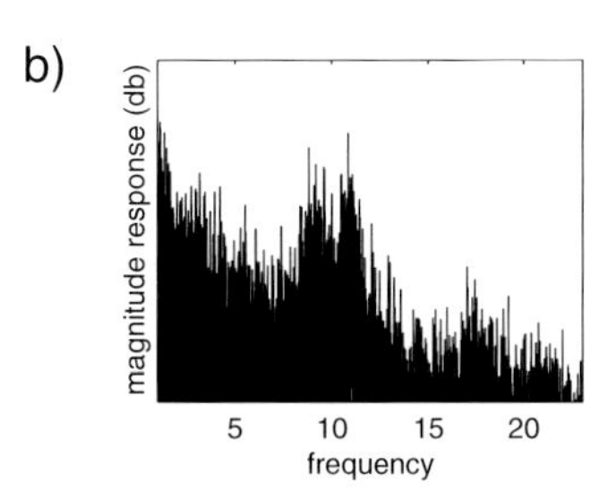

Fig. 4. Power spectra of EMG (a) and MEG signal (B, channel over the contralateral sensorimotor cortex). The burst activity of the EMG gives rise to the broadened part of the power spectrum ranging up to over 100 Hz.

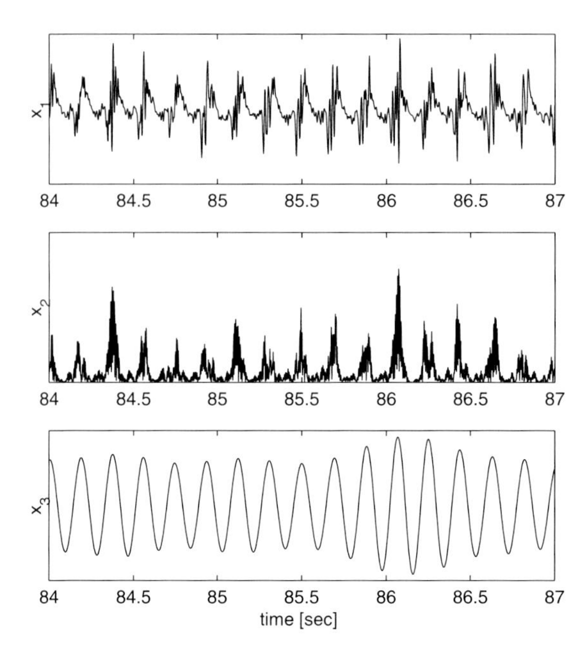

Fig. 5. High-pass filtering (> 60 Hz) and rectifying the EMG signal (x_1) yields x_2 which corresponds to the burst activity. By bandpass filtering (5 Hz to 7 Hz) x_2 one obtains x_3. The latter reflects the time course of the bursts' center of mass.

extracted. Each maximum of this signal coincides with the corresponding burst's center of mass (Fig. 5).

2. *Pre-processing of the MEG:* The MEG is bandpass filtered with the same bandpass as the EMG (from 5 Hz to 7 Hz for the data presented here) and, additionally, with a bandpass corresponding to the first harmonic (from 10 Hz to 13 Hz for the data presented here). In both frequency bands we observe peaks in the power spectra of MEG signals, in particular, from channels over the contralateral sensorimotor cortex (Fig. 4).

Our results are robust against changes of the parameters of both high-pass and bandpass filters. We will discuss this below.

6.2 Spatio-Temporal Patterns of Tremor Locked Cerebral Activity

Figures 6 and 7 show the time course of the 1:1 and the 1:2 synchronization index (7) between the EMG of the right flexor digitorum muscle and all MEG signals, and between the EMG of the right extensor indicis muscle and all MEG signals. The length of the time window $[t - \Delta t, t + \Delta t]$ was 10 seconds.

1:2 *synchronization between MEG and EMG signals* (Fig. 6): Preprocessed EMG signals were bandpass filtered between 5 Hz and 7 Hz, whereas the band edges of the MEG signals were 10 Hz and 13 Hz. Hence, the cerebral rhythm is twice as fast as the principal EMG activity. The main focus of the 1:2 synchronization is located over the contralateral sensorimotor area. In this context contralateral means contralateral to the tremor, i.e. left. Additionally, we observe 1:2 locked activity over (a) the temporal cortex (predominantly contralateral), (b) the parietal cortex (more pronounced contralaterally), (c) the premotor and frontal cortex (more pronounced contralaterally). In particular for times $t > 220$ sec synchronization between flexor EMG and corresponding MEG signals is stronger than synchronization between extensor EMG and MEG signals.

1:1 *synchronization between MEG and EMG signals* (Fig. 7): Band edges of both EMG and MEG were 5 Hz and 7 Hz. In contrast to the 1:2 synchronization the 1:1 locking is weaker. The latter is predominantly located over (a) the contralateral sensorimotor cortex, (b) a contralateral parieto-occipital area, (c) a contralateral frontal area.

The onset of both 1:2 and 1:1 synchronization coincides with the onset of tremor activity in flexor and extensor muscles as will be illustrated below in detail.

It is important to note that the results revealed by our approach are robust against changes of the parameters of both high-pass and bandpass filters. For instance, by performing a bandpass filtering between 10 Hz and 14 Hz the time course of the 1:2 synchronization index is practically not canged, however, its amplitude is diminished. This indicates that comparing the synchronization behaviour for different band edges of the bandpass filter may be an appropriate approach for separating frequency ranges of distinct physiological rhythms.

6.3 Synchronization Among Separate Cerebral Areas

According to Figs. 6 and 7 tremor locked activity is not restricted to a single cortical area. Hence, the question suggests itself of how distinct brain areas interact with each other. Figures 8 and 9 display the time course of the 1:1 synchronization index between a particular MEG channel and all other MEG channels. In Fig. 8 a channel over the premotor cortex serves as reference channel, whereas in Fig. 9 a channel over the contralateral parietal cortex is chosen as reference channel. The synchronization behaviour was analysed for the higher frequency band (10 Hz ... 13 Hz) and for the lower (5 Hz ... 7 Hz) frequency band. According to Fig. 8 (a) the parietal reference channel is 1:1 locked with contralateral sensorimotor and premotor areas, whereas according to Fig. 9 (a) the premotor reference channel is 1:1 locked to frontal, premotor, contralateral sensorimotor and (less pronounced) contralateral parietal areas. Thus, in the higher frequency band (10 Hz ... 13 Hz) all areas which are 1:2 locked to the tremor (cf. Fig. 6) are 1:1 locked to each other. Note that the onset of the strong synchronization between remote areas reflects the onset of the tremor.

Fig. 6. Time course of the 1:2 synchronization index $\rho_{1,2}(t)$ (7) between the EMG of the right flexor digitorum muscle and all MEG signals (a), and between the EMG of the right extensor indicis muscle and all MEG signals (b). Data from a 36 years old patient suffering from an early-stage ideopathic Parkinsonian's disease almost exclusively restricted to the right side. Bandpass filter between 5 Hz and 7 Hz for rectified EMG data and between 10 Hz and 13 Hz for MEG data. Time axis runs from 0 to 320 sec. Arrangement of plots corresponds to the SQUIDs' arrangement. At each measurement location a pair of SQUIDs determines the two orthogonal derivatives of the component of the magnetic field perpendicular to the helmet surface. As indicated by the schematic head in the right upper corner frontal and occipital sensors are located at the top and at the bottom, respectively, . 'L' and 'R' denote left and right side. Plots in the right lower corner show the EMG of the right flexor digitorum muscle (a) and the right extensor indicis muscle (b) plotted over time. Single plots are linearly scaled between 0 and 0.45 in (a) and between 0 and 0.38 in (b). 1:2 synchronization index was calculated from the normalized relative phase $\psi_{1,2}/4\pi$ (cf. Fig. 10).

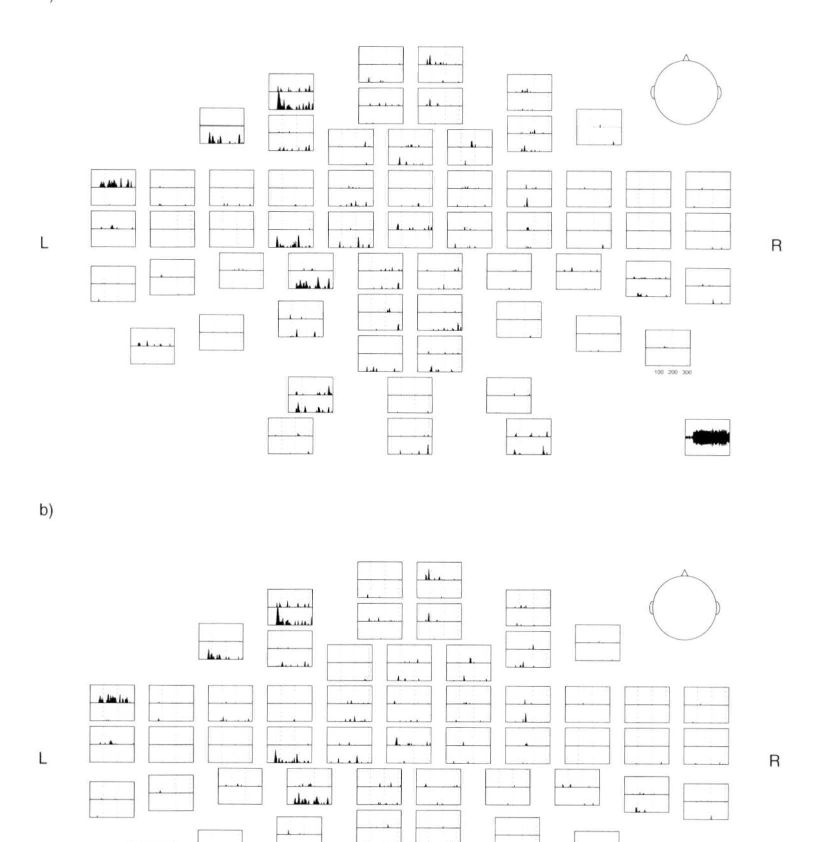

Fig. 7. 1:1 synchronization index $\rho_{1,1}(t)$ (7) between the right flexor digitorum muscle and all MEG signals (a), and between the right extensor indicis muscle and all MEG signals (b). Same format and data as in Fig. 7. Plots in the right lower corner show the EMG of the right flexor digitorum muscle (a) and the right extensor indicis muscle (b) plotted over time. Bandpass filter between 5 Hz and 7 Hz for MEG data and rectified EMG data. Single plots linearly are scaled between 0 and 0.23 in (a) and between 0 and 0.21 in (b).

Fig. 8. Time course of the 1:1 synchronization index $\rho_{1,1}(t)$ (7) between an MEG channel above the left premotor cortex (indicated by the gray rectangle) and all MEG channels. Same format and data as in Fig. 7. Plots in the right lower corner show the EMG of the right flexor digitorum muscle. Bandpass filter between 10 Hz and 13 Hz (*a*) and between 5 Hz and 7 Hz (*b*). Single plots linearly scaled between 0 and 0.19 in (*a*) and between 0 and 0.15 in (*b*).

Fig. 9. Time course of the 1:1 synchronization index $\rho_{1,1}(t)$ (7) between an MEG channel above the left parietal cortex (indicated by the gray rectangle) and all MEG channels. Same format and data as in Fig. 7. Plots in the right lower corner show the EMG of the right flexor digitorum muscle. Bandpass filter between 10 Hz and 13 Hz (a) and between 5 Hz and 7 Hz (b). Single plots linearly scaled between 0 and 0.19 in (a) and between 0 and 0.1 in (b).

In contrast, in the lower frequency band (5 Hz ... 7 Hz) the reference channels are synchronized almost exclusively with neighbouring channels. In this context we must keep in mind that the composition of the signals detected by neighbouring sensors overlap to a certain extent. In particular, there is no strong 1:1 synchronization between remote areas in the lower frequency range.

No 1:2 synchronization was found between areas exhibiting tremor locked activity.

6.4 Peripheral and Central Synchronization Patterns

Apart from the localization and strength of the n:m synchronization the relative phase $\psi_{n,m}$ provides us with additional information that is important, in particular, with regard to the temporal relationship of synchronized activities. This is illustrated in Fig. 10 showing the time course of the flexor's and extensor's burst activity, the corresponding distributions of relative phases and the related synchronization indexes.

After 60 seconds the tremor starts with an alternating flexor/extensor activation, i.e. flexor and extensor muscles are synchronized in antiphase (Figs. 10a and b). During each tremor cycle of flexor muscle (Fig. 10c) and extensor muscle (Fig. 10d) we observe two maxima of the MEG signal over the contralateral sensorimotor cortex. Note that within one tremor cycle an increasing relative phase $\psi_{1,2}/4\pi \bmod 1$ (as defined according to Fig. 10) corresponds to an advancing MEG signal. Concerning the timing of the central and peripheral rhythmic activity it is important to remark that the first prefered value of the relative phase (at ≈ 0.2) corresponds to the timing of the tremor-related field 1 (TRF 1) described by Volkmann et al (1996). TRF 1 reflects the synchronized activity of the pyramidal cells in the motor cortex (M1) which drives the tremor bursts via the spinal motor neuron pool.

After 220 seconds, the 1:2 locking between flexor muscle and, in particular, extensor muscle on the one hand and contralateral sensorimotor cortex on the other hand is attenuated (Fig. 10d). This attenuation is accompanied by a decrease of the 1:1 synchronization between flexor and extensor muscles (Fig. 10b). Hence, the time course of peripheral coordination patterns reflects the time course of synchronization processes between central and peripheral activity.

6.5 Summary and Interpretation

Let us summarize our main results. We observe a 1:2 synchronization between the tremor EMG and different cortical areas. The main focus of the 1:2 synchronization is located over the contralateral sensorimotor cortex. Additionally, this type of synchronization is observed over premotor, frontal, contralateral parietal and contralateral temporal areas. In contrast to the 1:2

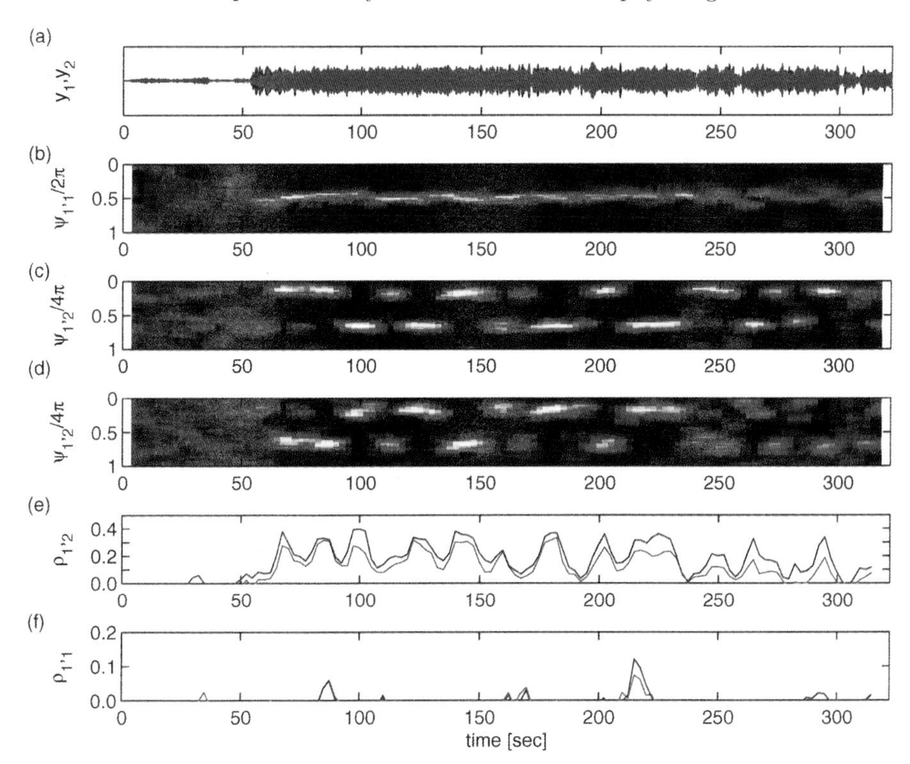

Fig. 10. Plot shows the time course of peripheral and central synchronization patterns. (*a*): EMG of right flexor digitorum (blue) and right extensor indicis (red). (*b*), (*c*) and (*d*) show the time course of the distribution of relative phases. To this end for every timing point t the distribution of the relative phase within the window $[t - 10\text{ sec}, t + 10\text{ sec}]$ is colour coded displayed (colour scale: min. = black, red, yellow, white = max.). (*b*) 1:1 synchronization between both EMG signals with relative phase $\psi_{1,1}(t)/2\pi$, where ϕ_1 and ϕ_2 are the phases of right extensor and flexor EMG, respectively (cf. (1)). (*c*) and (*d*): 1:2 synchronization between an MEG channel over the left sensorimotor cortex and both EMG signals: The relative phase is $\psi_{1,2}(t)/4\pi$, where ϕ_1 is the phase of the MEG signal, whereas ϕ_2 is the phase of the flexor (*c*) and extensor (*d*) EMG, respectively. Due to the normalization (i.e. division by 4π) one period of the relative phase corresponds to one tremor cycle.

locking, the 1:1 synchronization is much weaker, and it is observed over contralateral sensorimotor, parieto-occipital and frontal areas. All areas which are 1:2 locked with the tremor are 1:1 locked among each other. On the contrary the areas with 1:1 tremor locked activity are not locked among each other. Our analysis of the phase relationship between tremor activity and MEG signals above the sensorimotor area confirms the sequential activation of motor cortex (M1) and tremor burst revealed by Volkmann et al. (1996)

and, thus, supports the hypothesis of a central oscillator driving parkinsonian tremor.

So, considering time course and locations of the 1:2 locking compared to the 1:1 locking clearly shows that the synchronized thalamic (VLa) activity induces coherent activity running twice as fast as the tremor within a network located in several cortical areas as listed above. In particular, our results suggest that already in premotor areas (PMC/SMA) a rhythm is generated which is twice as fast as the principal tremor rhythm.

In this context it should be mentioned that the pattern of spindle discharges in parkinsonian tremor is similar to that observed during repetitive voluntary movements (Hagbarth et al. 1975). Therefore Hagbarth et al. (1975) suggested that parkinsonian resting tremor is caused by an involuntary activation of a central mechanism which is normally used for producing rapid voluntary movements. This suggestion was supported by the study of Volkmann et al. (1996) which showed that the sources of the tremor locked activity had the same localization as the generators known from voluntary movements. Additional support comes from MEG recordings showing that the μ-rhythm is suppressed during parkinsonian tremor (Maekelae et al. 1993), where μ-rhythmicity in EEG and MEG is typically attenuated during periods of voluntary movements (Lado et al. 1992). The suggestion of Hagbarth et al. (1975) is also confirmed by our findings: We observe the 1:2 synchronization between the EMG and the contralateral sensorimotor cortex during parkinsonian resting tremor as well as during voluntary repetitive movements in healthy subjects. Further details will be reported in a forthcoming article.

The length of the time window $[t-\Delta t, t+\Delta t]$ chosen in this study was 10 seconds. We performed the same analysis with the length of the time window ranging from 2 seconds to 30 seconds. Qualitatively all investigations yield the same results. The main difference was that with decreasing Δt the time course of the synchronization indices becomes more fine-grained. For this reason a window with length of 10 seconds turned out to be appropriate, as the dynamics of the synchronization process under consideration obviously acts on a time scale of several seconds. For the analysis of other cerebral activity, e.g., α-, β- or γ-activity suitable window lengths have to be chosen. The latter depends on the frequency range as well as the time scale of the investigated synchronization processes.

7 Discussion and Comparison with other Methods

Based on the concept of phase synchronization we have developed a method for detecting $n{:}m$ phase locking between MEG signals and related signals such as the EMG (Tass et al. 1998). In a patient suffering from Parkinson's disease with our approach we revealed

(a) 1:1 and, in particular, 1:2 phase locking between the EMG of flexor and extensor muscles and MEG signals from several areas, and

(b) 1:1 phase locking between MEG signals from areas exhibiting tremor locked activity.

The results revealed by our method (Sect. 6) agree with the findings of previous (Sect. 2.1) studies concerning the localization of areas exhibiting tremor locked activity. In contrast to the other techniques mentioned in Sect. 2.1 our method additionally enables us to analyse (a) the time course of synchronization processes and (b) dynamical processes which are not tightly locked to an event used as a trigger for an averaging procedure (e.g., the burst onset). For this reason our single run analysis method has good prospects for the investigation of the fundamental issues mentioned in Sect. 2.2.

Acknowledgement

This study was supported by grants from the Deutsche Forschungsgemeinschaft (SFB 194, A5 and Z2) and from the Max-Planck-Gesellschaft.

References

Ahonen, A.I., Hämäläinen, M.S., Kajola, M.J., Knuutila, J.E.T., Lounasmaa, O.V., Simola, J.T., Tesche, C.D., Vilkman, V.A. (1991): Multichannel SQUID systems for brain research. IEEE Trans. Magn. **27**, 2786–2792

Alberts, W.W., Wright, E.J., Feinstein, B. (1969): Cortical potentials and parkinsonian tremor. Nature **221**, 670–672

Andronov, A., Vitt, A., Khaykin, S. (1966): *Theory of Oscillations.* (Pergamon Press, Oxford)

Blekhman, I.I. (1988): *Synchronization in Science and Technology.* (ASME Press, New York)

Blekhman, I.I., Landa, P.S., Rosenblum, M.G. (1995): Synchronization and Chaotization in Interacting Dynamical Systems. Applied Mechanics Reviews **48**, 733–752

Cohen, D. (1972): Magnetoencephalography: Detection of the brain's electrical activity with a superconducting magnetometer. Science **175**, 664–666

Eckhorn, R., Bauer, R., Jordan, W., Brosch, M., Kruse, W., Munk, M.,] Reitboeck, H.J. (1988): Coherent oscillations: a mechanism of feature linking in the visual cortex? Biol. Cybern. **60**, 121–130

Eckhorn, R., Dicke, P., Kruse, W., Reitboeck, H.-J. (1990): Stimulus-related facilitation and synchronization among visual cortical areas: experiment and models. In: Schuster, H.G., Singer, W. (Eds.): *Nonlinear Dynamics and Neural Networks* (VCH, Mannheim)

Elble, R.J., Koller, W.C. (1990): *Tremor* (John Hopkins University, Baltimore)

Freeman, W.J. (1975): *Mass action in the nervous system* (Academic Press, New York)

Gabor, D. (1946): Theory of communication. J. IEE London, **93**, 429–457

Gray, C.M., Singer, W. (1987): Stimulus specific neuronal oscillations in the cat visual cortex: a cortical function unit. Soc. Neurosci. **404**, 3

Gray, C.M., Singer, W. (1989): Stimulus-specific neuronal oscillations in orientation columns of cat visual cortex. Proc. Natl. Acad. Sci. USA **86**, 1698–1702

Hagbarth, K.E., Wallin, G., Lofstedt, L., Aquilonius, S.M. (1975): Muscle spindle activity in alternating tremor of parkinsonism and in the clonus. J. Neurol. Neurosurg. Psychiatry **38**, 636–641

Haken, H. (1983): *Advanced Synergetics.* (Springer, Berlin)

Hämäläinen, M., Hari, R., Ilmoniemi, R.J.,Knuutila, J., Lounasmaa, O.V. (1993): Magnetoencephalography – theory, instrumentation, and applications to noninvasive studies of the working human brain. Rev. Mod. Phys. **65**, 413

Hayashi, C. (1964): *Nonlinear Oscillations in Physical Systems.* (McGraw-Hill, New York)

Lado, F., Ribary, U., Ioannides, A. et al. (1992): Coherent oscillations in primary motor cortex and sensory cortices detected using MEG and MFT. Soc. Neurosci. Abstr. **18**, 848

Lamarre, Y., Joffroy, A.J. (1979): Experimental tremor in monkey: activity of thalamic on precentral cortical neurons in the absence of peripheral feedback. Adv. Neurol. **24**, 109–122

Landa, P.S. (1996): *Nonlinear Oscillations and Waves in Dynamical Systems.* (Kluwer Academic Publishers, Dordrecht–Boston–London)

Lenz, F.A., Kwan, H.C., Martin, R.L., Tasker, R.R., Dostrovsk, J.O., Lenz, Y.E. (1994): Single unit analysis of the human ventral thalamic nuclear group. Tremor-related activity in functionally identfied cells. Brain **117**, 531–543

MacKay, W.A. (1997): Synchronized neuronal oscillations and their role in motor processes. Trends in Cognitive Sciences **1**, 176–183

Maekelae, J.P., Hari, R., Karhu, J., Salmelin, R., Teraevaeinen, H. (1993): Suppression of magnetic mu rhythm during parkinsonian tremor. Brain res. **617**, 189–193

Otnes, R.K., Enochson, L. (1972): *Digital time series analysis.* (John Wiley & Sons, New York)

Panter, P. *Modulation, Noise, and Spectral Analysis* (McGraw–Hill, New York, 1965)

Parker, F., Tzourio, N., Blond, S., Petit, H., Mazoyer, B. (1992): Evidence for a common network of brain structures involved in parkinsonian tremor and voluntary repetitive movements. Brain. Res. **584**, 11–17

Pikovsky, A.S., Rosenblum, M.G., Osipov, G.V., Kurths, J. (1997): Phase synchronization of chaotic oscillators by external driving. Physica D, **104**, 219–238

Pikovsky, A.S, Osipov, G.V., Rosenblum, M.G., Zaks, M.A., Kurths, J. (1997): Attractor–repeller collision and eyelet intermittency at the transition to phase synchronization. Phys. Rev. Lett. **79**, 47–50

Pompe, B. (1993): Measuring statistical dependencies in a time series. J. Stat. Phys. **73**, 587–610

Rényi, A. (1970): *Probability Theory.* (Akadémiai Kiadó, Budapest)

Roelfsema, P.R., Engel, A.K., König, P., Singer, W. (1997): Visuomotor integration is associated with zero time-lag synchronization among cortical areas, Nature **385**, 157–161

Rosenblum, M.G., Pikovsky, A.S., Kurths J. (1996): Phase synchronization of chaotic oscillators. Phys. Rev. Lett. **76**, 1804–1807

Rosenblum, M.G., Pikovsky, A.S., Kurths, J. (1997): From Phase to Lag Synchronization in Coupled Chaotic Oscillators. Phys. Rev. Lett. **78**, 4193–4196

Rosenblum, M.G., Pikovsky, A.S., Kurths, J. (1997): Phase synchronization in driven and coupled chaotic oscillators. IEEE Trans. On Circuits and Systems–I **44**, 874–881

Rosenblum, M.G., Kurths, J. Analysing synchronization phenomena from bivariate data by means of the Hilbert transform. In: Kantz, H., Kurths, J., Mayer-Kress, G. (Eds) *Nonlinear Analysis of Physiological Data*, (Springer Series for Synergetics. Springer, 1998).

Rosenblum, M.G., Kurths, J., Pikovsky, A., Schäfer, C., Tass, P., Abel, H.-H. (1998): Synchronization in Noisy Systems and Cardiorespiratory Interaction. IEEE, in press

Schäfer, C., Rosenblum, M.G., Kurths, J., Abel, H.-H. (1998): Synchronization in human cardiorespiratory system. Nature, in press.

Singer, W., Gray, C.M. (1995): Visual feature integration and the temporal correlation hypothesis. Annu. Rev. Neurosci. **18**, 555–586

Soikkeli, R., Partanen, J., Soininen, H., Pääkkönen, A., Riekkinen, P. (1991): Slowing of EEg in Parkinson's disease. Electroencephalogr. Clin. Neurophysiol. **79**, 159–165

Steriade, H., Jones, E.G., Llinás, R.R. (1988): *Thalamic Oscillations and Signaling.* (John Wiley & Sons, New York)

Stratonovich, R.L. (1963): *Topics in the Theory of Random Noise.* (Gordon and Breach, New York)

Tass, P., Kurths, J., Rosenblum, M.G., Guasti, G., Hefter, H. (1996): Delay-induced transitions in visually guided movements. Phys. Rev. E **54**, R2224–R2227

Tass, P., Weule, J., Rosenblum, M.G., Kurths, J., Pikovsky, A., Volkmann, J., Schnitzler, A., Freund, H.-J.: Detection of n:m phase-locking in noisy data – application to magnetoencephalography. Phys. Rev. Lett., submitted

Volkmann, J., Joliot, M., Mogilner, A., Ioannides, A.A., Lado, F., Fazzini, E., Ribary, U., Llinás, R. (1996): Central motor loop oscillations in parkinsonian resting tremor revealed by magnetoencephalography. Neurology **46**, 1359–1370

Voss, H., Kurths, J. (1997): Reconstruction of nonlinear time delay models from data by the use of optimal transformations. Phys. Lett. A **234**, 336–344

Zimmermann, J.E. (1977): SQUID instruments and shielding for low-level magnetic measurements. J. Appl. Phys. **48**, 702–710

Spatio-Temporal Modeling Based on Dynamical Systems Theory

Christian Uhl[1], and Rudolf Friedrich[2]

[1] Max-Planck-Institute of Cognitive Neuroscience, Stephanstrasse 1a, D-04103 Leipzig, Germany
[2] Institute of Theoretical Physics and Synergetics, Universität Stuttgart, Pfaffenwaldring 57/IV, D-70550 Stuttgart, Germany

1 Introduction

In the preceding contributions of Part II "Methods & Applications" different approaches of analyzing neurophysiological data have been presented. They represent different perspectives of investigating EEG/MEG data. The methods concerning source modeling and the ones concerning synchronization analyses substantially differ in two aspects:

1. **Type of data sets under investigation:** The basis of source modeling are averaged data sets, whereas in the case of coherence and synchronization analysis raw data are investigated. Source modeling is mainly concerned with event related potentials, which are based on stimuli triggering an event. Averaging with respect to the different stimulus classes is performed to cancel out fluctuations uncorrelated with the stimuli, leading to the general response of measured potentials/ fields with respect to the stimuli and the connected tasks the subjects have to perform. On the other hand, by averaging, information is also lost about underlying processes. Methods to investigate synchronization phenomena circumvent this problem by analyzing raw ("single-trial") data, leading – compared to source modeling – to a completely different description of brain functioning.
2. **Type of parameters estimated from the data:** In the case of source modeling, parameters are estimated from the data which describe spatial characteristics of the signal. The temporal evolution is described either by amplitudes corresponding to the obtained spatial distributions or by estimating for every time step new spatial configurations. In the case of synchronization analysis parameters are estimated describing temporal aspects of the data sets. The spatial characteristics are investigated thereafter. That is, in both approaches one focuses first on either spatial or temporal aspects, the second aspect is considered in a subsequent step.

The aim of our contribution is to add an additional perspective for an investigation of neurophysiological data. Our method aims at simultaneously estimating parameters, which describe both spatial and temporal characteristics of the data sets. This approach is based on concepts of synergetics and its application to brain functioning, which have been discussed in Part I "Models & Concepts" of this book.

In the following section we will briefly sketch the theory of synergetics and its relation to a concept of brain function. For a more elaborate discussion we refer the reader to the chapter "What can Synergetics Contribute to the Understanding of Brain Functioning?" by Hermann Haken. In Sect. 3, we will present an algorithm to extract equations from averaged event-related data sets to describe dynamic interactions of human brain information processing. We will demonstrate this approach by its application to two different event-related potential studies (Sect. 5). In Sect. 6 we will propose an approach to investigate raw data of EEG/MEG signals, which – in addition to deterministic features – also includes stochastic temporal fluctuations. In our opinion, by that, we may bridge the gap between the mentioned differences of source modeling and analyses of coherences and synchronization phenomena.

2 Brain Functioning and Complex Dynamical Systems

2.1 Synergetics

Synergetics (Haken 1983, 1987) provides general concepts describing complex self-organizing systems. It allows us to describe the emergence of macroscopic patterns of a system consisting of a large number of non-linearly interacting elements. These elements may be considered as the microscopic level of the complex system.

If the evolution equations on this microscopic level are known, the formation of macroscopic patterns can be derived by a bottom-up approach. This leads to an enormous reduction of degrees of freedom: on the microscopic level we deal with high-dimensional dynamics, whereas on the macroscopic level with low-dimensional mode interactions. Formally this reduction of degrees of freedom is achieved by the slaving principle in the vicinity of critical values of control parameters. Assuming the system is described by a state vector $\mathbf{d}(t)$, the macroscopic patterns are represented by the order parameters, $x_i(t)$, and the corresponding spatial modes, $\mathbf{v}_i$, of the system:

$$\mathbf{d}(t) \simeq \sum_i x_i(t)\ \mathbf{v}_i \ . \tag{1}$$

If the evolution equations of the interacting elements are known, the order parameter equations can be derived analytically:

$$\frac{\mathrm{d}}{\mathrm{d}t}\mathbf{x} = \mathbf{N}[\mathbf{x}] + \mathbf{F}(t) \ , \tag{2}$$

with the vector field $\mathbf{N}[\mathbf{x}]$ representing the deterministic contribution and $\mathbf{F}(t)$ time-dependent fluctuating forces.

To describe the stochastic process defined by (2) one can introduce a probability density function $p(\mathbf{x}, t)$ and one can derive the time evolution of this function, leading in the case of white noise forces $\mathbf{F}(\mathbf{t})$ to the Fokker–Planck equation:

$$\frac{\partial}{\partial t} p(\mathbf{x}, t) = \left\{ -\sum_i \frac{\partial}{\partial x_i} N_i[\mathbf{x}] + \sum_{ij} \frac{\partial^2}{\partial x_i \partial x_j} Q_{ij} \right\} p(\mathbf{x}, t) \tag{3}$$

So far we have sketched the *bottom-up* approach, based on microscopic evolution equations. In the next subsection we will discuss human brain information processing in the context of complex dynamical systems and we will motivate a *top-down approach*, which will be presented in Sects. 3–6.

2.2 Neurophysiological Data

From a physical and neurophysiological point of view, the brain represents a very complex system with interactions on various scales: interaction of cortical areas and modules, neuronal cell assemblies and single neurons. As outlined in Part I of this book (and in the books of H. Haken 1996, S. Kelso 1995 and P. Nunez 1995) concepts of the physical theory of complex dynamical systems can be applied to brain dynamics.

Of course, since the evolution equations of the neurons and their interaction is not exactly known, EEG/MEG signals cannot rigorously be described by a bottom-up approach as sketched in Sect. 2.1. However, attempts in this direction show very promising results, as outlined in the chapters of Hermann Haken, Paul Nunez and Viktor Jirsa. To link these results with observed macroscopic patterns (EEG/MEG signals or behavioral measurements) it is important – as discussed in the chapters by Hermann Haken and Scott Kelso – to extract a spatio-temporal model from the signal.

Here, we will present such a top-down approach, i.e., an approach to obtain from a measured EEG/MEG signal a decomposition into state vectors and time-depending amplitudes, and to obtain the corresponding dynamics of the amplitudes. The decomposition (1) will be performed on the level of potentials/fields on the surface of the head: Instead of conceiving $\mathbf{d}(t)$ as a state vector of cerebral components, we conceive $\mathbf{d}(t)$ as the resulting potential or field due to cerebral interactions.

Thereby we will focus first on averaged data sets and their deterministic dynamics. Of course, averaging EEG/MEG data leads to information losses, however, in our opinion, this is a good starting point for investigating dynamic interactions. By averaging, uncorrelated fluctuations cancel out – as it is the basis of event related potential/field studies for the last at least two decades. Results of our approach can also easily be compared with results of conventional ERP/F analyses. In Sect. 6 we will then propose a method dealing with an extraction of both deterministic and stochastic features from spatio-temporal signals, which may yield a useful tool for analyzing single-trial EEG/MEG data.

The top-down approach of decomposing an EEG/MEG signal *and* describing the dynamics is closely related to answer the following questions concerning brain functioning:

1. Which regions in the brain do interact?
2. How do these regions interact?

Compared to conventional source modeling, the novelty of our approach is based on the following: We constrain our search for spatial field distributions by searching modes which interact in terms of dynamical systems. Thus, we take advantage of the high temporal resolution, which is one of the major advantages of EEG/MEG measurements, to draw conclusions on interacting spatial field distributions. So far our method only leads to field distributions on the surface of the head, further work may incorporate spatial sources into the framework of constraining the associated dynamics to differential equations.

To avoid notional misunderstandings we want to mention that *spatio-temporal source localization* techniques (see chapter "Source Modeling") aim at solving the neuroelectromagnetic inverse problem by *fitting spatial parameters* to the data and *describing the temporal evolution* by amplitudes. Our *spatio-temporal* approach aims at fitting *spatial and temporal* parameters to the dataset on the level of potential fields on the scalp surface to find characteristics of underlying field distributions and their dynamics, i.e. *to explain the temporal evolution* of the amplitudes.

3 Deterministic Spatio-Temporal Models

3.1 Starting Point

We consider datasets $\mathbf{d}(t)$, with vector components $d_i(t)$ representing the signal of channel i with $0 < i \leq n$. Summarizing the concept of synergetics and its application to brain functioning, as described in Sect. 2, the goal of spatio-temporal modeling is to find a decomposition of the signal,

$$\mathbf{d}(t) = \sum_{i=1}^{N} x_i(t)\, \mathbf{v}_i \quad , \tag{4}$$

and a model of the evolution equations. For averaged datasets and data of strongly coherent processes, as epileptic seizures, one can neglect fluctuating components and consider a set of deterministic differential equations to describe the evolution of the data:

$$\frac{\mathrm{d}}{\mathrm{d}t} x_i = f_i[x_j] \quad . \tag{5}$$

Since a set of biorthogonal modes $\mathbf{v}_i^+$ can be introduced, i.e.,

$$\mathbf{v}_i \cdot \mathbf{v}_j^+ = \delta_{ij} \quad \Rightarrow \quad x_i(t) = \mathbf{v}_i^+ \cdot \mathbf{d}(t) \ , \tag{6}$$

the model is complete, if one specifies the spatial modes, $\mathbf{v}_i$, the biorthogonal modes, $\mathbf{v}_i^+$ and the functions f_i.

In the next two sections we will present a method to obtain these parameters from a given dataset and an interpretation of the obtained model in terms of human brain information processing units.

3.2 Detection of Interacting Modes

To obtain a model we consider a variational principle. Therefore we define a cost function depending on spatial parameters, $\mathbf{v}_i$ and $\mathbf{v}_i^+$, and on coefficients of functions f_i describing the temporal evolution of the system. The parameter space can be reduced to a dependence on $\mathbf{v}_i^+$ only, and further reduction of parameters is achieved by considering constraints to fix the scaling of the modes. Finally, by numerical minimization of the resulting non-linear cost function, a model of the spatio-temporal signal can be obtained.

Definition of a Cost Function
Let us first introduce brackets, $\langle \ldots \rangle$, to denote time averages:

$$\langle F(t) \rangle = \frac{1}{T} \int_{t_0}^{t_0+T} \mathrm{d}t F(t) \ . \tag{7}$$

A common cost function, S, to decompose a signal is given by the principal component analysis (PCA),

$$S[\mathbf{v}_i] = \frac{\langle (\mathbf{d}(t) - \sum_i x_i(t)\, \mathbf{v}_i)^2 \rangle}{\langle (\mathbf{d}(t))^2 \rangle} \ , \tag{8}$$

which leads to a linear eigenvalue problem of the correlation matrix. This represents a spatial approach, since the modes, $\mathbf{v}_i$, are optimal for signal representation (with respect to the L_2-norm), but are not necessarily optimal for representing the dynamics of the signal. To avoid this problem, we proposed a cost function in (Uhl et al. 1993, 1995) to simultaneously fit spatial modes and temporal coefficients to a given dataset:

$$D[\mathbf{v}_i^+, f_i] = \sum_{i=1}^{N} \frac{\langle (\dot{x}_i - f_i[x_j])^2 \rangle}{\langle (\dot{x}_i)^2 \rangle} \ . \tag{9}$$

This cost function aims at finding a dynamically relevant subspace, spanned by the spatial modes, $\mathbf{v}_i^+$, with dynamics described by the functions f_i.

A similar cost function, based on a methodology first pointed out by Hasselmann (1988), is used in (Kwasniok 1996, 1997) to obtain a reduction of complex dynamical systems given by partial differential equations.

In the following we will discuss a generalized cost function, C, which represents a combination of both of the above mentioned functions:

$$C[\mathbf{v}_i, \mathbf{v}_i^+, f_i] = (1-p) \cdot D + p \cdot S + \text{constraints} \ . \tag{10}$$

The parameter $p \in [0,1]$ determines the influence of the cost function D, considering the dynamics, and of the cost function S, considering the signal representation. $p = 0$ represents the pure dynamics cost function, $p = 1$ the pure signal representation cost function.

The functions $f_i[x_j]$ describing the interaction of the modes $\mathbf{v}_i$ can be expressed by low-dimensional polynomials,

$$f_i = \sum_{j=1}^{N} a_{i,j}\, x_j + \sum_{\substack{j,k=1\\ k \geq j}}^{N} a_{i,jk}\, x_j x_k + \sum_{\substack{j,k,l=1\\ l \geq k \geq j}}^{N} a_{i,jkl}\, x_j x_k x_l \ , \tag{11}$$

since they can capture most of the generic cases of dynamics. If one knows more about the underlying dynamics other classes of functions can be considered to model the dynamics.

As constraints to fix the scaling of the vectors and functions we require,

$$\langle x_i x_j \rangle \ = \ \langle (\mathbf{d}\mathbf{v}_i^+)(\mathbf{d}\mathbf{v}_j^+) \rangle \ = \ c\, \delta_{ij} \ , \tag{12}$$

to deal with comparable amplitudes and therefore comparable coefficients $a_{i,j}, a_{i,jk}, a_{i,jkl}$.

Reduction of the Parameter Space

The cost function C depends on $\mathbf{v}_i^+$, $\mathbf{v}_i$, $a_{i,j}$, $a_{i,jk}$, $a_{i,jkl}$:

$$C = C[\mathbf{v}_i^+, \mathbf{v}_i, a_{i,j}, a_{i,jk}, a_{i,jkl}] \ . \tag{13}$$

Variation with respect to $\mathbf{v}_i$ and with respect to $a_{i,j}, a_{i,jk}, a_{i,jkl}$ leads to an elimination of these parameters, since they can be expressed as functions of $\mathbf{v}_i^+$ (see Appendix A for details):

$$\mathbf{v}_i = \mathbf{v}_i[\mathbf{v}_j^+] \tag{14}$$

$$a_{i,j}, a_{i,jk}, a_{i,jkl} = a_{i,j}, a_{i,jk}, a_{i,jkl}[\mathbf{v}_m^+] \ . \tag{15}$$

Inserting these expressions into the cost function, (10), leads to

$$C[\mathbf{v}_i^+] = (1-p) \cdot D[\mathbf{v}_i^+] + p \cdot S[\mathbf{v}_i^+] \ , \tag{16}$$

depending on spatial modes $\mathbf{v}_i^+$ only.

Writing the biorthogonal modes $\mathbf{v}_i^+$ as

$$\mathbf{v}_i^+ = \left(v_{i,1}^+, \ldots, v_{i,n}^+\right)^T \ , \tag{17}$$

the constraints, (12), can be written as

$$\langle x_i x_j \rangle = \sum_{\alpha,\beta=1}^{n} (M_c)_{\alpha\beta} v^+_{i,\alpha} v^+_{j,\beta} \ , \tag{18}$$

with $(M_c)_{\alpha\beta} = \langle d_\alpha d_\beta \rangle$ representing the correlation matrix of the signal. Transformation of the signal $\mathbf{d}$ by principal component analysis (PCA) leads to a diagonal correlation matrix $(M_c)_{\alpha\beta} = \lambda_\alpha \delta_{\alpha\beta}$ and therefore to the following constraint:

$$\tilde{\mathbf{v}}^+_i \cdot \tilde{\mathbf{v}}^+_j = \delta_{ij} \ , \ \text{with} \ \tilde{v}^+_{i,\alpha} = v^+_{i,\alpha} / \sqrt{\lambda_\alpha} \ . \tag{19}$$

To fulfill these constraints, one can use generalized elliptic coordinates with angles $\phi^{(j)}_i$, and one ends up with the following set of free parameters:

$$\begin{aligned} \tilde{\mathbf{v}}^+_1 &= \tilde{\mathbf{v}}^+_1[\phi^{(1)}_1, \ldots, \ \phi^{(n-N)}_1, \ \ldots, \phi^{(n-2)}_1, \phi^{(n-1)}_1] \\ \tilde{\mathbf{v}}^+_2 &= \tilde{\mathbf{v}}^+_2[\phi^{(1)}_2, \ldots, \ \phi^{(n-N)}_2, \ \ldots, \phi^{(n-2)}_2] \\ &\vdots \\ \tilde{\mathbf{v}}^+_N &= \tilde{\mathbf{v}}^+_N[\phi^{(1)}_N, \ldots, \phi^{(n-N)}_N] \ , \end{aligned} \tag{20}$$

and a cost function depending on these angles:

$$C = C[\phi^{(1)}_1, \ldots, \phi^{(n-1)}_1, \phi^{(1)}_2, \ldots, \phi^{(n-2)}_2, \phi^{(1)}_N, \ldots, \phi^{(n-N)}_N] \ . \tag{21}$$

The dimensionality of parameter space is therefore given by

$$\dim = N \cdot n - \frac{N \cdot (N+1)}{2} \ . \tag{22}$$

In Appendix B, we outline the procedure of introducing generalized elliptic coordinates as an example for the case of three-mode ($N = 3$) interaction. This can be extended in a straight forward but tedious manner to N-mode ($N > 3$) interaction.

We also would like to emphasize, that the dimensionality of the problem can be reduced considerably: The above mentioned transformation of the signal $\mathbf{d}$ by PCA does not need to be complete: it is sufficient to consider the first few PCA modes for projection of the signal into a subspace and to perform the search for the spatio-temporal model in that subspace.

Minimization

Because of the non-linear dependence of the resulting cost function on the angles $\phi^{(j)}_i$ numerical methods have to be considered for further calculation. Our analysis of EEG data of petit-mal epileptic seizures (1996) was performed with gradient based techniques. A recent study of ERP datasets (Uhl et al. (1998)) was based on a genetic algorithm (Holland (1987)) which is – for the typical order of dimensionality of parameter space ($n = 5 - 10$, $N = 2 - 5$ $\rightarrow \dim = 7 - 35$) – a more robust method for minimizing the cost function C. However, any choice of optimization method is acceptable if it is able to deal with local minima in a 10–50 dimensional solution space.

The global minimum of the cost function (21) represents the best choice to model the signal with respect to a decomposition (4) and a description of the dynamics in terms of differential equations (5).

3.3 Interpretation of Dynamics

As a result of the obtained model, the EEG/MEG signal can be conceived as a trajectory in a subspace, spanned by the modes, $\mathbf{v}_i$. The dynamics in that subspace are described by the functions f_i, which represent the basis for an interpretation of the signal in terms of information processing units. Since any linear coordinate transformation of the modes $\mathbf{v}_i$ yields the same subspace, it is difficult to interpret the modes themselves. It is therefore necessary to find linear combinations of modes, which represent characteristic "states" for the underlying dynamics.

From a mathematical point of view, one can characterize dynamical systems by discussing sets of points, which are invariant under the vector field, $\dot{x}_i = f_i[x_j]$, and by discussing the behavior of the trajectory in the vicinity of these invariant manifolds. Fixed points are an example of invariant points: These are points, $\mathbf{x}^{(0)}$, which obey the equation,

$$\frac{\mathrm{d}}{\mathrm{d}t}\mathbf{x}^{(0)} = 0 \ , \quad \text{i.e. } \mathbf{f}[\mathbf{x}^{(0)}] = 0 \ . \tag{23}$$

With respect to these fixed points a trajectory of a dynamical system can be interpreted as being attracted and repelled in the course of time by different "states".

Considering such an interpretation for the analysis of EEG/MEG data, occurring fixed points may be interpreted as processing units of the underlying mental processes. This approach of describing neurophysiological signals in terms of interaction of different fixed points representing human brain information processing units is linked both to spatial microstate analysis and to conventional ERP/ERF component identification:

1. Spatial microstate analysis (SMA), introduced by Lehmann (1987), is a statistical approach to obtain quasistable field maps on the scalp surface. The signal is described by stepping from one state to another, and the states are interpreted as processing units of brain functioning. Our approach leads to brain dynamics which is governed by changing attractions of different states, instead of stepping from one state to the next one. The states of our analysis are not really occupied, as in SMA, but the trajectory is attracted by these states and therefore moves into the vicinity of these states.
2. Conventional ERP/ERF component identification (see, e.g., Roesler (1982)) is based on minima and maxima of the time series of a single channel i, which is given by $\frac{\mathrm{d}}{\mathrm{d}t}d_i = 0$. Our "definition" of processing units in terms of fixed points is similar: The fixed point is a state in

which *all* channels would obey the "definition" of a component: $\frac{\mathrm{d}}{\mathrm{d}t}d_i = 0$. It is never reached by the trajectory, since it would stay there forever, but it is attracted and repelled by them. This is closely related to minima and maxima, which occur in different channels at slightly different latencies but belong to the same component. This may allow to detect components in noisy datasets which show minima and maxima in single channels not all of them being correlated to brain processing of a cognitive task. Therefore our approach may lead to a more reliable identification components.

With respect to source localization, the fixed points in the phase portrait represent potential fields, $\mathbf{v}^{(0)}$, on the surface of the scalp,

$$\mathbf{v}^{(0)} = \sum_{i=1}^{N} x_i^{(0)} \mathbf{v}_i \ , \tag{24}$$

which can be modeled by any source localization technique: Either one considers time intervals close to fixed points as basis for source modeling, or one considers the fixed point itself as potential field to be represented by neural sources in the brain.

Studies of EEG data of cognitive processes, which are discussed in Sect. 5, will further demonstrate our approach and outline the interpretation of the obtained models.

4 EEG Data of Petit-Mal Epileptic Seizures

In previous studies we investigated EEG data of petit mal epileptic seizures with a similar algorithm as described above. We could reconstruct the spatio-temporal signals by a three-mode interaction, with the underlying dynamics showing Shilnikov type of behavior: The three-dimensional set of differential equations possesses a fixed point of saddle-focus stability and fulfills Shilnikov's condition (Šil'nikov (1965)). According to the theorem of Shilnikov, the systems then possesses in the vicinity of the homoclinic orbit an infinite number of unstable periodic orbits and therefore the trajectory can show chaotic but low-dimensional behavior. For details we refer the reader to our paper (1996).

5 Examples of the Application on Event-Related Potentials (ERP)

In the following we present two ERP studies with the results and interpretation of the obtained deterministic spatio-temporal models. In both cases we restrict ourselves to two-dimensional models,

$$\mathbf{d}^{(s)}(t) = x_1^{(s)}(t) \cdot \mathbf{v}_1 + x_2^{(s)}(t) \cdot \mathbf{v}_2 \ , \tag{25}$$

to allow a transparent presentation of the results and an interpretation based on two-dimensional phase portraits. To allow a comparison of the different stimulus conditions (s), we fitted simultaneously the same biorthogonal modes, $\mathbf{v}_1^+$ and $\mathbf{v}_2^+$, to the different datasets $\mathbf{d}^{(s)}$. To characterize the dynamic interactions, we assumed polynomial functions considering all linear, quadratic and cubic terms:

$$\frac{\mathrm{d}}{\mathrm{d}t}\begin{pmatrix} x_1 \\ x_2 \end{pmatrix} = \begin{pmatrix} f_1(x_1, x_2) \\ f_2(x_1, x_2) \end{pmatrix} \tag{26}$$

with

$$\begin{aligned} f_i = & \, a_{i,1}x_1 + a_{i,2}x_2 + a_{i,11}x_1^2 + a_{i,12}x_1x_2 + a_{i,22}x_2^2 \\ & + a_{i,111}x_1^3 + a_{i,112}x_1^2x_2 + a_{i,122}x_1x_2^2 + a_{i,222}x_2^3 \ . \end{aligned} \tag{27}$$

In both cases we chose the parameter p of the cost function (10)) as $p = 0.5$ to equally consider dynamics and signal representation (compare Sect. 3.2).

5.1 An Auditory ERP Study

The first example is an auditory ERP study of B. Opitz and A. Mecklinger (1997). The results of the dynamical systems based spatio-temporal analysis are published in (Uhl et al. (1998)).

Experiment

The experiment was performed to examine the neuronal and functional characteristics of processing deviancy and novelty in auditory stimuli. Three different stimuli (with presentation probability p) were presented to 20 volunteers:

- standard (s): tone of 600 Hz ($p = 0.8$)
- deviant (d): tone of 660 Hz, to be counted (attend condition, $p = 0.1$)
- novel (n): an identifiable (meaningful) novel sound selected from a class of unique sounds ($p = 0.1$)

During the experiment 128 channel EEG recordings were carried out. The datasets were averaged for the different stimuli after artifact elimination and bandpass (1–20 Hz) filtering. In the following we discuss the datasets of deviant and novel condition in the time interval [0 ms, 540 ms] after stimulus presentation.

Figure 1 shows the time series of electrode Fz for the two different conditions and Fig. 2 the spatio-temporal signals.

One observes a negative peak at around 100 ms (the so-called N100 component) in both cases. They show both a strong positive peak between 250

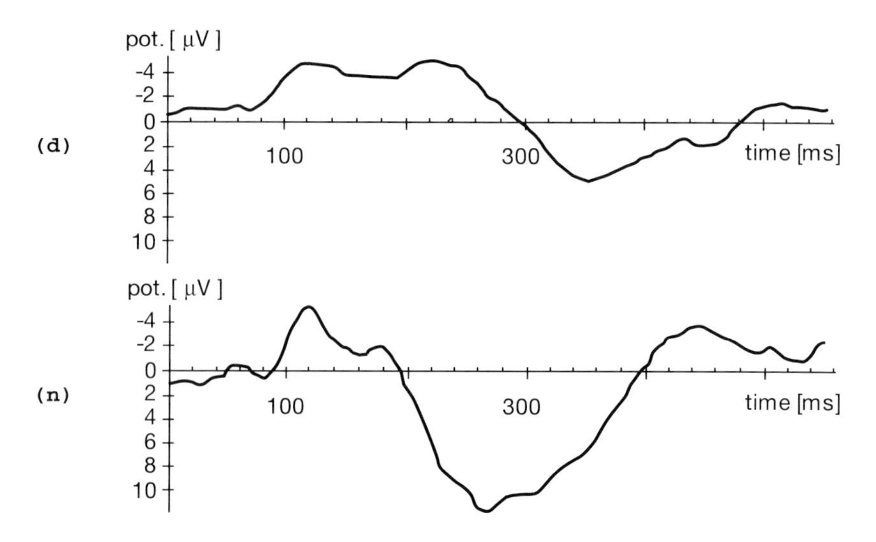

Fig. 1. Time series of the auditory ERP experiment at the Fz-electrode: (d) deviant condition and (n) novel condition

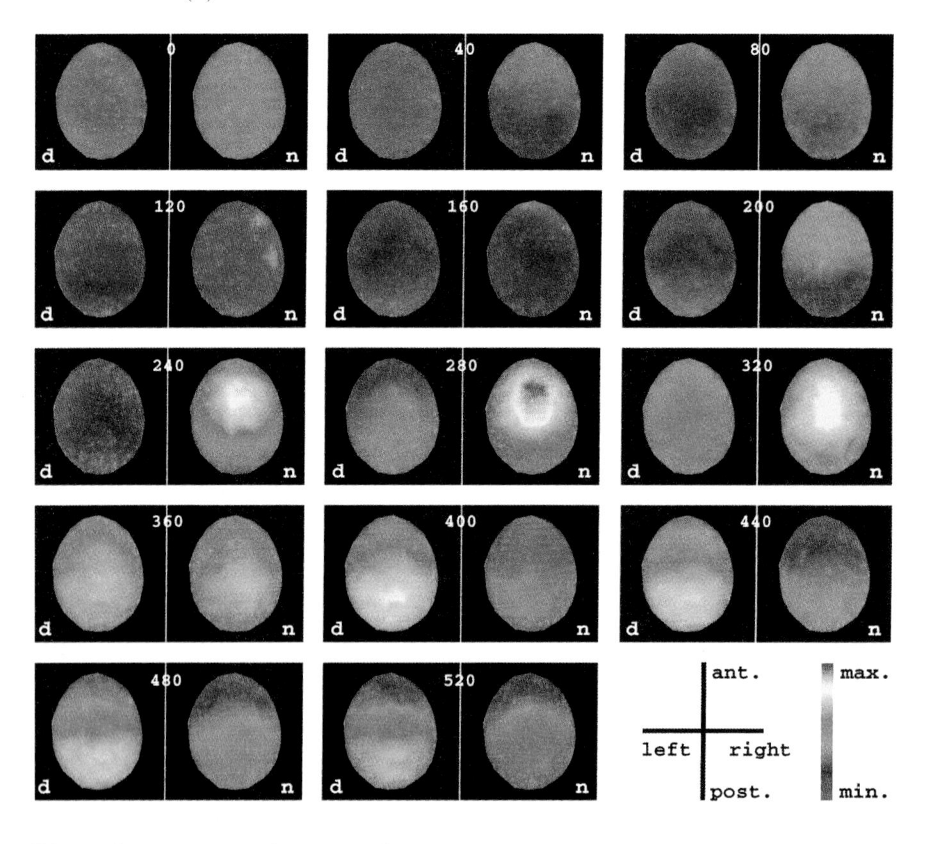

Fig. 2. Spatio-temporal signals of the auditory ERP study, deviant condition (d) and novel condition (n)

and 400 ms (P300) but clear differences with respect to the spatial distributions. The goal of the analysis of the signals is to investigate the differences in processing deviant and novel stimuli in this time range.

Spatio-Temporal Analysis

The global minimum of the cost function considering least-square-error functions of eqs. (25) and (27) was obtained by the genetic algorithm and led to a signal representation of 85 % (deviant condition) and 84 % (novel condition) with respect to the L_2 norm.

Figure 3 shows the spatial modes $\mathbf{v}_1$ and $\mathbf{v}_2$ corresponding to the global minimum of the cost function. They differ clearly and span the space of the relevant dynamics.

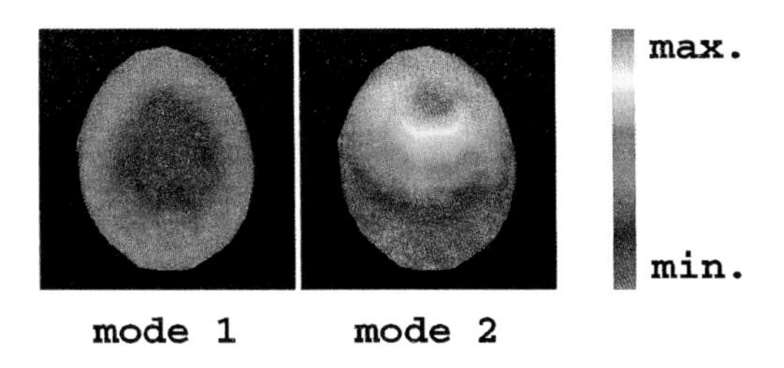

Fig. 3. The obtained modes of the spatio-temporal analysis

Figure 4 presents the two-dimensional phase-portrait for deviant (d) and novel (n) condition of the obtained amplitudes $x_1^{(d,n)}(t)$ and $x_2^{(d,n)}(t)$ corresponding to the EEG signal $\mathbf{d}^{(d,n)}(t)$ (solid pink line, with time markers every 40 ms). The underlying dynamics are visualized by plotting the absolute values of the vector field color coded: red areas represent small absolute values of the vector field, $|\mathbf{f}(x_1, x_2)| \simeq 0$, blue regions large values, $|\mathbf{f}(x_1, x_2)| \gg 0$. The axes of the phase-portraits, x_1, x_2, represent the amplitudes of the corresponding spatial modes, $\mathbf{v}_1$ and $\mathbf{v}_2$. Each point in that space represents a different spatial field distribution: a linear combination of $\mathbf{v}_1$ and $\mathbf{v}_2$ weighted by the corresponding x_1 and x_2 values:

$$\mathbf{d} = x_1 \mathbf{v}_1 + x_2 \mathbf{v}_2 \ . \tag{28}$$

The crossing of the trajectories is from a theoretical point of view incorrect and can be avoided by modeling the signal with more than two modes. However, for a more transparent presentation of our method, a two-dimensional projection was preferred, as these two-dimensional phase portraits represent approximations of the skeletons of the underlying dynamics. Finally we have plotted approximated fixed points (fp) calculated from the set of differential equations. They are marked by the letters A–E.

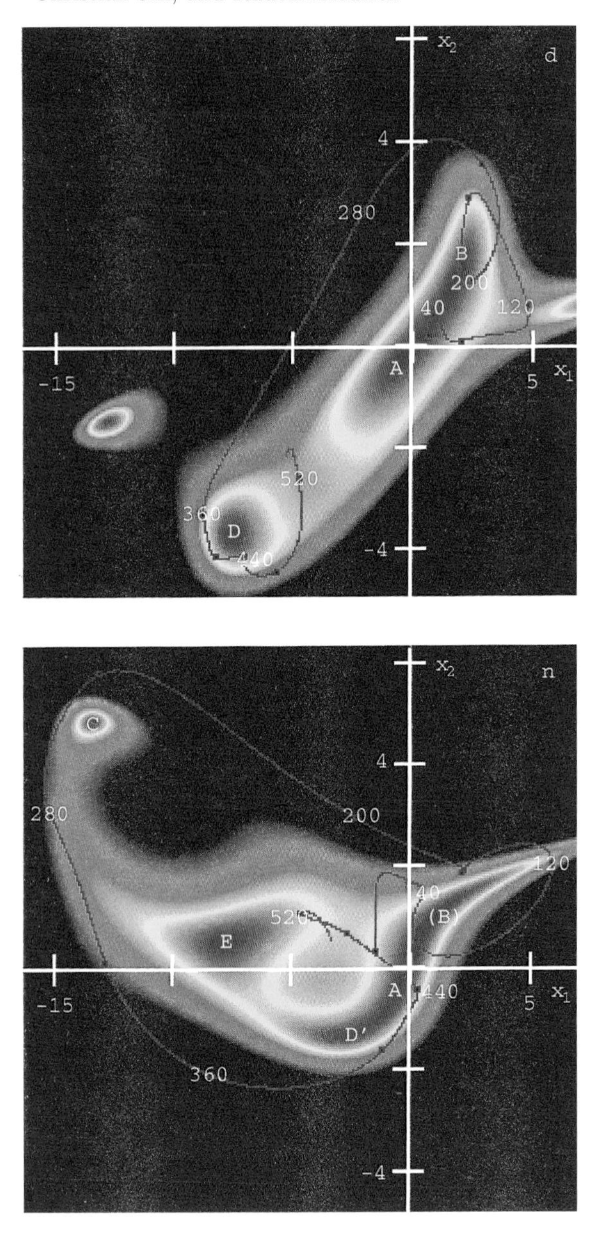

Fig. 4. Phase portraits of the obtained dynamics, see text for details

Interpretation of Dynamics
In both conditions the trajectory is first repelled by fixed point A and then influenced by fixed point B. The time range of the influence of B varies from 0–160 ms (novel condition) up to 0–240 ms (deviant condition) and reveals the N100 component. (As already mentioned the fixed points are approximations: $|\mathbf{f}(\mathbf{x}_{\mathrm{fp}})| \simeq 0$. In the case of fp B we observe $|\mathbf{f}^{(d)}(\mathbf{x}_{\mathrm{B}})| < |\mathbf{f}^{(n)}(\mathbf{x}_{\mathrm{B}})|$. Therefore, we marked the fixed point in the novel case with "(B)". The observation of a more distinct fixed point in the deviant case might reflect the so-called change detection process leading to mismatch negativity, see, e.g., Schröger (1997) for an overview.)

In the case of the deviant condition the trajectory after 240 ms leaves the region of fp B and is strongly attracted by fp D. This region is reached after 340 ms and the trajectory stays there up to 520 ms. This might reflect the processing of deviancy, with the corresponding component called target P3 (or P3b). The trajectory of the novel condition shows a different behavior: After 160 ms the trajectory is attracted by fp C, reaches the region of that fixed point in the interval 230–270 ms, representing the so-called novel P3, and is then influenced by fp D′, passing at 400 ms. Since the fixed point D′ lies approximately on the line of origin and fp D, it represents the same spatial distribution as fp D (see also Fig. 5), with decreased value of amplitude. This suggests that also in the novel condition brain functions represented by fp D play a role in human brain information processing, but are less important than in the deviant case, an observation also found by Mecklinger and Müller, 1996. Finally the trajectory is attracted by fp E at 480–520 ms, which might be associated with semantic aspects (Mecklinger et al., 1997).

Figure 5 (bottom rows) presents the spatial field distributions corresponding to the occurring fixed points for each condition. The amplitude of these distributions is given by the distance of the fixed point to the origin. Therefore, the spatial information given by fp A can be neglected, the amplitude is zero. Fp B corresponds to the N100 potential maps. Fp C and fp D differ clearly, indicating that different brain regions may be active. The potential maps corresponding to fp D and D′ are quite similar and reflect the already mentioned observation that the same processing may occur for deviant and novel stimuli with lower impact in the novel case. The spatial distributions associated with fixed points fp C and fp E are quite similar as well, both might represent novelty processing, the latter one might correspond to the verbal representation of the novel sound.

Another way of interpreting spatial potential field distributions is to investigate differences between different fixed points, i.e., to investigate, which potential fields (or corresponding sources) are necessary to get from one fixed point to another one. In terms of conventional analysis this would represent source localization of a process with respect to an earlier (or another) process. Thereby, one could overcome the problem of baseline corrections – a major problem of non-stationary signals. In the top rows of Fig. 5 the spatial distri-

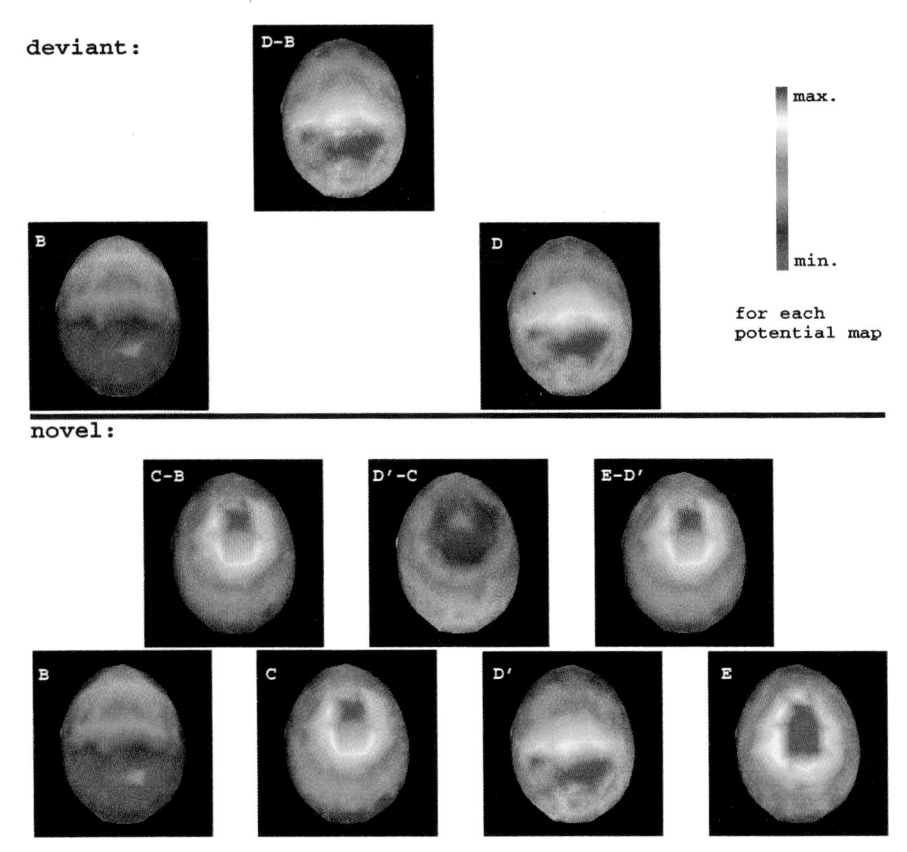

Fig. 5. Spatial field distributions of fixed points and their differences

butions of the transitions from the occurring fixed points are presented. The similarities of the processing corresponding to fp C and fp E are even more distinct considering the transitions from fp B to fp C (C–B) and from fp D′ to fp E (E–D′). Further studies will focus on the validation and significance of studying differences between process components.

Figure 6 shows the corresponding dipole fits of the spatial field distributions of Fig. 5. The dipole fits are based on spherical head models and were performed with ASA (from A.N.T. Software b.v., Hengelo, Netherlands). For each field map, one dipole is sufficient to represent the potential maps with a residual error less than 3%. These dipole fits demonstrate that the fixed points and their differences may be described by dipole models, although our approach originally does not aim at dipole modeling. Furthermore, do our dipole models correspond to findings of Opitz and Mecklinger (1997), where conventional component identification and dipole modeling lead to quite similar results. The positions of the obtained dipoles are from a neuroanatomic point of view not interpretable, since only one central dipole is sufficient to represent the spatial field distribution quite well. More sophisticated source localization models should be incorporated to possibly allow a neuroanatomic interpretation.

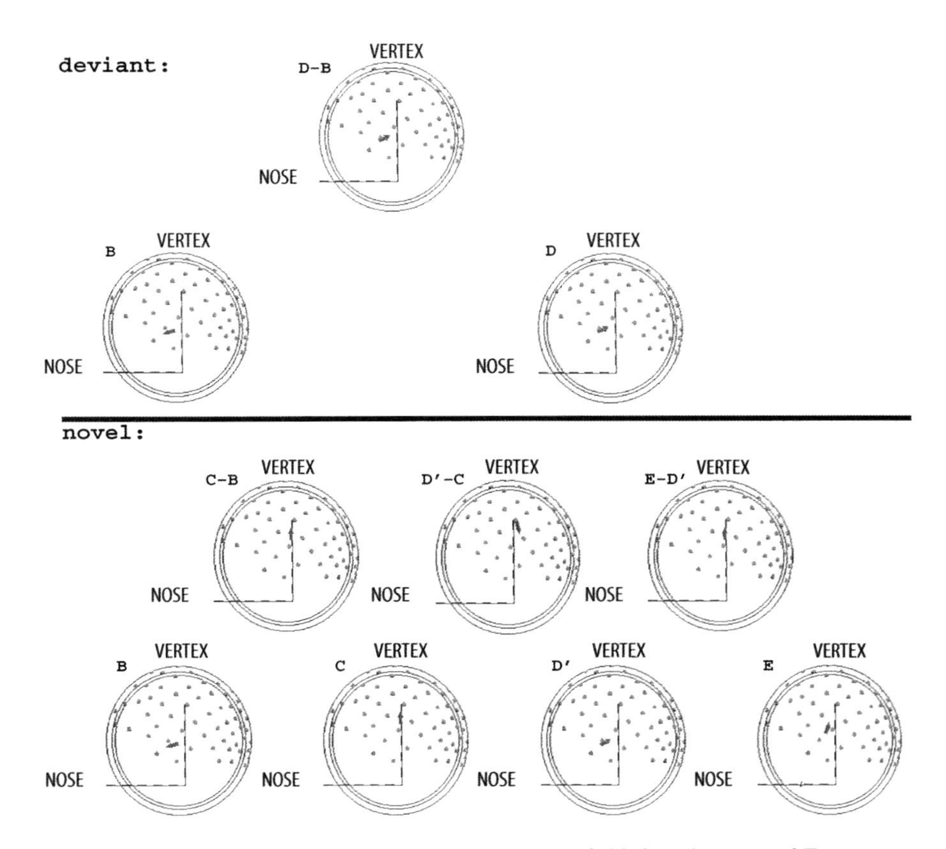

Fig. 6. Dipole models corresponding to the spatial field distributions of Fig. 5

Summarizing this example, we were able to represent the two different spatio-temporal signals in the same two-dimensional phase portrait, and we could characterize the signals by differential equations. An interpretation of the dynamics in terms of occurring fixed points with different spatial distributions suggest that different cortical regions in processing novelty and deviancy are active. Compared with processing of the deviant condition additional processes are activated in the case of the novel condition: fixed point C and E. Fixed point C might represent a process which is active before "processing deviancy" (fixed point D) and fixed point E might reflect access to conceptual-semantic processes of identifiable novel sounds.

5.2 A Psycholinguistic ERP Study

The second example we would like to present is based on a language related ERP study of A. Friederici and coworkers (1998). It was aimed at investigating parsing strategies in temporarily ambiguous sentences.

Experiment

Ten volunteers with tested high memory span capacity were asked to read, among other types of sentences, German subject and object first constructions of both relative and complement clauses, which are ambiguous until the last word. These sentences were presented on a PC monitor in six chunks of either one or two words. Two seconds after sentence presentation they were asked a question about the content of the sentence which had to be answered by a button press. Here, four different conditions are investigated (the temporarily ambiguity is due to German being a verb-final language, therefore a word-by-word English translation is added):

- Subject relative clause (SR):
  Das ist die *Professorin*, die die Studentinnen gesucht *hat.*
  This is the *professor* that the students sought *has.*
- Object relative clause (OR):
  Das ist die Professorin, die die *Studentinnen* gesucht *haben.*
  This is the professor that the *students* sought *have.*
- Subject complement clause (SC):
  Er wußte, daß die *Professorin* die Studentinnen gesucht *hat.*
  He knew that the *professor* the students sought *has.*
- Object complement clause (OC):
  Er wußte, daß die Professorin die *Studentinnen* gesucht *haben.*
  He knew that the professor the *students* sought *have.*

During the experiment 25-channel EEG recording was obtained with the onset of the last word representing "zero" on the time scale. The signals were averaged over all trials and subjects for the different stimulus conditions after artifact elimination and low pass filtering.

In Fig. 7 the time series of the Fz electrode are presented for the four different stimulus types SR, OR, SC and OC and in Fig. 8 the spatio-temporal signals are shown. The signals of the four different conditions show similar behavior in the interval [0 ms , 250 ms] as represented by the so-called N100 and P200 components. In the time range of 300–900 ms the signals differ slightly and the aim of the study was to investigate these differences.

This language related ERP experiment does not show such distinct differences of the spatio-temporal signals as in the above presented auditory experiment. We have actually chosen such an experiment to demonstrate

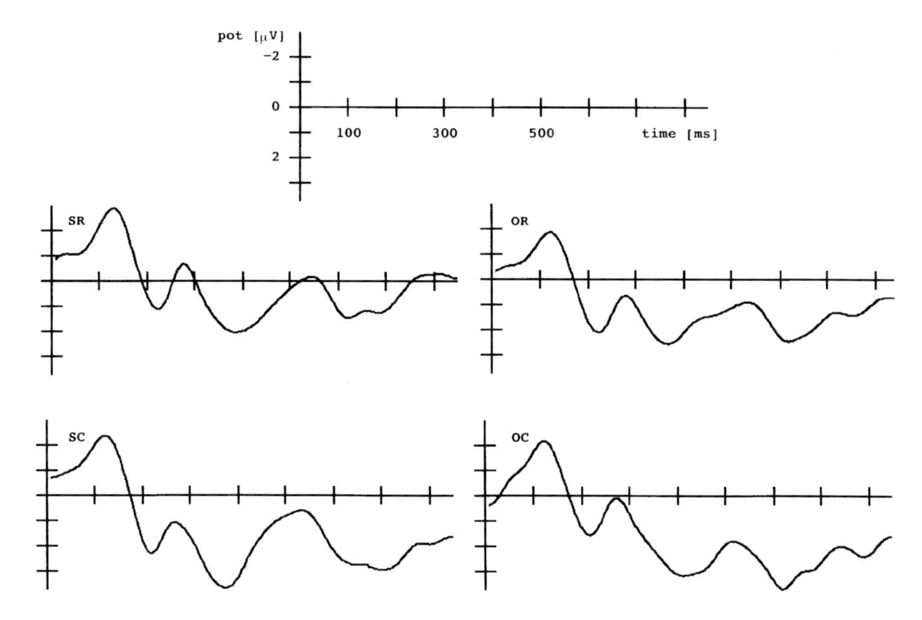

Fig. 7. Time series of Fz electrode of the language related ERP study for the four different conditions: subject relative (SR), object relative (OR), subject complement (SC) and object complement (OC)

the advantage of visualizing similar ERP datasets in phase portraits after detection of the dynamically relevant subspace.

Spatio-Temporal Analysis

Again the global minimum of the cost function is obtained by the genetic algorithm. This leads to a signal representation with respect to the L_2-norm of 75% for the subject relative clauses (SR), 90% for the object relative (OR), 89% for the subject complement (SC), and 95% for the object complement (OC) clauses.

Figure 9 shows the spatial modes $\mathbf{v}_1$ and $-\mathbf{v}_2$ corresponding to the global minimum of the cost function. (We present the negative of spatial mode $\mathbf{v}_2$ to better grasp the differences of $\mathbf{v}_1$ and $\mathbf{v}_2$ as they are quite similar.)

Figure 10 presents the two-dimensional phase-portraits for the four different conditions (SR, OR, SC and OC). Again the amplitudes $x_1(t)$ and $x_2(t)$ corresponding to the different EEG signals $\mathbf{d}(t)$ are presented as solid pink line, with time markers every 40 ms. The underlying dynamics are visualized color coded as in the phase portraits of the previous auditory ERP study: Red areas represent small values of $|\mathbf{f}(x_1, x_2)|$ the centers of which representing the fixed points of the dynamical systems.

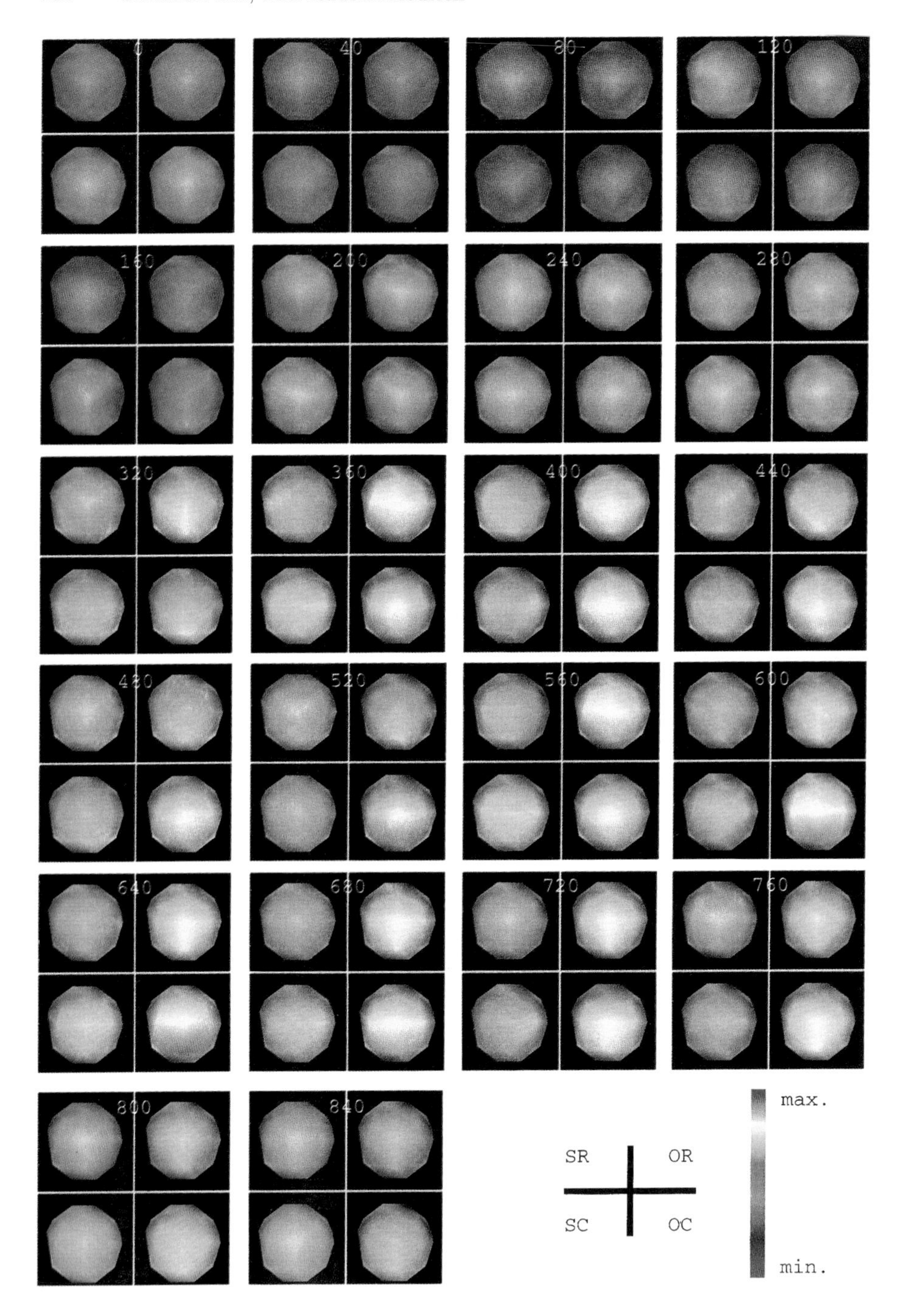

Fig. 8. Spatio-temporal ERP signals of subject relative (SR), object relative (OR), subject complement (SC) and object complement (OC) clauses

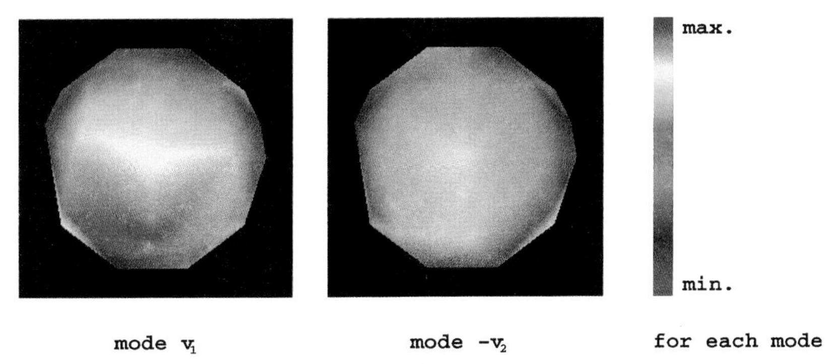

Fig. 9. The obtained modes of the spatio-temporal analysis

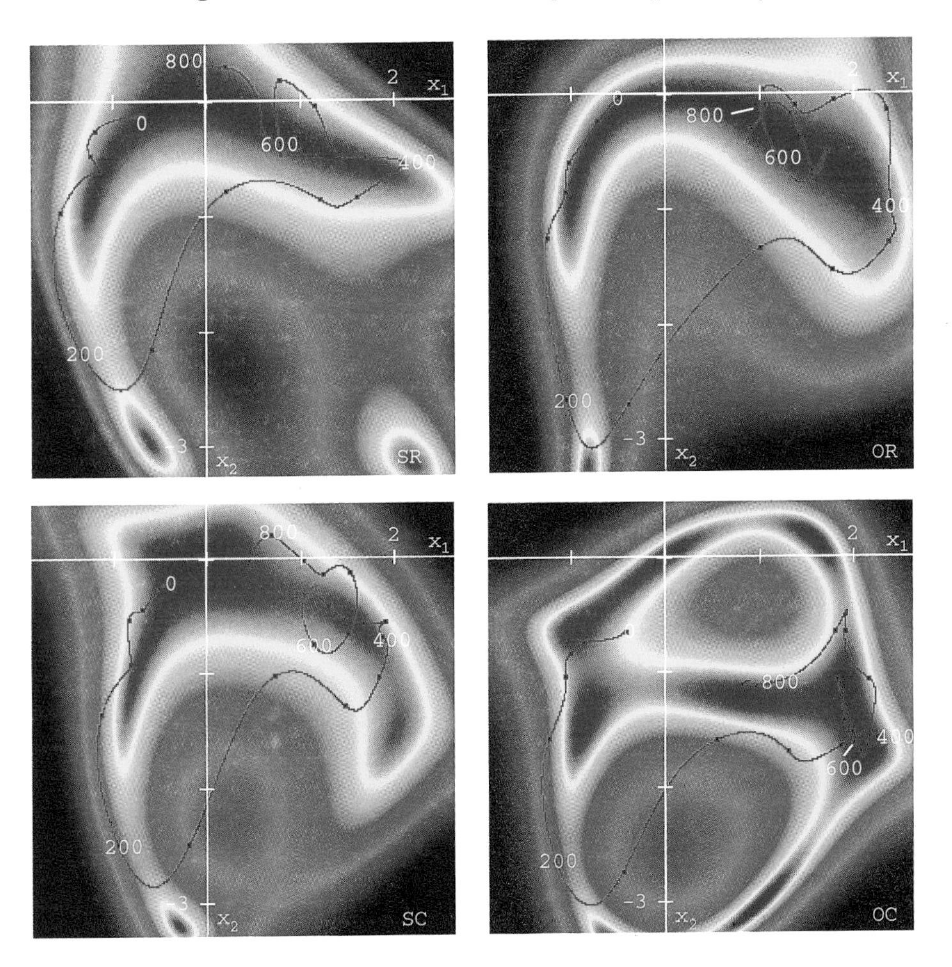

Fig. 10. Phase portraits of the obtained dynamics, see text for details

Interpretation
The phase portraits show clearly the similarities of the measured data sets. Comparing the object first with the subject first clauses, one observes two major differences: In the subject first construction (SR, SC) the trajectories at the end of the observed measurement are much closer to the origin as in the object first structures (OR, OC), an observation which may be interpreted as a "rechecking process" which is still in progress at 800 ms in the case of object first clauses. The second finding is the distinct influence of a fixed point in the time interval of 360 ms – 640 ms in the case of object first complement clause. Compared to the subject first complement clause this influence seems to be much stronger and for a longer temporal endurance. In the relative clauses, this fixed point seems also to influence the object first structure, but compared to the complement clauses for a shorter period of time at approximately 400 ms. This visually confirms the hypothesis of a syntactic reanalysis process with a latency of 400 ms (P400) in the object relative condition and the same process for the object first complement clause, but longer in duration (360 ms – 640 ms). Further studies will conclude a statistical test (by a multivariate ANOVA with spatial and temporal factors) of these observed phenomena.

Summarizing this example, we were able to represent the spatio-temporal signals of the four different conditions in the same phase portrait by different dynamics and thereby visualize common and diverse underlying processes. This suggests that our spatio-temporal approach based on dynamical systems theory might be a useful tool for visualizing similar datasets and for generating hypotheses of the underlying processes.

6 Stochastic Approach

Up to now we have not taken into account that the time signals of complex systems like EEG- and MEG-data contain significant temporal fluctuations, especially for the case of single trial measurements. Usually, averaging methods are applied to extract deterministic features of the spatio-temporal signal. However, it would be highly desirable to obtain information on both the deterministic part as well as on the noisy part of the signal. In the following we shall propose a method which, in our opinion, may turn out to be quite useful for the analysis of single trial EEG- and MEG-recordings. However, the method has not yet been applied to real EEG- and MEG-data sets. Nevertheless, we shall verify our method for several, synthetically generated data sets of stochastic systems.

The presence of noise in signals of self-organizing systems is a well-known fact (Haken 1983, 1987). The signatures of intrinsic noise become important especially close to phase transitions. In the context of biological systems it has been explicitly demonstrated in the Haken–Kelso–Bunz experiment. There, a phase transition in a coordination experiment has been examined. This underlines the necessity to extend the procedures for analyzing signals of complex systems described in the previous sections in order to take care of the noise.

For the following we assume that the system under consideration is governed by the Langevin equation:

$$\dot{\mathbf{x}}(t) = \mathbf{N}[\mathbf{x}] + \mathbf{F}(t) \tag{29}$$

Here, the vector $\mathbf{x}(t)$ consists of the collective variables. In fact, we assume that they are the order parameters. Our task is to determine the deterministic part of this equation, $\mathbf{N}[\mathbf{x}]$, as well as the properties of the fluctuating forces, $\mathbf{F}(t)$. We assume that these forces can be taken to be white noise Gaussian forces:

$$\langle F_i(t)F_j(t')\rangle = Q_{ij}\delta(t-t') \tag{30}$$

For the description of the statistical properties of the stochastic process generated by the Langevin equation one can introduce the probability density $p(\mathbf{x},t)$. It is well-known that this probability density obeys the following Fokker–Planck equation:

$$\frac{\partial}{\partial t}p(\mathbf{x},t) = \left\{ -\sum_i \frac{\partial}{\partial x_i} N_i[\mathbf{x}] + \sum_{ij} \frac{\partial^2}{\partial x_i \partial x_j} Q_{ij} \right\} p(\mathbf{x},t) \tag{31}$$

If the drift term and the diffusion term of this equation is known, the probability density can be calculated.

Let us now investigate the inverse problem. A complete analysis of a complex system exhibiting noisy signals, therefore, should yield the drift term $\mathbf{N}[\mathbf{x}]$ as well as the correlations of the noise term, Q_{ij}. Based on the maximum information principle, Hermann Haken (1988) suggested an approach to derive these terms from measured data. Lisa Borland (1993) further developed this method and applied it to model systems. In the following we will discuss an approach to determine drift and noise term directly from the signal. In an analysis of the statistical properties of a turbulent fluid motion a method has been devised which allows to extract these quantities directly from the data (Peinke and Friedrich (1997)).

The Langevin equation generates a Markovian process which is entirely determined by the conditional probability densities $p(\mathbf{x}',t+\tau|\mathbf{x},t)$,

$$p(\mathbf{x}',t+\tau|\mathbf{x},t) = p(\mathbf{x}',t+\tau;\mathbf{x},t)/p(\mathbf{x},t) \tag{32}$$

Here, $p(\mathbf{x}',t+\tau;\mathbf{x},t)$ is a joint probability density. The conditional probability density is a transition probability from state $\mathbf{x}$ at time t to the state $\mathbf{x}'$ at time $t+\tau$. The derivation of the Fokker–Planck equation from the Langevin equation yields operational expressions for drift and diffusion terms in terms of the conditional probability densities:

$$
\begin{aligned}
N_i[\mathbf{x}] &= \lim_{\tau \to 0} \frac{1}{\tau} \int \mathrm{d}\mathbf{x}'(x_i' - x_i) p(\mathbf{x}', t+\tau|\mathbf{x}, t) \\
Q_{ij} &= \lim_{\tau \to 0} \frac{1}{\tau} \int \mathrm{d}\mathbf{x}(x_i' - x_i)(x_j' - x_j) p(\mathbf{x}', t+\tau|\mathbf{x}, t)
\end{aligned}
\tag{33}
$$

Therefore, if the conditional probability distribution has been determined, the expressions for the drift and diffusion terms can be evaluated. However, the probability density can be estimated from the experimental data by evaluating histograms.

Let us add some remarks. If one considers methods which have been designed to fit a set of differential equations to experimental data, one immediately recognizes that one has to establish an ansatz for the vector field $\mathbf{N}(\mathbf{x})$, e.g. as superposition of polynomials. In order to obtain reasonable results one has to make guesses on the functional of $\mathbf{N}(\mathbf{x})$ on $\mathbf{x}$. The present evaluation of $\mathbf{N}(\mathbf{x})$ is not based on such type of ansatz: The drift term is obtained numerically as a data set. Subsequently it may be approximated by analytical formulas.

6.1 Verification of the Method for Synthetic Data Sets

In order to demonstrate that the present method leads to a successful solution of the problem we consider various synthetically determined data sets. Let us first resort to the case of one-dimensional systems.

Noisy Pitchfork Bifurcation

We start by considering the following Langevin equation:

$$
\dot{q}(t) = \epsilon q(t) - q(t)^3 + F(t) \tag{34}
$$

This equation applies for systems exhibiting noisy pitchfork bifurcations. Figure 11 exhibits a part of the calculated noisy time series. Figures 12 and 13 exhibit the reconstructed drift term as well as the diffusion term in comparison to the exact values.

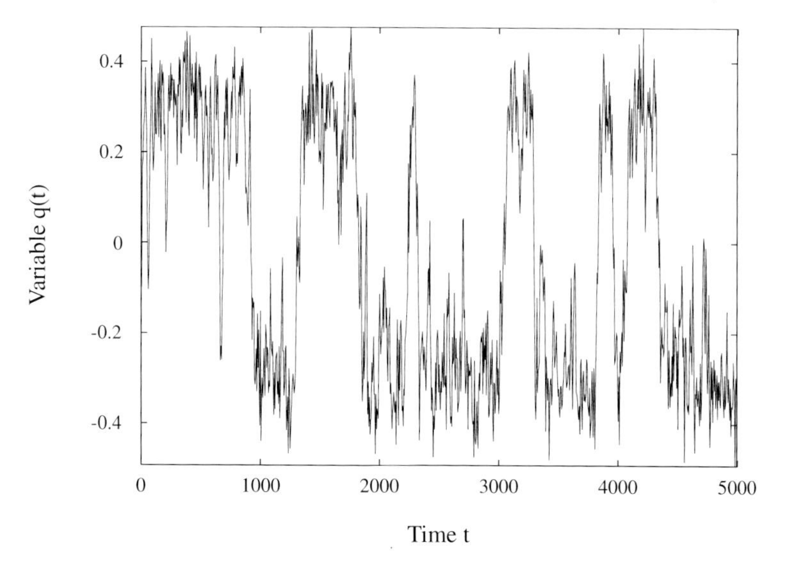

Fig. 11. Realization $q(t)$ of the Langevin equation (34).

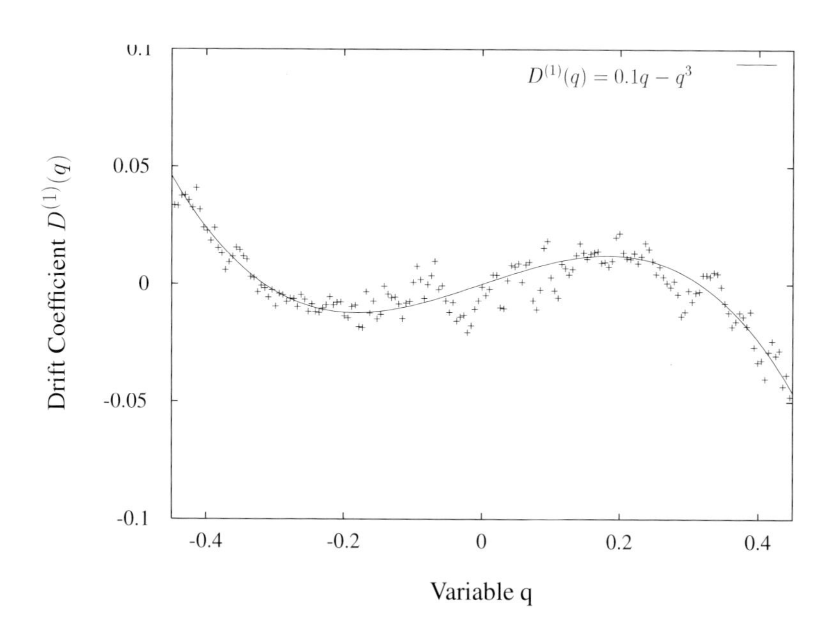

Fig. 12. Reconstruction of the drift term, $N(q) = \epsilon q - q^3$ of the Langevin equation (34).

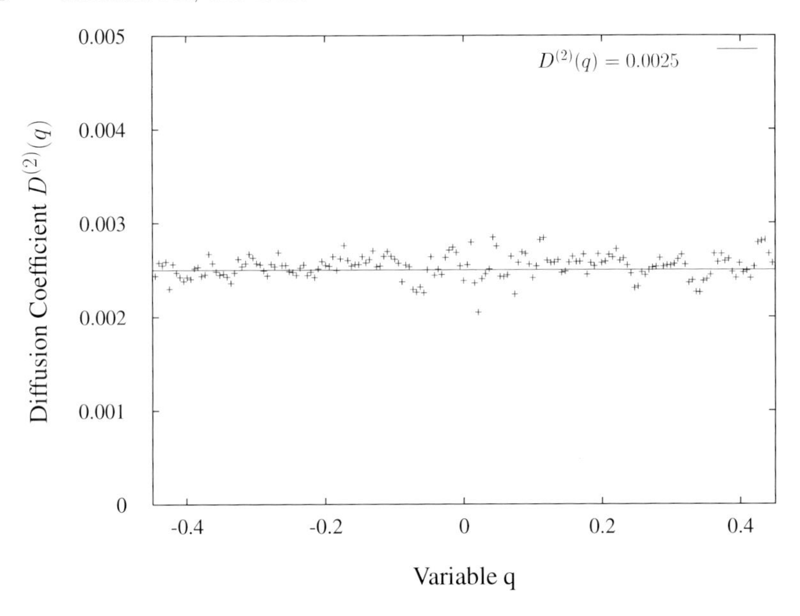

Fig. 13. Reconstruction of the diffusion coefficient of the Langevin equation (34).

Analysis of Synchronization Phenomena
Our next example is concerned with the equation

$$\dot{\Phi} = \omega + h(\Phi) + F(t) \tag{35}$$

where Φ is a 2π-periodic function in Φ. In the present case we have taken $h(\Phi) = \omega_0 + \sin \Phi$, $\omega = 0.2$, $Q = 0.6$. This type of equation describes the dynamics of the phase difference $\Phi = \varphi_1 - \varphi_2$ of the phases of two coupled nonlinear oscillators. For $\omega > 0$ we obtain running solutions, as exhibited in Fig. 14. Figure 15 exhibits the drift term obtained by the present method. Equation (35) applies to a variety of synchronization phenomena. Therefore, we expect that the present method will turn out to be quite useful for examining the synchronization of different channels in EEG- and MEG-signals. To this end one has to determine suitable phase variables φ_i for the different channels and may then proceed as in our example.

Finally, let us mention that the present method has already been successfully applied to various two-dimensional Langevin equations (Siegert et al. (1998)).

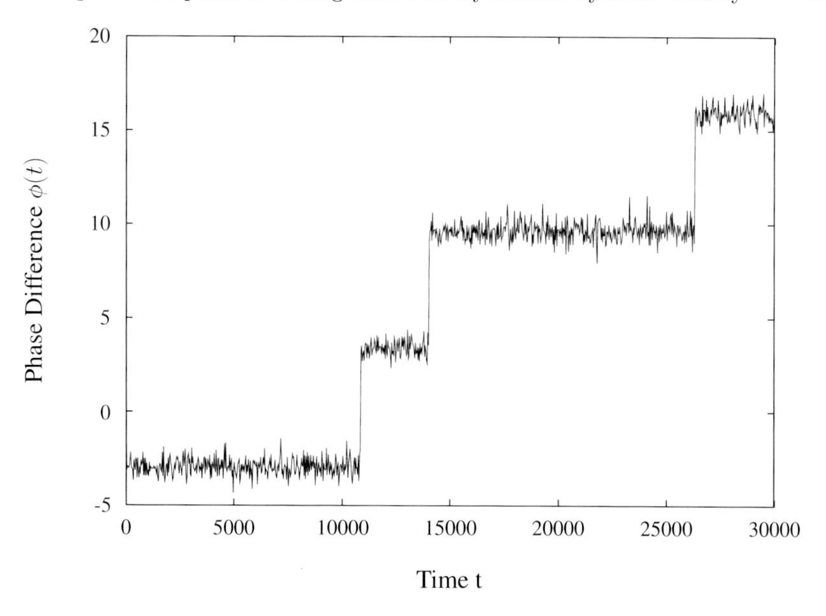

Fig. 14. Realization $\Phi(t)$ of the Langevin equation (35).

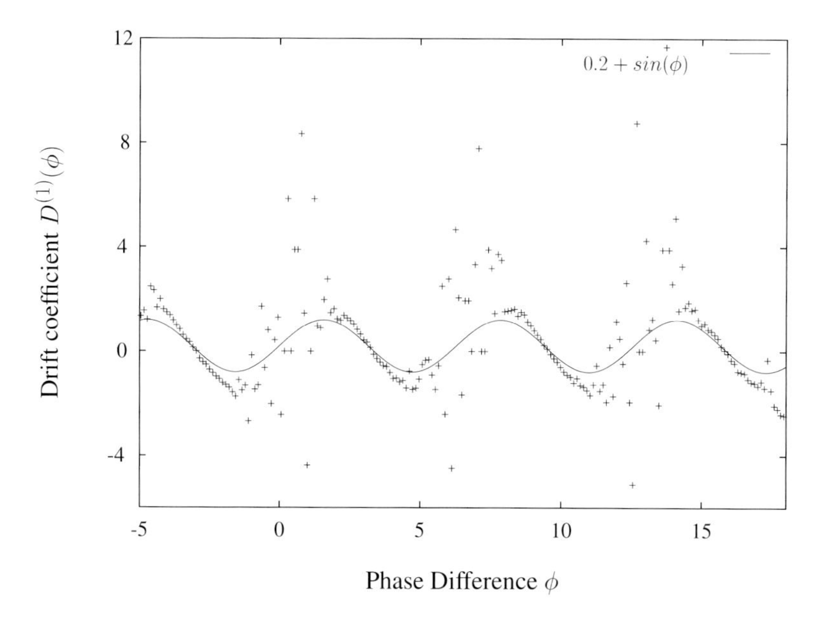

Fig. 15. Reconstruction of the drift term of the Langevin equation (35).

7 Conclusions

We have presented an algorithm to determine deterministic models of spatio-temporal EEG/MEG signals and we have linked its interpretation to the description of event-related potentials/fields in terms of components and their influence to the signals. By this shift from the discussion of components to the discussion of dynamical systems, in our opinion, we open a wide field of possible quantitative descriptions of brain functioning.

In the second part of this chapter we have discussed an approach, which allows for an extraction of deterministic as well as stochastic features from the data. This approach may lead to a helpful tool for investigating single-trial EEG/MEG data and may lead to an understanding of brain functioning combing concepts of source modeling and synchronization.

Acknowledgments. We gratefully acknowledge stimulating and encouraging discussions concerning synergetics and brain functioning with Hermann Haken, D. Yves von Cramon and Frithjof Kruggel. We thank Marjan Admiraal, Axel Hutt and Silke Siegert for their support and for preparing some of the figures. We also appreciate helpful discussions and support concerning the ERP studies by Angela D. Friederici, Axel Mecklinger, Bertram Opitz, Karsten Steinhauer and Martin Meyer.

Appendix A

Variation of C with respect to $\mathbf{v}_j$ yields

$$\frac{\partial C}{\partial \mathbf{v}_j} = \frac{\partial S}{\partial \mathbf{v}_j} = 0 \quad \Rightarrow \quad \sum_{i=1}^{N} \langle x_i x_j \rangle \mathbf{v}_j = \langle \mathbf{d}\, x_j \rangle \ , \tag{36}$$

which can be solved by

$$\mathbf{v}_j[\mathbf{v}_i^+] = \sum_{i=1}^{N} \langle \mathbf{d}\, x_i \rangle M_{ij}^{-1} \ , \tag{37}$$

with $M_{ij} = \langle x_i x_j \rangle = \langle (\mathbf{d} \cdot \mathbf{v}_i^+)(\mathbf{d} \cdot \mathbf{v}_j^+) \rangle$. Inserting (37) back into (8) leads to

$$S[\mathbf{v}_i^+] = 1 - \sum_{i,j=1}^{N} \langle \mathbf{d}\,(\mathbf{d} \cdot \mathbf{v}_i^+) \rangle M_{ij}^{-1} \langle (\mathbf{d} \cdot \mathbf{v}_j^+)\, \mathbf{d} \rangle / \langle \mathbf{d}^2 \rangle \ . \tag{38}$$

Note, that in the case of $\langle x_i x_j \rangle = c\delta_{ij}$, (37) simplifies to

$$\mathbf{v}_j[\mathbf{v}_i^+] = \langle \mathbf{d}\, x_j \rangle / c \ , \tag{39}$$

and (38) to

$$S[\mathbf{v}_i^+] = 1 - \sum_{i=1}^{N} \langle \mathbf{d}\,(\mathbf{d}\cdot\mathbf{v}_i^+)\rangle^2/(c\langle\mathbf{d}^2\rangle) \ . \tag{40}$$

The polynomial function $f_i[x_j]$ (11) can be written as:

$$f_i[x_j] = \sum_{\alpha=1}^{M} a_{i,\alpha}\,\xi_{i,\alpha} \ , \tag{41}$$

after the introduction of abbreviations

$$\begin{aligned}\{a_{i,\alpha}\}_{\alpha=1,\dots,M} &= \{\ \{a_{i,j}\}_{j=1,\dots,N}, \quad \{a_{i,jk}\}_{j=1,\dots,N;k=j,\dots,N}, \\ &\qquad \{a_{i,jkl}\}_{j=1,\dots,N;k=j,\dots,N;l=k,\dots,N} \quad \} \\ \{\xi_{i,\alpha}\}_{\alpha=1,\dots,M} &= \{\ \{x_j\}_{j=1,\dots,N}, \quad \{x_j x_k\}_{j=1,\dots,N;k=j,\dots,N}, \\ &\qquad \{x_j x_k x_l\}_{j=1,\dots,N;k=j,\dots,N;l=k,\dots,N} \quad \} \ .\end{aligned} \tag{42}$$

The cost function D then reads

$$D = \sum_{i=1}^{N} \frac{\langle(\dot{x}_i - \sum_{\alpha=1}^{M} a_{i,\alpha}\xi_{i,\alpha})^2\rangle}{\langle\dot{x}_i^2\rangle} \ . \tag{43}$$

Variation with respect to $a_{j,\beta}$ yields

$$\frac{\partial C}{\partial a_{j,\beta}} = \frac{\partial D}{\partial a_{j,\beta}} = 0 \quad \Rightarrow \quad \sum_{\alpha=1}^{M} \langle \xi_{j,\alpha}\xi_{j,\beta}\rangle a_{j,\alpha} = \langle \dot{x}_j \xi_\beta\rangle \ , \tag{44}$$

which can be solved by

$$\mathbf{a}_i = Q_i^{-1}\mathbf{b}_i \ , \tag{45}$$

$$\text{with} \quad (\mathbf{a}_i)_\alpha = a_{i,\alpha}, \quad (Q_i)_{\alpha\beta} = \langle \xi_{i,\alpha}\xi_{i,\beta}\rangle, \quad (\mathbf{b}_i)_\beta = \langle \dot{x}_i \xi_{i,\beta}\rangle \ . \tag{46}$$

Inserting this expression back into D leads to

$$D[\mathbf{v}_i^+] = N - \sum_{i=1}^{N} \mathbf{b}_i Q_i^{-1}\mathbf{b}_i/\langle\dot{x}_i^2\rangle \ . \tag{47}$$

Because of (42) and (6) this depends on the biorthogonal modes $\mathbf{v}_i^+$ only.

Appendix B

We will outline the derivation of free parameters of the cost function for the cases of two-mode and three-mode interaction.

For simplicity of notation we write

$$\mathbf{u} = \tilde{\mathbf{v}}_1^+, \quad \mathbf{v} = \tilde{\mathbf{v}}_2^+, \quad \mathbf{w} = \tilde{\mathbf{v}}_3^+ \ . \tag{48}$$

We fulfill the constraints step by step:

1. Constraint $\mathbf{u} \cdot \mathbf{u} = 1$:
   This constraint can be fulfilled by introducing elliptic coordinates,

$$u_1 = \cos\phi_1^{(1)} \quad , \quad u_n = \prod_{k=1}^{n-1} \sin\phi_1^{(k)} \tag{49}$$

$$u_i = \cos\phi_1^{(i)} \prod_{k=1}^{i-1} \sin\phi_1^{(k)} \quad \text{if} \quad i \neq 1, n \ . \tag{50}$$

2. Constraints $\mathbf{u} \cdot \mathbf{v} = 0, \quad \mathbf{v} \cdot \mathbf{v} = 1$:
   The orthogonality constraint leads to

$$\sum_{i=1}^{n} u_i v_i = 0 \Rightarrow v_n = c_n \sum_{i=1}^{n-1} u_i v_i, \quad c_n = -\frac{1}{u_n} \ . \tag{51}$$

Without loss of generality we assumed $u_n \neq 0$.
Inserting this expression into the normalization constraint, we obtain

$$\sum_{i=1}^{n-1} v_i^2 + c_n^2 \sum_{i,j=1}^{n-1} u_i v_i u_j v_j = 1 \ , \tag{52}$$

which can be written as

$$\mathbf{v} \cdot \mathbf{v} = \sum_{i,j=1}^{n-1} v_i A_{ij} v_j, \quad A_{ij} = \delta_{ij} + c_n^2 u_i u_j \ . \tag{53}$$

In index-free notation ($\mathbf{v}'$ denoting the vector $\mathbf{v}$ without the component v_n) this reads

$$\mathbf{v}'^T A \mathbf{v}' = 1 \ . \tag{54}$$

By a coordinate transformation $\mathbf{v}' = K\tilde{\mathbf{v}}'$ the expression can be diagonalized with eigenvalues a_i:

$$\tilde{\mathbf{v}}'^T K^T A K \tilde{\mathbf{v}}' = \sum_{i=1}^{n-1} a_i (\tilde{v}_i')^2 = 1 \ . \tag{55}$$

Again this can be solved by elliptic coordinates,

$$\tilde{v}_1' = r^{(1)} \cos\phi_2^{(1)} \quad , \quad \tilde{v}_{n-1}' = r^{(n-1)} \prod_{k=1}^{n-2} \sin\phi_2^{(k)} \tag{56}$$

$$\tilde{v}_i' = r^{(i)} \cos\phi_2^{(i)} \prod_{k=1}^{i-1} \sin\phi_2^{(k)} \quad \text{if} \quad i \neq 1, n-1 \ ,$$

with radii $r^{(i)} = 1/\sqrt{a_i}$.

3. Constraints $\mathbf{u} \cdot \mathbf{w} = \mathbf{v} \cdot \mathbf{w} = 0, \quad \mathbf{w} \cdot \mathbf{w} = 1$:
   Analogous to the calculation above, the first orthogonality constraint leads to

$$\sum_{i=1}^{n} u_i w_i = 0 \Rightarrow w_n = d_n \sum_{i=1}^{n-1} u_i w_i, \quad d_n = -\frac{1}{u_n} \ . \tag{57}$$

Inserting this into the second orthogonality constraint we obtain

$$\sum_{i=1}^{n} v_i w_i = \sum_{i=1}^{n-1} \left(v_i + v_n d_n u_i\right) w_i = 0 \tag{58}$$

$$\Rightarrow \quad w_{n-1} = d_{n-1} \sum_{i=1}^{n-2} \left(v_i + v_n d_n u_i\right) w_i \tag{59}$$

with

$$d_{n-1} = \frac{u_n}{v_n u_{n-1} - v_{n-1} u_n} \ . \tag{60}$$

Inserting (59) back into (57), this yields

$$w_n = d_n \sum_{i=1}^{n-2} \left(u_i + u_{n-1} d_{n-1} \left(v_i + v_n d_n u_i\right)\right) w_i \ . \tag{61}$$

Summarizing eqs. (59,61) we obtain the following expressions

$$w_{n-1} = d_{n-1} \sum_{i=1}^{n-2} g_i w_i \quad , \qquad w_n = d_n \sum_{i=1}^{n-2} h_i w_i \ , \tag{62}$$

$$g_i = v_i + d_n v_n u_i \quad , \qquad h_i = u_i + d_{n-1} u_{n-1} g_i \ . \tag{63}$$

Inserting these expressions into the normalization constraint, we obtain

$$\sum_{i=1}^{n-2} w_i^2 + d_{n-1}^2 \sum_{i,j=1}^{n-2} g_i w_i g_j w_j + d_n^2 \sum_{i,j=1}^{n-2} h_i w_j h_j w_j = 1 \ , \tag{64}$$

which can be written as

$$\mathbf{w} \cdot \mathbf{w} = \sum_{i,j=1}^{n-2} w_i B_{ij} w_j, \quad B_{ij} = \delta_{ij} + d_{n-1}^2 g_i g_j + d_n^2 h_i h_j \ . \tag{65}$$

With $\mathbf{w}'$ denoting the vector $\mathbf{w}$ dropping the components w_{n-1}, w_n we obtain,

$$\mathbf{w}'^T B \mathbf{w}' = 1 \ , \tag{66}$$

which can be diagonalized by a coordinate transformation:

$$\tilde{\mathbf{w}}'^T L^T B L \tilde{\mathbf{w}}' = \sum_{i=1}^{n-2} b_i (\tilde{w}_i')^2 = 1 \ . \tag{67}$$

And again this can be solved by elliptic coordinates,

$$\tilde{w}_1' = \tau^{(1)} \cos\phi_3^{(1)} \quad , \quad \tilde{w}_{n-2}' = \tau^{(n-2)} \prod_{k=1}^{n-3} \sin\phi_3^{(k)} \tag{68}$$

$$\tilde{w}_i' = \tau^{(i)} \cos\phi_3^{(i)} \prod_{k=1}^{i-1} \sin\phi_3^{(k)} \quad \text{if} \quad i \neq 1, n-2 \ , \tag{69}$$

with radii $\tau^{(i)} = 1/\sqrt{b_i}$.

References

Borland, L. (1993): Ein Verfahren zur Bestimmung der Dynamik stochastischer Prozesse, Thesis Stuttgart (1993). Learning the Dynamics of Two-Dimensional Stochastic Markov Processes, Open Systems & Information Dynamics, Vol. 1, No. 3 (1992). On the constraints necessary for macroscopic prediction of time-dependent stochastic processes, Reports on Mathematical Physics, Vol. 33, No. 1/2 (1993)

Friederici, A. D., Steinhauer, K., Mecklinger, A., Meyer, M. (1998): Working Memory Constraints on Syntactic Ambiguity Resolution as Revealed by Electrical Brain Responses, Biological Psychology 47, 193–221

Friedrich, R., Uhl, C. (1996): Spatio-temporal analysis of human electroencephalograms: petit-mal epilepsy, Physica D 98, 171–182

Haken, H. (1983): *Synergetics. An Introduction.* Springer, Berlin, Heidelberg, 3rd edition

Haken, H. (1987): *Advanced Synergetics.* Springer, Berlin, Heidelberg, 2nd edition

Haken, H. (1988): *Information and self-organization.* Springer, Berlin, Heidelberg

Haken, H. (1996): *Principles of Brain Functioning.* Springer, Berlin, Heidelberg

Hasselmann, K. (1988): PIPs and POPs: The Reduction of Complex Dynamical Systems Using Principal Interaction and Oscillation Patterns, J. Geophys. Res. 93, D9, 11015–11021

Holland, J.H. (1975): *Adaption in Natural and Artificial Systems* Michigan University Press, Ann Arbor

Kelso, J.A.S. (1995): *Dynamic Patterns. The self-organization of brain and behavior.* MIT Press, Cambridge

Kwasniok, F. (1996): The reduction of complex dynamical systems using principal interaction patterns, Physica D 92, 28–60

Kwasniok, F. (1997): Optimal Galerkin Approximations of Partial Differential Equations Using Principal Interaction Patterns, Phys. Rev. E 55, 5365–5375

Lehmann, D. (1987): Principles of spatial analysis. In: Gevins, A.S., Remond, A. (eds): *Methods of Brain Electrical and Magnetic Signals. Handbook of Electroencephalography and Clinical Neurophysiology, revised series, Volume 1* Elsevier, Amsterdam

Mecklinger, A., Müller (1996): Dissociations in the processing of "what" and "where" information in working memory: an event-related potential analysis, Journal of Cognitive Neuroscience, 8:5, 453–473

Mecklinger, A., Opitz, B., Friederici, A., D. (1997): Semantic aspects of novelty detection in humans, Neuroscience Letters 235, 65–68

Nunez, P.L. (1995): *Neocortical Dynamics and Human EEG Rhythms* Oxford University Press, Oxford

Peinke, J., Friedrich, R. (1997): Statistical properties of a turbulent cascade, Physica D 102, 147

Opitz, B., Mecklinger, A. (1997): What are the similarities in hearing cars and tones ?, Fourth Annual Meeting of the Cognitive Neuroscience Society, March 23 - 25, Boston, Massachusetts

Rösler, F. (1982): *Hirnelektrische Korrelate Kognitiver Prozesse* Springer, Berlin, Heidelberg

Schröger, E. (1997): Measurement and Interpretation of the Mismatch Negativity, submitted to Behavior Research Methods, Instruments, & Computers

Siegert, S., Friedrich, R., Peinke, J. (1998): Analysis of data sets of stochastic systems, Preprint

Šil'nikov, L.P. (1965): A case of the existence of a countable number of periodic motions, Sov. Math. Dokl. 6, 163–166 (1965)

Uhl, C., Friedrich, R., Haken, H. (1993): Reconstruction of spatio-temporal signals of complex systems, Z. Phys. B 92, 211–219

Uhl, C., Friedrich, R., Haken, H. (1995): Analysis of spatio-temporal signals of complex systems, Phys. Rev. E 51, 3890–3900

Uhl, C., Friedrich, R. (1996): Spatio-temporal Signal Analysis with Applications to EEG-Analysis, NeuroImage, 3, Suppl., 260

Uhl, C., Kruggel, F., Opitz, B., von Cramon, D.Y. (1997): A New Concept for EEG/MEG – Signal Analysis: Detection of Interacting Spatial Modes, Human Brain Mapping 6:137–149

Subject Index

Springer and the environment

At Springer we firmly believe that an international science publisher has a special obligation to the environment, and our corporate policies consistently reflect this conviction.

We also expect our business partners – paper mills, printers, packaging manufacturers, etc. – to commit themselves to using materials and production processes that do not harm the environment. The paper in this book is made from low- or no-chlorine pulp and is acid free, in conformance with international standards for paper permanency.

Printing: Mercedesdruck, Berlin
Binding: Buchbinderei Lüderitz & Bauer, Berlin